Cryochemical Technology of Advanced Materials

Powder Technology Series

EDITED BY

BRIAN SCARLETT	and	GENJI JIMBO
Delft University of Technology		*Chubu Powtech Plaza Lab*
The Netherlands		*Japan*

Many materials exist in the form of a disperse system, for example powders, pastes, slurries, emulsions and aerosols. The study of such systems necessarily arises in many technologies but may alternatively be regarded as a separate subject which is concerned with the manufacture, characterization and manipulation of such systems. Chapman & Hall were one of the first publishers to recognize the basic importance of the subject, going on to instigate this series of books. The series does not aspire to define and confine the subject without duplication, but rather to provide a good home for any book which has a contribution to make to the record of both the theory and the application of the subject. We hope that all engineers and scientists who concern themselves with disperse systems will use these books and that those who become expert will contribute further to the series.

Particle Size Measurement
Terence Allen
5th edn, 2 vols, hardback
(0 412 75350 2), 552 and 272 pages

Chemistry of Powder Production
Yasuo Arai
Hardback (0 412 39540 1), 292 pages

Particle Size Analysis
Claus Bernhardt
Translated by H. Finken
Hardback (0 412 55880 7), 428 pages

Particle Classification
K. Heiskanen
Hardback (0 412 49300 4), 330 pages

Powder Surface Area and Porosity
S. Lowell and Joan E. Shields
3rd edn, hardback (0 412 39690 4), 256 pages

Principles of Flow in Disperse Systems
O. Molerus
Hardback (0 412 40630 6), 314 pages

Processing of Particulate Solids
J.P.K. Seville, U. Tüzan and R. Clift
Hardback (0 751 40376 8), 384 pages

Pneumatic Conveying of Solids
A theoretical and practical approach
G.E. Klinzing, R.D. Marcus, F. Rizk and L.S. Leung
2nd edn, hardback (0 412 72440 5), 610 pages

Cryochemical Technology of Advanced Materials

Y.D. TRETYAKOV
Higher School of Materials Science, Moscow State University, Russia

N.N. OLEYNIKOV
Department of Chemistry, Moscow State University, Russia

and

O.A. SHLYAKHTIN
Department of Chemistry, Moscow State University, Russia

CHAPMAN & HALL
London · Weinheim · New York · Tokyo · Melbourne · Madras

Published by Chapman & Hall,
2–6 Boundary Row, London SE1 8HN

Chapman & Hall, 2–6 Boundary Row, London SE1 8HN, UK

Chapman & Hall GmbH, Pappelallee 3, 69469 Weinheim, Germany

Chapman & Hall USA, 115 Fifth Avenue, New York, NY 10003, USA

Chapman & Hall Japan, ITP-Japan, Kyowa Building, 3F, 2-2-1 Hirakawacho, Chiyoda-ku, Tokyo 102, Japan

Chapman & Hall Australia, 102 Dodds Street, South Melbourne, Victoria 3205, Australia

Chapman & Hall India, R. Seshadri, 32 Second Main Road, CIT East, Madras 600 035, India

First edition 1997

Typeset in 10/12 Times Roman by Tradespools Ltd, Frome, Somerset
Printed in Great Britain by St Edmundsbury Press, Bury St Edmunds

ISBN 0 412 63980 7

A catalogue record for this book is available from the British Library

∞ Printed on permanent acid-free text paper, manufactured in accordance with ANSI/NISO Z39.48-1992 and ANSI/NISO Z39.48-1984 (Permanence of Paper)

Contents

Preface

The global problem of the gap between ever-increasing human needs (food, clothes, habitation, health, security, spiritual requirements, and social contacts) and ever-decreasing traditional resources (mineral and energetic raw materials and pure environment) can only be solved by developing new generations of technologies and materials. About one half of the gross national product of the developed countries is related to processing natural and manufacturing artificial materials. The demand for new, promising materials grows continuously. Usually, this demand is satisfied through the search for new chemical compositions and the development of multicomponent and multiphase materials.

The principles of chemical, structural, and phase complication, when applied successfully, make it possible to achieve progress, as is visually demonstrated by new generations of ceramic superconductors and various fullerene derivatives. However, in pursuing this approach, one encounters obstacles on the way to improved materials. First, physical response is not a linear function of the chemical complexity of a system and passes through a maximum. Secondly, the more complex the material, the more difficult it is to reproduce its properties. This is especially true for the group of so-called 'frontier' materials whose properties depend on the behaviour of grain boundaries in polycrystalline structures, and of the group of functional materials used in microelectronics.

The causes of irreproducibility of the properties of materials are various. Thus, the majority of inorganic solid-state materials are non-stoichiometric compounds, and even in single-phase products, there occur composition variations that lead to variations of structure-sensitive properties. These variations are inevitable when conventional solid-state processes requiring high temperatures and prolonged syntheses are used. For instance, in the preparation of high-permeability magnetic materials based on ferrospinels by the reaction

$$Fe_2O_3 + (1 - x)ZnO + xMnCO_3 = Mn_xZn_{1-x}Fe_2O_{4+x}$$

perfect bulk homogeneity cannot be achieved under moderate heating because of diffusion limitations and at high temperatures because of selective evaporation of oxides as a result of self-dissociation or the dissociation of intermediate products formed in interaction with water followed by chemical transport. This is the reason why it was suggested that solid-state materials with maximum homogeneity and reproducible properties be prepared from high-homogeneity systems (true solutions, melts and gaseous mixtures) as precursors of multicomponent materials. Even then, however, success depends on finding precursor processing methods that make it possible to preserve initial homogeneity in the final product. This is a fairly difficult task, considering the 'incongruence' of the behaviour of the majority of multicomponent systems and technological requirements for optimum combinations of thermodynamic and kinetic factors. Fulfilling this task is the ultimate goal of cryochemical technology, the foundations of which were laid about 30 years ago in Moscow State University and have since been consistently developed at the Laboratory of Cryochemical Technology created by one of the authors of this book.

Cryochemical technology is based on a well-balanced combination of cold and heat. Cold is used to prevent undesirable changes of intermediate and final products and to deliberately control their properties. Cryochemical technology uses various non-traditional (original) processes such as cryocrystallization, sublimation drying (freeze-drying), cryoprecipitation, cryoextraction, cryo-impregnation, and cryogrinding. The fundamental and applied aspects of these processes were studied with the participation of the present authors, who in part generalized the results in *Fundamentals of Cryochemical Technology*, edited in 1987 by Vysshaya Shkola Publishers, Moscow. The authors rejected the proposal to translate the book into English because, among other reasons, of its educational character and excessive universality, which was the consequence of the desire to cover various problems of cryochemistry and cryotechnology equally thoroughly.

In the present book, the authors have concentrated their attention on the special problems of cryochemical technology of advanced inorganic materials that have been the object of their activities for many years. For various reasons, a major part of the scientific information obtained by the authors and their colleagues has not yet been published or has been published in very limited editions even within the former USSR (DSc and PhD theses, MSc projects, annual thematic reports, and proceedings of conferences).

Today, when many of the former restrictions are eliminated, writing this new monograph is an excellent opportunity to present the results of interesting investigations performed in the former USSR. It is interesting to note that the most recent review on cryochemical technology was published over ten years ago, but thanks to great interest in ceramic superconductors and other popular materials, some interesting works have recently been published in English.

The present book, which comprises four chapters and a Conclusion, is not only a summary of the studies in which the present authors participated but also a short synopsis of the major part of the published cryochemical works aimed at solving the problems of materials science by cryochemical means, mainly by freeze-drying.

Modern cryochemical technology is a peculiar conglomerate of low- and high-temperature chemical and technological approaches. We therefore considered it expedient to preface the systematic treatment of the problem with an Introduction (Chapter 1) which contains brief characteristics of techniques for obtaining low temperatures, application of cold in the technologies of pure and ultra-pure substances, the peculiarities of chemical reactions at low temperatures, and the special features of the behaviour of materials under conditions of natural or artificially created cold.

Chapter 2 describes the principal processes of cryochemical technology in the manufacture of materials, beginning with the preparation of precursors homogeneous at the atomic level and ending with transition to homogeneous solid-state precursors of the type of solid solutions, amorphous salt phases, or high-dispersity multiphase mixtures. Processing of cryochemical products, including thermolysis, phase formation, forming and sintering, are also considered.

In Chapter 3 the special features of cryochemical technologies of such advanced materials as metals, alloys, ferrites, ferroelectrics, solid electrolytes, ceramic superconductors, electrode materials, bioceramics, drugs etc. are discussed.

The problems of safety and hazards in cryochemical technology of advanced materials are discussed in Chapter 4.

Acknowledgments

The authors are deeply indebted to their colleagues and graduate and post-graduate students of the Faculty of Chemistry, Moscow State University, who at various stages and at different times contributed to the development of cryochemical technology. Special thanks are due to Drs N.V. Anastasyuk, A.V. Golovchansky, G.N. Fedotov, A.P. Mozhaev, S.V. Naumov, V.I. Pershin, R.N. Pyagai, L.A. Savchenko, V.P. Shabatin and F.Yu. Sharikov.

The authors are especially grateful to Dr Renée G. Ford and Prof. N.G. Eror (University of Pittsburgh) who initiated work on this book.

Sincere gratitude should be given to Prof. D. Klvana (Ecole Politechnique, Montreal), Prof. M. Ristich and Dr V. Vulichevich (Serbian Academy of Sciences and Art) for their kind help in gathering the literature on the cryochemical studies and to Prof. M. Jansen (University of Bonn) for his valuable support and supplying the database facilities.

We thank Dr Yu.G. Metlin, Dr V.A. Sipachev, Dr E.A. Goodilin, Mr D.I. Grigorashev, Dr S.R. Lee and Mr V.V. Ischenko, who helped us at all stages of the preparation of the manuscript.

1

Introduction

1.1 TECHNIQUES FOR ATTAINING LOW AND EXTRA-LOW TEMPERATURES

1.1.1 Laboratory methods for obtaining low temperatures

The concepts of low temperature and cold are relative. In chemical technology, the term 'cold' refers to temperatures below 273 K. Sometimes (conventionally) a distinction is drawn between moderate cold (273–120 K) and deep cold (<120 K). In low-temperature physics, the term 'deep cold' is applied to temperatures corresponding to relict radioradiation of the universe (<3 K).

To attain low temperatures in any physicochemical system, it is necessary to transfer heat to another system, which is the source of cold. The simplest natural source of cold is ice, which melts when brought into contact with a warmer body and absorbs 333 kJ of heat per 1 kg of molten ice. Clearly, with pure ice, the limiting degree of cooling corresponds to 273 K, which is close to the temperature of the invariant ice–water–vapour equilibrium.

To attain lower temperatures, ice–salt mixtures have long been used. These mixtures melt at different temperatures depending on the concentration and the nature of the salt. For any ice–salt binary system, the lowest melting temperature (eutectic) is attained at a fixed water-to-salt ratio determined by the type of the phase diagram. For instance, the eutectic point coordinates in the NH_4NO_3–H_2O system are 41.2 wt.% of NH_4NO_3 and 257 K (Figure 1.1). This means that frozen aqueous solutions of NH_4NO_3 can be used as a source of cold for cooling objects to 257 K. The compositions of several common cooling mixtures and the limiting temperatures attainable with these mixtures are listed in Tables 1.1 and 1.2.

One of the lowest temperatures attained with the help of ice-containing cooling mixtures was 210 K (Lomonosov and Braun, 1759). This mixture consisted of ice and $CaCl_2*6H_2O$.

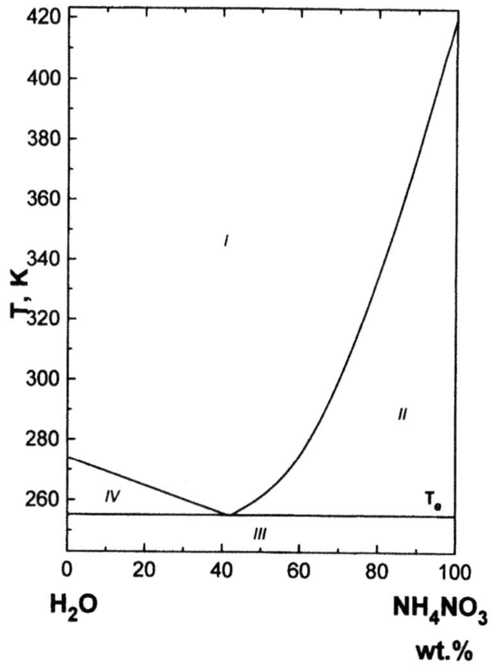

FIGURE 1.1 Phase diagram for system NH_4NO_3–H_2O: (I) aqueous solution; (II) aqueous solution + NH_4NO_3; (III) ice + NH_4NO_3; (IV) ice + aqueous solution.

TABLE 1.1 Cooling mixtures: ice + salts.

Salt	*Weight of salt per 100 g of ice*	*T*, K
$Na_2S_2O_3*H_2O$	67.5	262
KCl	30	261.9
NH_4Cl	25	257.2
NH_4NO_3	45	255.7
$NaNO_3$	59	254.5
$(NH_4)_2SO_4$	62	254
NaCl	33	251.8
$CaCl_2*6H_2O$	143	218

Still lower temperatures (200 K) can be reached using carbon dioxide ice (dry ice) mixed with alcohols or ethers. For example, a mixture of dry ice with ethanol can be used for cooling to 201 K, a similar mixture with ethyl ester gives 196 K, and mixtures with chloroform, acetone and trichloroethylene have temperatures below 196 K.

TABLE 1.2 Cooling mixtures: ice + two salts.

Salt	*Weight of salt per 100 g of ice*	*Salt*	*Weight of salt per 100 g of ice*	*T*, K
KNO_3	13.5	NH_4Cl	26	255.2
NH_4NO_3	52	$NaNO_3$	55	247.2
KNO_3	9	NH_4SCN	67	244.8
NH_4Cl	20	NaCl	40	243
NH_4NO_3	32	NH_4SCN	59	242.4
NH_4Cl	13	$NaNO_3$	37.5	242.3
KNO_3	2	KSCN	112	238.9
NH_4SCN	39.5	$NaNO_3$	54.5	235.6
NH_4NO_3	41.6	NaCl	41.6	233

Low and extra-low temperatures can be obtained in expansion of a compressed gas when work is done by the gas throttling or by using the thermoelectric phenomenon (Peltier effect), and the adiabatic demagnetization process (Rose-Innes, 1973).

1.1.2 Cooling cycles

J.P. Joule and W. Thomson have discovered that adiabatic throttling, when gas pressure decreases as a result of its passage through a porous membrane, a diaphragm, or a throttle without heat exchange with the environment, changes the temperature of the gas. The Joule–Thomson effect is considered positive if gas temperature decreases in adiabatic throttling and negative if gas temperature increases. The Joule–Thomson effect determined by intermolecular interaction forces is absent in ideal gases only. Throttling liquids rather than gases results in even more effective cooling, because additional heat is absorbed in liquid vaporization.

Cryogenic devices, i.e. equipment for attaining deep cold (<120 K), use complex cooling cycles, in particular, including evaporation and condensation processes. These cycles can be classified into three groups:

1. cycles based on the Joule–Thomson effect without doing any external work (Linde cycle);
2. cycles with adiabatic expansion of the working substance in which external work is done (Claude cycle improved by P.L. Kapitsa);
3. combined cycles involving various combinations of cycles (the above mentioned ones).

The Linde technique for obtaining low temperatures is used, for instance, for the liquefaction of air. A schematic of the unit is shown in Figure 1.2a, and the

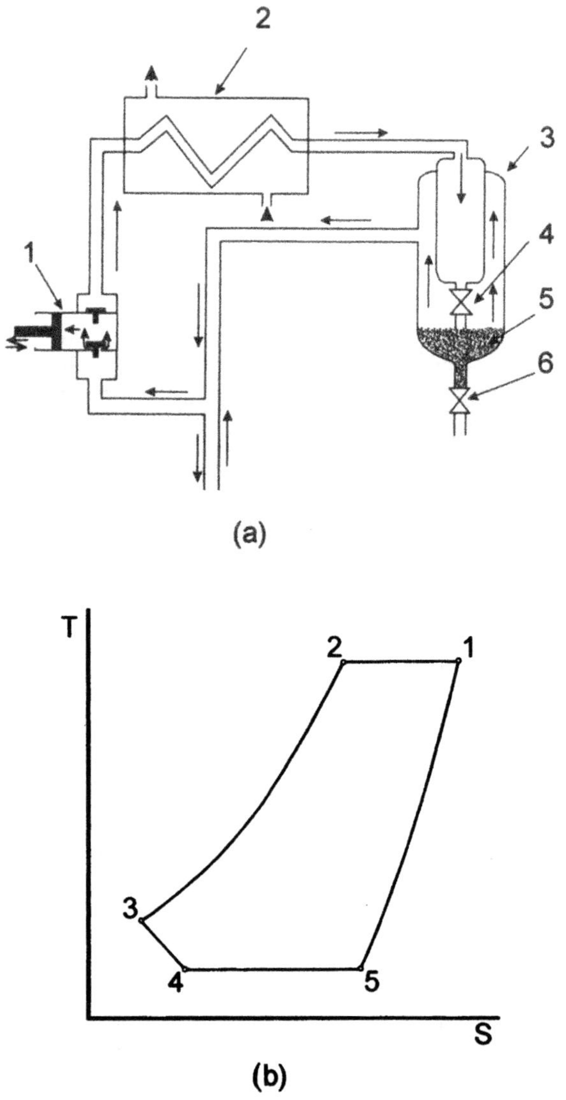

FIGURE 1.2 Linde cycle for *liquefaction* of gases: (1) compressor; (2), (3) heat exchangers; (4) throttle valve; (5) collector; (6) valve for removal of liquefied gas.

corresponding thermodynamic cycle temperature *T*–entropy *S* is illustrated by Figure 1.2b.

A gas compressed in compressor 1 is cooled by water in heat exchanger 2 (line 1–2 in Figure 1.2b) and transferred to countercurrent regenerative heat exchanger 3 where it is cooled further with the cold of counterflowing gases (curve 2–3).

Throttling the compressed gas through throttle 4 leads to its partial condensation (line 3–4). The uncondensed part of the gas is returned from heat exchanger 3 to compressor 1 (branch 5–1) or discharged in the atmosphere. The liquid air is removed from collector 5 through valve 6.

The Claude cycle, in addition, involves gas expansion in a piston expander where the gas does work and therefore cools. A schematic of the unit and the diagram of the process are shown in Figures 1.3a and 1.3b.

The Claude cycle differs from the Linde cycle in that the compressed gas in heat exchanger 3 is divided into two fluxes (point 2 in Figure 1.3b). The major

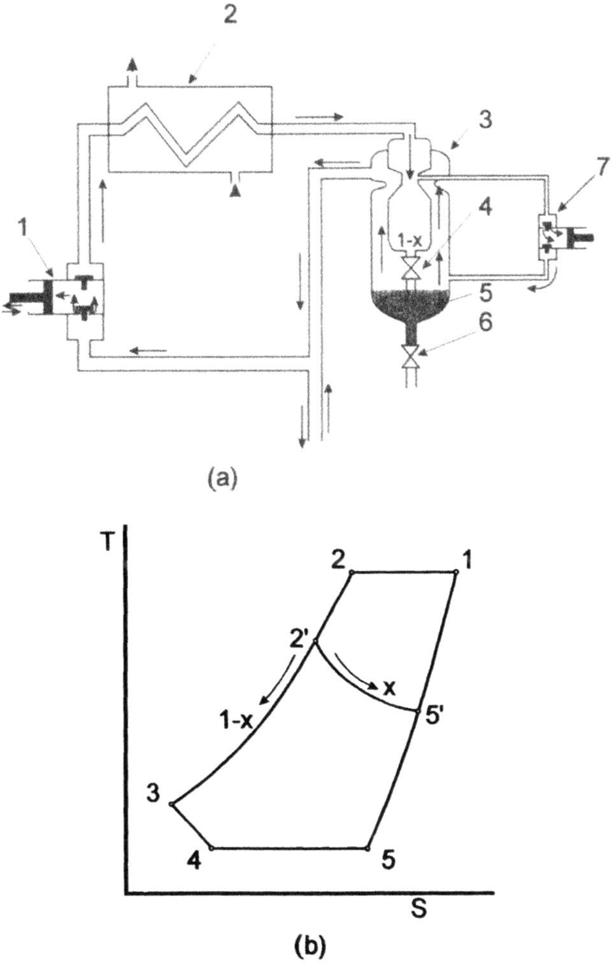

FIGURE 1.3 Claude cycle for *liquefaction* of gases: (1) compressor; (2), (3) heat exchangers; (4) throttle valve; (5) collector; (6) valve for removal of liquefied gas; (7) piston expander.

flux (up to 70%) is directed to the expander, where the gas does work and cools (line 2′–5′). The minor flux (30%) is throttled, condensed, and collected in collector 5.

P.L. Kapitsa replaced the piston expander by a turbine (turbodetander). A cooling engine with a turbodetander of the same efficiency as an engine with a piston expander is approximately five times smaller (see Kapitsa, 1964–86).

The energetic characteristics of deep-cooling cycles can only be compared under certain conditions. In large units, it is expedient to use preliminary ammonia cooling. In high-productivity units, liquid nitrogen and oxygen are obtained in a cycle with double throttling and ammonia cooling. In medium-productivity units, single-throttling cycles are used for reasons of simplicity and convenience of handling.

1.1.3 Cooling agents and cold carriers

Working fluids of cooling engines, called cooling agents or coolants, are selected by considering their physical and chemical properties and ecological and economical conditions. Vapour pressures at coolant boiling and condensation temperatures should not be lower than the atmospheric pressure, for otherwise air inflow can occur. At the same time, the pressure should not be very high, for this would require more metal to build the unit and cause coolant losses. Increasing the heat of vaporization decreases the required amount of the coolant involved in the cooling cycle. The freezing point of the coolant should be much lower than the working temperature of coolant boiling in the vaporizer. This determines the choice of the coolant depending on the temperature range of unit operation. The use of coolants with low viscosities and densities at working temperatures decreases energy expenditures for pumping coolants in a cooling or cryogenic engine. It is also necessary to take into account that many cooling agents are toxic.

The first working fluids used in cold production were water and air. At present, there exist many coolants with radically different compositions and properties.

To obtain deep cold in cryogenic techniques, liquid gases are extensively used. Table 1.3 (Malkov, 1985) lists the physical properties of several coolants.

Bearing in mind that liquefied gases are used not only as coolants for cooling engines but also as sources of cold in laboratories, we will briefly discuss problems of their storage. Liquid gases are kept in Dewar flasks, which are glass vessels with double walls coated with silver. The residual pressure between the walls is 10^{-2}–10^{-4} Pa. Less costly are vessels with a 1 Pa residual pressure and metal grain reflectors (vacuum–powder insulation) or polymeric films coated by aluminium (vacuum–laminar insulation). Vessels made of foam plastics (foam polystyrene) are also used. For simultaneously storing different gases, common reservoirs made as cylinders mounted one within another are used. The central

TABLE 1.3 Properties of liquid gases, utilized as cooling agents.

	Cooling agents					
Properties	He	H_2	N_2	O_2	Ar	NH_3
Boiling temperature, K	4.2	20.4	77.3	90.2	87.2	239.8
Temperature of triple point, K	—	14.0	62.2	54.4	83.8	—
Critical temperature, K	5.2	33.2	126.0	154.3	150.8	405.4
Critical pressure, MPa	0.23	1.3	3.5	5.0	4.9	11.5
Heat of evaporation, J/g	20.7	442.7	200.8	213.4	158.6	1368
Density, g/cm^3	0.12	0.07	0.80	1.14	—	—

cylinder is for the liquefied gas with the lowest boiling point. Gases that evaporate from interior cylinders are directed to cool exterior ones. This is especially necessary when vessels made of metals are used. It should be borne in mind that the heat of vaporization of liquid helium is an order of magnitude lower than the heats of vaporization of the other liquefied gases. Handling helium therefore requires minimizing heat inflow from the outside. For this purpose, Dewar flasks with double vacuum jackets are commonly used. Liquid helium is the source of many experimental difficulties. For instance, when air gets into Dewar vessel inlet and outlet tubes, it forms solid locks clogging the tubes. This leads to explosion of glass flasks. The experimenters working with liquid helium (or hydrogen, for which the explosion hazard is even larger than with helium) should therefore be well aware of safety regulations.

Technological objects are cooled by coolants proper or cold carriers. Cold carriers are intermediate agents carrying heat from an object to be cooled to a coolant. Cold carriers must have fairly low freezing points, high boiling points, high flare points, and the highest possible heat capacity and conductivity values. A low viscosity of a carrier at working temperatures is prerequisite for its free motion in the cooling system. As with coolants, it is important that cold carriers be stable and have low corrosivity towards the materials of the apparatus. In Table 1.4, some carriers commonly used in chemical industry and their freezing temperatures are listed.

The most commonly used carriers are aqueous or non-aqueous solutions of eutectic compositions. Aqueous solutions of ethylene glycol, commonly known as antifreezes, are extensively used as cold carriers down to 208 K. To decrease their corrosivity, additives such as triethanolaminophosphate are used. The weak point of antifreezes is a considerable increase in viscosity at low temperatures. Clearly, some coolants can also be used as cold carriers. For example, Freon-30 and Freon-11 are extensively used as cold-use carriers down to 183 and 168 K, respectively. Their advantage in comparison with alcohols used in the same temperature range is their non-explosive and non-toxic character, and their

TABLE 1.4 Freezing points of some cooling agents, utilized in technology.

Cooling agent	*T*, K
Water	273.2
Ethylenglycol	255.7
Water solution of NaCl (30.4 wt.%)	250.0
Water solution of glycerin (70 wt.%)	233.2
Water solution of $CaCl_2$ (29.9 wt.%)	218.2
Methyl alcohol	175.4
Ethyl alcohol	156.2

disadvantages are high penetrability through fittings and valves and hygroscopic properties, and their effects on the ozone layer.

1.1.4 Cooling based on Peltier effect

Among the physical phenomena that depend on interrelation between thermal and electric processes, the only one that has been used for attaining low temperatures is the Peltier effect. If a current is passed in a circuit consisting of two different metals or a metal and a semiconductor, heat is liberated at one junction and absorbed at the other.

Thermoelectric elements using the Peltier effect are widely used in Thian–Calvet microcalorimeters for compensating measured thermal effects (Calvet, 1963). At present, the sources of cold based on the Peltier effect cannot compete with compression or absorption devices either in efficiency or in temperature drop values, which do not exceed several dozen degrees. In recent years, however, interest in such cooling devices has increased considerably. They have obvious advantages, in the first place, as sources of cold in long duration experiments. Their important disadvantage is the impossibility of attaining temperatures below 220–230 K even with the help of two-stage thermoelectric batteries. Even now, however, there exist four-stage microcoolers ensuring temperature drops of 150 K and operating at 130–350 K. We can hope that, thanks to a further improvement in thermoelectric batteries and the development of new semiconductivity materials, these cooling devices will find wide applications in low-temperature chemistry.

1.2 CHEMICAL REACTIONS AT LOW TEMPERATURES (CRYOCHEMISTRY)

1.2.1 Spontaneous reactions

It is well known that increasing temperature usually increases the rate of chemical reactions. According to S. Arrhenius, the influence of heating resulted

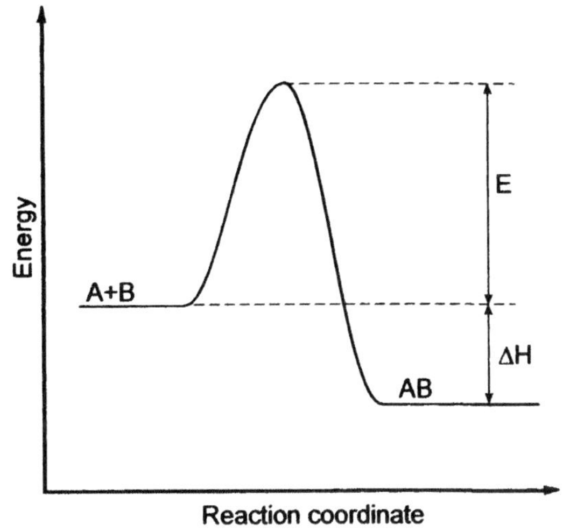

FIGURE 1.4 Schematic energy diagram of chemical reaction.

in increasing the number of active particles (molecules and radicals), i.e. particles whose collisions lead to the formation of products. The fraction of effective collisions equal to the ratio of the number of effective collisions (n_{eff}) to the total number of collisions (n) changes with temperature according to the equation

$$a = n_{\text{eff}}/n = e^{-E/RT}$$

where E is the activation energy. The difference between the E value and the thermal effect of the reaction ΔH is illustrated by the scheme given in Figure 1.4.

The activation energy corresponds to the energy barrier that should be overcome by colliding molecules for a redistribution of chemical bonds to occur. This barrier owes its existence to electrostatic repulsion between the electronic shells of the reacting molecules. As the reaction rate (v) is proportional to the number of effective collisions per unit time, we have

$$v = ke^{-E/RT}$$

where k is the rate constant for the reaction.

If a reaction does not take place at room temperature (for example, in a mixture of solid reagents) it may be initiated by heating. However, this is not always so. More than a century ago, chemical reactions were discovered, the rates of which, contrary to common views, increased on cooling rather than on heating. This implies that the E value in the Arrhenius equation is negative. Negative E values are at variance with common sense if, as previously, these values are treated as the energies of activation of molecules, which should be positive by definition.

This anomaly was given an explanation when it was found that low-temperature initiation of processes is related to the formation of thermally unstable complexes facilitating reactions in given directions. Decreasing temperature differently affects the mechanisms of concurrent reactions. At low temperatures, the most probable processes are those with the lowest activation energies. Decreasing temperature can therefore simultaneously give two desirable results. First, thanks to a change of the mechanism of the formation of the major product, its accumulation via low-temperature molecular complexes is facilitated. Secondly, side processes with (as a rule) higher activation energies are suppressed. As the end result, a high-selectivity chemical process becomes predominant.

The earliest studies of low-temperature chemical reactions were performed by J. Dewar who investigated low-temperature reactions of alkaline metals, hydrogen sulphide, and hydrogen iodide with liquid oxygen. And although much work has since been done in this direction, only few chemical reactions that occur at low temperatures spontaneously without special initiation have been discovered. In the first place, these are reactions with the participation of molecular fluorine. One example is the reaction that goes at temperatures below 85 K

$$2Na + F_2 = 2NaF$$

Another example is the interaction

$$2NO + O_2 = N_2O_4$$

The rate of this reaction increases sharply as temperature decreases in the range 83–90 K. This increase is explained on the assumption that the rate-determining step of the reaction is interaction between oxygen molecules and $(NO)_2$ dimers which are formed from nitrogen(II) oxide with evolution of heat. The process can be represented by the scheme

$$2NO = (NO)_2$$

$$(NO)_2 + O_2 = N_2O_4$$

Naturally, the concentration of dimeric molecules from nitrogen(II) oxide increases with a decreasing temperature. If the rate of this process is higher than the rate at which the number of collisions decreases with a decrease in temperature, the overall temperature coefficient will be negative, that is, the reaction will formally have a negative activation energy.

The most important feature of these low-temperature spontaneous reactions is that the energy for process initiation is taken from the system itself rather than from an external source, as is characteristic of induced reactions.

A reaction can be initiated by spontaneous energy exchange in collisions of molecules. A relevant example of a high reactivity at low temperatures is molecular fluorine characterized by an anomalously low atomization energy.

Another type of initiation is related to changes of chemical bonds in reagent molecules (N_2O_4 and N_2O_5) as a result of changes in their states caused, in turn, by the lowering temperature. An example of activation of molecules because of strengthening of interatomic interaction in binary systems is the formation of molecular complexes or charge-transfer complexes observed in the hydrohalogenation of olefins.

Self-initiation reactions are rare, and to extend the possibilities of low-temperature actions, it is necessary to use special techniques for the preparation of reaction mixtures or for reaction initiation.

In the absence of such techniques, the efficiency of low-temperature actions can nevertheless be substantially increased by combining them with high-temperature ones. For this purpose, the walls of reactors are cooled to low temperatures, which makes it possible to fix instantaneously metastable reaction products that are formed at high temperatures.

Another technique for initiating gas-phase reactions, which is frequently used in combinations with low-temperature stabilization of high-reactivity products, is electric discharge (Table 1.5).

1.2.2 Low-temperature polymerization

Low-temperature solid-state polymerization can be initiated either by irradiation or by special initiators introduced into the monomer. The first technique does not require using special experimental procedures applied in cryochemical syntheses for the preparation of the initial condensate. The monomer is first cooled, frozen, and then subjected to irradiation. To initiate solid-state polymerization, ionizing radiation with high penetrating ability through large monomer volumes is used most frequently. The action of ionizing radiation is of a very universal character because of the formation of a wide spectrum of secondary particles (radicals, ions and secondary electrons). This decreases process selectivity and sometimes makes chemists use photochemical initiation with a much lower penetrating ability.

The principal special feature of low-temperature polymerization is the presence of initial monomers in the solid crystalline or vitreous state. The mobility of molecules is reduced in these states. In addition, the crystalline state is characterized by a high degree of molecular ordering (long-range order) which is preserved in the vitreous state (short-range order). The existence of molecular

TABLE 1.5 Characteristics of the reactions initiated by electrical discharge occurring in the gaseous phase.

Starting mixture	*Active atomic particles appearing under electrical discharge*	*Temperature of condenser*, K	*Products*	*Output*, %
Br_2, N_2, O_2	O	83	BrO_2, $3NO_2$ BrO_2, N_2O_3	—
Br_2, O_2	O	83	BrO_2	about 100
HCl, O_2	O	—	Cl_2, H_2O, ClO_2	—
H_2S, O_2	O	83	H_2SO_4	70
O_2	O	77	O_3	26
O_2	O	4.2	O_3	78
H_2, O_2	H	20	H_2O_2	82
H_2, O_3 (1)	H	—	H_2O_4	20
H_2, O_2	H	77	H_2O_4	—
C_6H_6, H_2	H	77	C_6H_5	—
OsF_6, F_2	F	77–173	OsF_7	output grows
ReF_6, F_2	F	77–193	ReF_7	with the temperature

ordering and reduced molecular mobility in solid-state systems experiencing low-temperature polymerization results in fundamental changes in the character and mechanisms of the processes.

The second approach to initiating low-temperature polymerization is through introduction of initiators. The corresponding procedure can involve co-condensation of monomers and initiators from molecular beams on a cooled surface. Polymerization induced by initiators ensures obtaining purer products not contaminated by side-products formed in radiolysis. Initiators can be metals, oxides and salts.

Initiators show considerable selectivity in polymerization of monomers. As with thermally and mechanically activated processes, reaction acceleration is related to the glass–crystal, crystal I–crystal II, and crystal–melt phase transitions. The current state of the art of low-temperature polymerization is reviewed in the monograph (Benderskii *et al.*, 1994).

1.2.3 Halogenation and hydrohalogenation of olefins

Among the low-temperature reactions of hydrohalogenation of olefins, the best-studied one is the reaction between ethylene and hydrogen bromide. Under irradiation by light, the reaction follows the chain mechanism:

$$HBr + h\nu = H + \dot{B}r$$

$$HBr + \dot{H} = H_2 + \dot{B}r$$

$$\dot{B}r + C_2H_4 = C_2H_4\dot{B}r$$

$$C_2H_4\dot{B}r + HBr = C_2H_5Br + \dot{B}r$$

$$C_2H_4\dot{B}r + \dot{B}r = C_2H_4Br_2$$

This mechanism was determined by layer-by-layer freezing on the initial substances followed by photoinitiation of pure components. It is clear from the scheme that the leading role in chain propagation in the condensed state is played by bromine.

Another important feature of the reaction between ethylene and hydrogen bromide is the existence of a certain limiting dose of irradiation by light

$$D_{expl} = I\tau$$

where I is the intensity, τ is the time, and D_{expl} is the absorbed dose starting with which the reaction develops in a fast explosive mode.

It follows that for an explosive chain reaction to occur at low temperatures, it is necessary that a certain concentration of active centres be reached. Naturally, the determining role is played by the irradiation dose rather than the irradiation intensity. Note in addition that no explosion is observed when the recombination rate of active bromine atoms exceeds the rate of their formation.

Halogenation and hydrohalogenation of more complex olefins follow the same pattern as the reaction in the C_2H_4–HBr system.

1. In binary systems, unstable molecular complexes of a 1 : 1 composition are formed. These complexes have electric dipole moments, which facilitate their solvation by olefin or halogen molecules. The structure of these solvated exothermic molecular complexes appears to be similar to the transition state.
2. The reaction temperature, as a rule, corresponds to phase transitions of the glass–crystal, crystal I–crystal II, and crystal–melt types, when molecules acquire mobilities necessary for overcoming activation barriers of restructuring.
3. Low-temperature halogenation of olefins goes in one direction and gives quantitative product yields. The presence of molecular complexes in the system leads to a predominant mutual orientation of component molecules. This, in turn, results in the formation of a single product. On the contrary, the same reactions in the gas and liquid phases give both addition and substitution products. It follows that at low temperatures, halogenation of olefins occurs as a selective reaction.
4. High selectivity of chemical reactions appears to be a special feature of low-temperature interactions. For instance, chlorine is easily added to ethylene but does not react with acetylene. The addition of hydrogen bromide to isobutylene occurs in the dark, whereas the reaction with propylene goes only under the action of light. A detailed analysis of the recent advances in low-temperature halogenation and hydrohalogenation is given in the monograph (Sergeev, 1981) and review (Barelko *et al.*, 1988).

1.2.4 Synthesis of organometallic compounds

Advances in practical applications of low-temperature chemical transformations in the condensed state with the participation of organometallic compounds are tightly bound up with studies of the peculiarities of the preparation and stabilization of elements in the atomic state. Practically, two techniques for obtaining atomic metal beams are extensively used:

1. the resistive technique of high-temperature evaporation of atoms from a metal surface heated by electric current, and
2. the arc, or electron-beam technique, when a directed flow of atoms appears at the surface of a metal under the action of a beam of accelerated electrons.

Note that when technique (1) is used, metal atoms leave the surface in the ground state, whereas with technique (2) they are formed in an excited state. Both techniques give high reactivity metal condensates. An interesting reaction is reduction of bisubstituted acetylenes by ammonia in a matrix of condensed magnesium prepared by resistive evaporation:

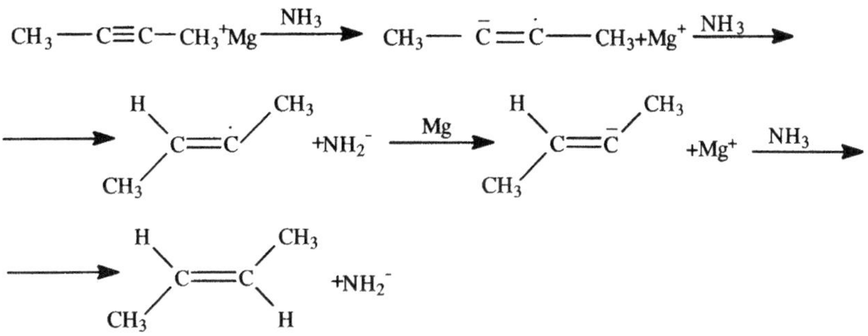

No matter what atomization technique is used, metal atoms that reach the surface of a condenser may undergo undesirable transformations, the most important of which is their 'polymerization'. To avoid this effect, metals M are condensed on a substrate together with organic ligands L at considerable dilutions [L] : [Me] = 10–100.

The degree of interaction depends on the preparation technique and the conditions of condensate treatment. At near room temperatures, the reaction of catalytic hydrogenation leads to the formation of the cis-isomer. Similar results were obtained for other metals. Thus, thermally obtained copper vapour (copper metal is the classic reagent for condensation of halides) does not react with compounds containing silicon–chlorine bonds, but copper vapour formed under bombardment by electrons does react to produce perchlorosilanes.

The differences in the mechanisms of low-temperature processes depending on the technique of the preparation of atomic metals are also observed in their interaction with alkyl halides.

Some of these reactions, especially low-temperature condensation reactions, can go in an explosive mode, if they are initiated mechanically. In other words, in [M] – [L] co-condensed systems, spontaneous chemical reactions can occur as a result of the development of an autowave, oscillating process in the system.

It was found that explosive reaction modes are favoured by a decrease of substrate heat capacity and an increase of the concentration of atomic metal in the co-condensate.

In recent years, syntheses of organometallic compounds have acquired great significance because of the possibility of wide applications of these compounds for pyrolytically obtaining films in electronics and general-purpose protective coatings. For this reason, in addition to the well-known syntheses of organometallic compounds with alkaline and alkaline-earth metals, magnesium, zinc, copper, silver, gold, aluminium and germanium, a considerable progress has been made in synthesizing transition metal compounds including iron, cobalt, nickel, chromium, molybdenum, tungsten, titanium, niobium and tantalum derivatives (Moskovits, 1989).

1.3 CRYOQUENCHING IN TECHNOLOGY OF METALLIC AND NON-METALLIC SUBSTANCES AND MATERIALS

1.3.1 Preparation of amorphous and ultradisperse crystalline materials

Cryoquenching is used for obtaining metallic and non-metallic substances and materials in two ways. One is bound up with the technology of amorphous and ultradisperse crystalline materials as target products (amorphous metals and alloys, oxides, chalcogenides, and salt compositions). The other is related to using active particles formed at high temperatures as cryochemical reagents for the preparation of organic and inorganic products with unusual properties.

At present, interest in the use of amorphous and ultradisperse materials of various types has increased greatly. Many such materials possess properties different from those of traditional substances and compositions. Amorphous materials obtained by cryoquenching may have unique semiconductivity, superconductivity, or high-resistivity properties and enhanced hardness and thermal stability. Moreover, cryoquenching techniques can be used to obtain materials as metastable solid solutions or mixtures of components and phases that do not show sufficient reciprocal solubility under equilibrium conditions and do not interact in the solid or even in the liquid state.

At present, there exist no unambiguous views on the structure of amorphous substances and glasses. However, the property common to all non-crystalline solids is the presence of short-range order in the arrangement of constituent particles (atoms, molecules and ions) and the absence of long-range order. The structure of such substances is considered in terms of two limiting models, the discrete model of microcrystallites and the continuum statistical model. According to the discrete model, an amorphous substance can be treated as a system of microscopic quasi-crystalline short-range order regions separated by intermediate more disordered regions. Such structures were observed, for instance, in $Al_{55}Cu_{45}$, $Cu_{35}Mg_{65}$, and $Ni_{77}P_{23}$ thin films.

Continuous statistical models are based on theoretical ideas according to which a regular arrangement of atoms, characteristic of a crystal lattice, is not always energetically favoured and thermodynamically stable and under certain conditions gives way to a statistical arrangement. There exists the opinion that statistical models are most suitable for describing amorphous substances with a complex chemical composition and different kinds of chemical bonds. It is obvious that the solution of the problem of the amorphous substances and materials structure is bound up with studies of the liquid state and with the development of the concept (M. Plank, Ya. Frenkel, M. Born) of a continuous transition from the liquid to the solid state.

The most important factors of amorphization of substances are those that decrease atomic or molecular mobility and prevent nucleation and growth of crystallites. In the first place, this is the character and strength of chemical bonds. It is known that substances with covalent bonds most easily form amorphous phases. Of importance are such factors as temperature, the degree of system deviation from the thermodynamic equilibrium state, gradients of concentrations and thermodynamic functions, and the presence of impurities that do not form solid solutions with the major component. The formation of amorphous substances by cryoquenching of liquids is facilitated by a high viscosity and a high activation energy of viscous flow in the supercooled state. The transition from the liquid to the amorphous state can most easily be effected for microquantities of substances (microdroplets), because this transition depends not only on kinetic, but also on thermodynamic factors. It is known that the melting point of droplets decreases with decreasing size. The influence of the size factor is especially strong when the diameter of droplets is less than 10 μm. For example, for drops with a 2 μm diameter, the melting point decreases by up to 20% of its value for the bulk sample.

At present, two procedures for obtaining amorphous materials as thin films are largely used. These are rapid quenching of small portions of a melt and quenching of condensed vapour. Other techniques include cathodic sputtering, hydrolysis of the vapour phase, combined action of high pressures and shear stresses, disordering under the action of shock waves or under neutron bombardment of crystals, and precipitation from organic and water solutions or in glow discharge. Cryoquenching is, however, the most popular technique because of its simplicity.

In obtaining amorphous phases by melt quenching, measures should be taken to remove heat as rapidly as possible from small melt portions placed on a metal surface cooled with a liquid coolant (most frequently, liquid nitrogen). There are several varieties of this technique. In the earliest studies, melt droplets were applied to the internal surface of a cooled metallic cylinder rotating at a high rate. The droplets were spread on the metal surface to produce a thin layer and rapidly solidified into an amorphous phase.

Higher cooling rates are achieved under the action of explosive waves, which shoot off a melt drop at a definite angle onto the internal surface of a rapidly rotating polished cylinder. The high velocity of the drop and the centrifugal force acting on the molten metal ensure a close thermal contact with the substrate. This technique gives cooling rates of 10^7 K/s. A schematic of a device for obtaining

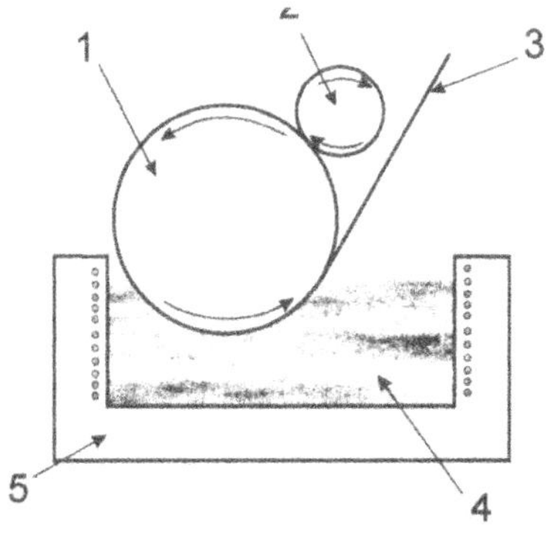

FIGURE 1.5 Device for preparation of powders and fibres of metals and alloys by high-speed melt solidification: (1) metal disk; (2) rotor; (3) fibre; (4) melt; (5) bath.

metal powders and fibres by subjecting melts to high-rate solidification is represented in Figure 1.5 (Tretyakov *et al.*, 1987).

A metal disc about 200 mm in diameter rotates at a 10^3–10^4 rpm rate. The lower part of the disc is immersed in a molten metal (or alloy) placed in bath 5. The temperature of the melt can reach 4300 K. The device works in air or under vacuum, or in an atmosphere of an inert gas to prevent melt oxidation. To increase the rate of melt cooling to 10^6–10^7 K/s, the disc is cooled by liquid nitrogen. Thanks to shrinkage, the solidified metal is spontaneously removed from the disc and transferred to a collector by the centrifugal force. The device can be used to obtain metal and alloy powders and fibres of a desired form and size (width 20–200 μm, thickness 10–100 μm, length from 500 μm to continuous fibre). The productivity of the device reaches 150 kg of powder or 350 kg of fibre per hour. Due to the high rate of cooling, metals and alloys can be obtained in the amorphous state. The high-rate solidification technique makes it possible to increase the solubility of doping elements by a factor of 10 to 100 in comparison with the solubility under equilibrium conditions.

Quenching of the vapour phase is a more effective technique for obtaining amorphous materials. For example, in the Cu–Ag system, quenching of the melt leads to the formation of ultradisperse crystalline phases only, and quenching from the vapour phase yields amorphous products. The properties of the phases obtained by this method can be changed by varying the condensation parameters, viz. the composition and the density of the molecular flow, the temperature of the substrate, the type of the substrate material, and the state of its surface. In addition, condensates can be annealed at various temperatures for various times. For example, in the Co–Cu system over a range of copper content of 12 to 96 atomic percent, equilibrium solid solutions are not formed at all temperatures, whereas metastable solid solutions can easily be obtained by co-condensing the components from vapour onto a cooled metallic surface.

An important requirement is temperature stability of cryoquenched materials. Among amorphous substances of various types, the highest temperature stability is exhibited by oxide and chalcogenide glasses. Amorphous metals and alloys show considerably lower stabilities. It is clear that the reason for this difference lies with the type of chemical bonds. Stability of non-crystalline materials increases in passing from metallic bonds to strong ionic or covalent bonds. The transition from the amorphous to the crystalline state is accompanied by a sharp decrease of the electrical resistance of the material. Thus, the resistance of metal films changes by a factor of 2 to 4, and the resistance of non-metallic films changes by several orders of magnitude. Oxide films can remain in an amorphous state after heating by several hundred degrees, whereas metal films crystallize at temperatures much lower than their melting points. For example, the crystallization temperatures of amorphous Fe, Ga, Bi and Sb films are 50, 20, 20 and 240 K, respectively, whereas the melting points of the metals are 1813, 303, 544 and 904 K. Note that the crystallization temperatures of amorphous materials cannot be treated as constants because their values depend not only on the nature of the material, but often also on the preparation procedure. Although temperature stabilities of elemental films are not high, these films are of great interest because they have elevated temperatures of transition into the superconducting state. For example, for amorphous films of Be, Ge, Bi, Al and Sn, the temperatures of transition to the superconducting state are 6–8.4(1), 8.4(1.1), 6.0(–), 5.1(1.14), 7.0(3.73) K, respectively. The transition temperatures for bulk crystalline samples are given in parentheses. (It should be emphasized that this correlates well with the well-known observation that metals with less ideal metallic order and stronger electron–phonon interaction tend to have higher T_cs.) Additional information about cryochemical syntheses and the properties of ultrafine powders can be found in a monograph (Veale, 1973) and a recent review (Sergeev, 1996).

1.3.2 Cryoquenching of active particles

Any synthesis in both preparative and matrix cryochemistry begins with obtaining active particles at high temperatures. These particles must satisfy at least three requirements. First, they must contain an atom or a group of atoms in a highly active (according to thermodynamic and kinetic criteria) state. The majority of such particles are formed at temperatures above 1273 K and pressures below 133 Pa under the conditions of thermodynamic equilibrium with the environment. Active chemical reagents can be obtained by non-thermal methods (in electric discharge or under irradiation of different types), but all these methods produce differently excited atoms and molecules. Next, for practical use of such particles, the rate of their formation should be fairly high (not less than several millimoles per hour).

At present, four classes of particles satisfying these requirements and used in preparative cryochemistry are known. These are:

- elements in the atomic state;
- molecular species containing elements in unusually low oxidation states (BF, BCl, CS, SiO, SiS, PbF_2, SiF_2, $SiCl_2$ and $SiBr_2$);
- coordination undersaturated molecular species ($FeCl_2$, CuCl, B_2O_2, ZnF_2, AlF_3, PN, TiO, KOH and KCN);
- sterically strained or unsaturated organic molecules.

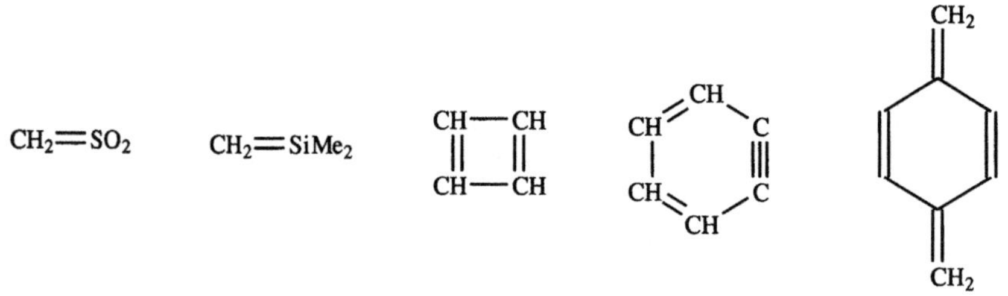

The first group of active particles, which is the most numerous one, includes atoms of chemical elements, which are more reactive than the elements themselves in the normal state. Because the condensation of isolated atoms on an inert surface at 77 K does not change their energy to any considerable degree, the atomization energy ΔH^{298}_{at} values can be used to analyse the energy characteristics of low-temperature condensation reactions. Excess energy of atoms compared with the elements in the normal state leads to changes in the character of reactions. There are no substantial changes in reaction mechanisms except when $\Delta H^{298}_{at} > 170$ kJ/mol, as, for instance, in interaction of alkaline metals with various reagents.

Changes in the reactivity of the elements caused by their atomization can be exemplified by the reactions of copper and sodium atoms with chlorine-containing compounds. The enthalpies of formation of NaCl and CuCl from the elements are −410 and −134 kJ/mol, respectively. It follows that under normal conditions, reactions of sodium with chlorine-containing reagents should be more exothermic than the corresponding reactions of copper. However, the energies of the dissociation of the Na–Cl and Cu–Cl bonds (410 and 364 kJ/mol, respectively) differ insignificantly because of the large enthalpy of formation of copper atoms in the vapour state. (The estimate of the dissociation energies of Na–Cl and Cu–Cl bonds can be done by calculations via thermochemical cycle using enthalpies of all constituting elementary processes. Enthalpy of Cu (met) evaporation (302 kJ/mole) is much higher than that for Na (90 kJ/mole) while most other enthalpies involved in the cycle calculation differ insignificantly. So, the difference of evaporation enthalpies compensates for most of the difference of enthalpies of formation from elements, and the final difference of dissociation energies becomes rather small.) Therefore, in co-condensation of copper atoms with chlorine-containing compounds on a cooled metal surface, these atoms can be as effective dechlorinating agents as sodium atoms.

The second large group of active particles formed at high temperatures comprises molecules containing elements in reduced valency states. Many of them were discovered in high-temperature spectroscopic studies of vapours of various substances. Molecular species must satisfy two requirements:

1. they should have strong chemical bonds hindering their dissociation at high temperatures;
2. they should appear in the system in ever-increasing amounts as pressure decreases and temperature increases.

Examples of the formation of such particles are given by the reactions

$$Si(s) + SiF_4(g) \xrightarrow{1523\,K,\,1.3\,Pa} 2SiF_2(g)$$

$$2B(s) + BF_3(g) \xrightarrow{1073\,K,\,133\,Pa} 3BF(g)$$

$$P_2F_4(g) \xrightarrow{1173\,K,\,0.1\,Pa} 2PF_2(g)$$

The synthetic potential of these particles under cryochemical conditions can be illustrated by the following transformations:

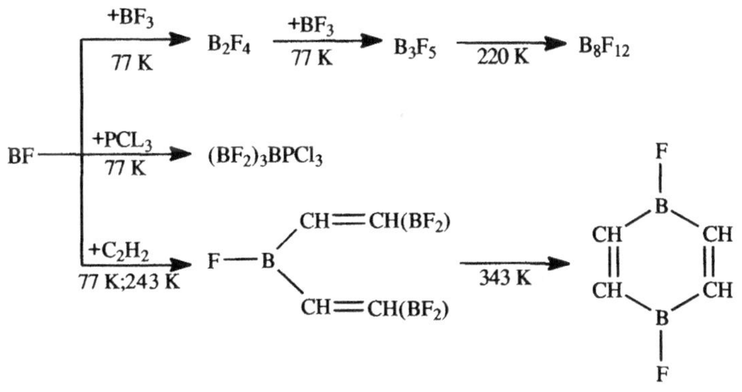

Boron monofluoride can easily be co-condensed with a large number of organic and inorganic compounds at a low temperature (77 K) to produce boron(III) compounds. These reactions compete successfully with polymerization of BF yielding $(BF)_n$. However, not all molecular particles containing elements in low oxidation states behave similarly. Many of them show a propensity for polymerization or the formation of compounds with unusual stoichiometries.

The third group of active particles includes high-temperature vapour products that contain coordination-undersaturated elements in their characteristic oxidation states (oxides, sulphides, and halides of some metals and non-metals). They are used comparatively rarely in cryochemical syntheses. There are, however, reasons to believe that their use can be effective if the difference of the Gibbs energies of the vapour and the polymerized condensed form is sufficiently large. For example, for zirconium dioxide, this difference equals 750 kJ/mol, and it can be expected that unlike crystalline zirconia, the product obtained by low-temperature condensation of $ZrO_2(g)$ will react with acids and alcohols.

The fourth group of active particles comprises organic molecules either sterically strained or characterized by high unsaturation degrees. Such molecules, usually obtained in fast pyrolysis of organic or organometallic compounds, are thermodynamically unstable, but can exist if rapidly removed from the high-temperature zone. Species of this type are also used fairly infrequently because of the diversity of pyrolysis products, low yields of required compounds, and considerable contents of side products.

Among the factors affecting low-temperature condensation reactions, an important role is played by the lifetime of particles in the gas phase. Lifetime variations (from milliseconds to several minutes) cause a variety of chemical interactions in the gas phase or in contact with a cooled surface. For example, $SiCl_2$ obtained from Si and $SiCl_4$ is a short-lived product in the gas phase with exceedingly high reactivity at low temperatures:

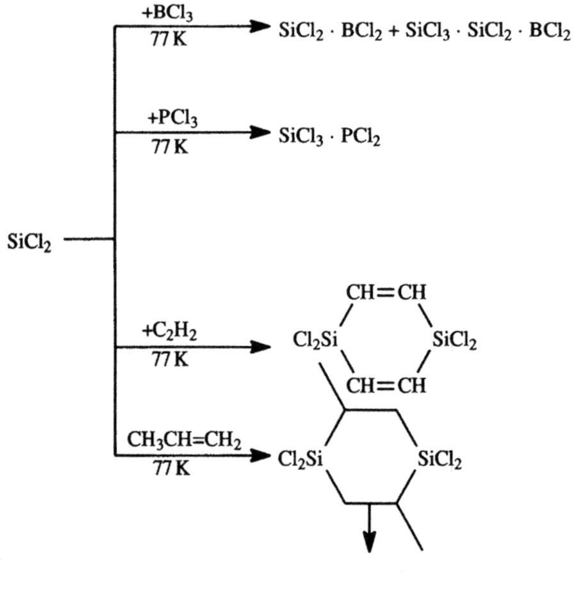

When brought in contact with solid surfaces, $SiCl_2$ readily yields $(SiCl_2)_n$. It rapidly reacts with $SiCl_4$ to produce perchlorosilanes. On the contrary, the SiF_2 molecule has a half-life of about 150 s at 13 Pa and only decomposes on a solid surface to produce $(SiF_2)_n$ but does not form perfluorosilanes.

Co-condensation on a cooled surface of high-temperature active particles and a stable chemical compound can result in the occurrence of reactions of three types.

1. Active particles polymerize and do not react chemically with the stable reagent.
2. Active particles react with the compound as monomers.
3. Active particles form dimers, trimers and various oligomers that interact with the compound more effectively than monomers.

The main factor hindering intense chemical interaction between active particles and the stable reagent co-condensed at a low temperature is polymerization. Very often, the rate of polymerization is high even at 20 K, although the reaction is diffusion-limited.

Sometimes high-temperature particles are co-condensed on a cooled surface with a large excess of an inert gas to isolate active particles in an inert matrix and slow down polymerization. After this, the second reagent is condensed on the same matrix, and the reaction occurs as a result of thermally induced diffusion of the reagents (the matrix isolation technique).

For obtaining active reagents, the following techniques can be used:

- high-temperature evaporation of liquid or solid substances; interaction in a mixture of two or several substances in the liquid or solid state at a high temperature;
- interaction of gases with solids;
- pulsed thermolysis of vapours or gases; and
- electric discharge.

Evaporation is performed from crucibles or from tungsten, tantalum or molybdenum spirals heated by electric current (the technique of a heated container). To protect heating elements and containers from interaction with vapours, various protecting coatings (for example, Al_2O_3) are used. The most frequently used procedures for obtaining gaseous elements are characterized in Table 1.6.

There are different techniques of container-free evaporation. For instance, a wire of the metal to be evaporated can be heated by electric current. There exist evaporators of various kinds based on the use of an electric arc, laser irradiation, and bombardment by electrons and ions. In bombardment by ions, the metal to be sprayed with a negative potential is bombarded by argon ions produced in an electric discharge.

TABLE 1.6 Methods for preparation of elements in the vapour state for utilizing it in cryochemical synthesis.

Element	*Method of preparation*
Aluminium	Evaporation from the TiB_2 crucibles or heated TiB_2–BN wire
Germanium	Evaporation from graphite crucibles or graphite wire heated by electrical current
Gold	Evaporation from Al_2O_3 crucibles
Silicon	Evaporation from BeO crucibles with outer heating element or rapid evaporation under electron bombardment
Lithium	Evaporation from the stainless steel crucibles
Molybdenum	Sublimation from the molybdenum wire, heated by electrical current
Nickel	Evaporation from Al_2O_3 crucibles
Titanium	Slow evaporation from the wire heated by electrical current
Carbon	Electric arc between carbon electrodes
Chromium	Sublimation from Mo, W or Al_2O_3 crucibles

For producing molecular particles by evaporating solids or liquids, or in interaction of substances in the condensed state, container techniques are used most frequently. If molecular species are formed in interaction between a solid and a gas, the reaction is performed by passing the gas over the solid heated up to temperatures above 1273 K. This technique is suitable for obtaining particles that are sufficiently stable and have low dissociation degrees, such as silicon dihalides.

In addition, techniques for obtaining active particles by high-temperature decomposition of substance vapours have been developed (for example, BCl is formed from B_2Cl_4 at 1273 K). The most frequently used techniques of the preparation of molecular species are included in Table 1.7.

The introduction of a stable reagent in the reactor, if the reagent is a solid non-volatile substance or if it decomposes during evaporation, can cause difficulties in cryochemical syntheses. Such a solid substance is dissolved in an inert and volatile solvent. The solution is then pulverized onto the walls of the reactor to produce a finely dispersed powder with a high reactivity. Non-volatile solvents with low vapour pressures at their freezing points can also be used. Active particles then interact with a solution of the reagent in the liquid or vitreous state.

At present, low-temperature co-condensation is frequently used for the preparation of many organic and organometallic compounds (Moskovits, 1989). This technique is often the only possible one, or it successfully competes with traditional techniques.

TABLE 1.7 Synthetic methods for the preparation of some molecular particles.

Active particles	*Temperature of synthesis*, K *(vacuum)*	*Synthetic method*
BF	2073	Flow of gaseous BF_3 under pressure 130 Pa over the boron placed in the graphite container
SiF_2, $SiCl_4$, $SiBr_2$	1523	Flow of the SiX_4 over the silicon, placed in the graphite or Al_3O_3 tubes under the pressure 130 Pa
SiC	1373	Heating of the Fe_1S and Si, taken over stoichiometry, in the Al_2O_3 crucibles
Metals halogenides, for example KCl, $MnCl_2$, LiF, ZnF_2	873–1273	Heating of the dry halogenides in the graphite, glassy graphite or Al_2O_3 crucibles

1.4 LOW-TEMPERATURE PROPERTIES OF SUBSTANCES AND MATERIALS

Passage to low and, especially, extra-low temperatures substantially changes the properties of many substances and materials. Some of the most visual manifestations of these changes are well known. A lead bell cooled by liquid nitrogen rings like a bronze one. A rubber sheet subjected to a similar treatment disintegrates into tiny pieces. Cooling is known to increase thermal and electrical conductivities of metals and decrease those of semiconductors. Liquid media become more viscous and turn into crystalline or vitreous solids. The loss of plasticity by metals in the cold creates serious problems in Arctic regions (cold-induced embrittlement of metals), but at the same time is beneficial for builders, because freezing of earth simplifies drilling operations considerably.

Substantial changes of the properties of solid substances and materials such as tin plague can be caused by phase transitions. (This phenomenon, well known from the early Middle Ages, involves the spontaneous break-up of tin goods and parts after continuous exposure at low temperatures (<0 °C). The reason for this is the phase transition of white (β) polymorph of tin, normally obtained by the casting of tin, into grey (α) polymorph with a diamond-like structure. The equilibrium temperature of this transition is +14–18 °C, but without activation by freezing the rate of this process is very low.) Much more dramatic changes are observed at the temperatures close to 0 K (superfluidity of liquid helium and superconductivity of some metals and alloys). These effects should be taken into account in organizing studies at low temperatures. It is worth noting that the differences of the thermophysical parameters of metals increase upon cooling to liquid helium temperatures in comparison with liquid nitrogen and, especially, room temperatures. For instance, the thermal conductivities of copper and stainless steel differ by a factor of 25 at room temperature and by three orders of magnitude at liquid helium temperature.

Let us consider patterns of cooling-induced construction property variations for substances and materials (Reed and Clark, 1983) in more detail.

1.4.1 Elastic properties

Cooling from ambient to near-zero temperatures changes the elastic constants of metals and alloys by up to 15%. Some non-metallics experience far larger changes. In all solids, stiffness (Young's moduli) increases, whereas elasticity and Poisson's ratios decrease during cooling. Temperature-induced changes of elastic constants are related to the basic elasticity of materials. Elasticity implies that atoms can move about their equilibrium positions periodically increasing and decreasing interatomic distances. In the absence of such lattice vibrations, there would be no such physical phenomena as thermal expansion and phonon thermal conductivity. Figure 1.6 illustrates the temperature behaviour of regular materials and Figures 1.7 and 1.8 show how Young's moduli and Poisson's ratios vary with temperature between 0 and 300 K for some typical metals.

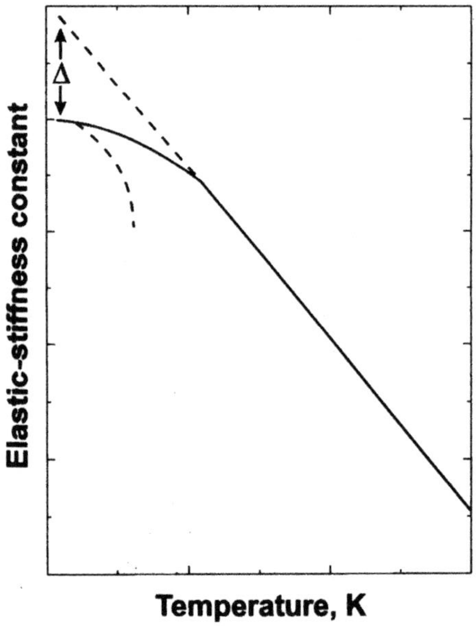

FIGURE 1.6 Regular behaviour of elastic-stiffness constant *vs.* temperature (scheme).

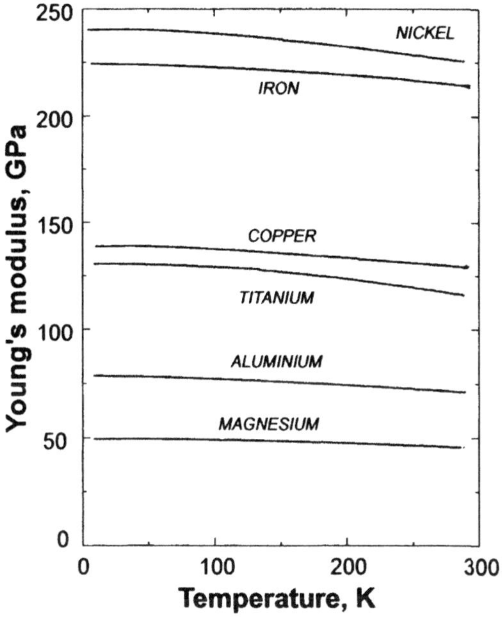

FIGURE 1.7 Temperature variation of Young's modulus for some metals.

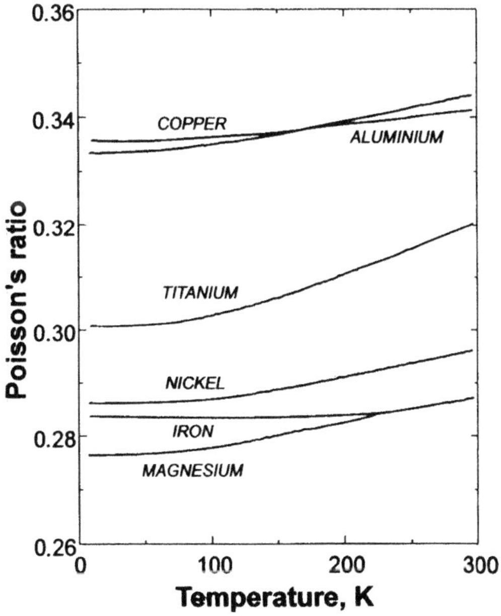

FIGURE 1.8 Temperature variation of Poisson's ratio for some metals.

1.4.2 Specific heat

Specific heat is a fundamental characteristic that relates the total heat per unit mass added to a system to the resulting change of the temperature of the system. This property has been studied for more than a century. Diverse needs have provided the impetus for this continuous study and development: specific heat experiments were used to verify Nernst's statement of the third law of thermodynamics (entropy changes approach zero at near-zero temperatures), to determine thermodynamic equilibria for chemical reactions, and for the purpose of calorimetric determinations of the energy states of substances. Specific heat is a permanent indicator of the validity of solid-state theories; these theories can be used to estimate the specific heat of a substance with varying degrees of success.

Specific heats are strong functions of temperature, particularly for $T < 200\,\mathrm{K}$ as shown in Figure 1.9. The specific heat for the semiconductor silicon, for example, changes by more than six orders of magnitude below 200 K. Typical behaviour for highly conducting metals, an amorphous material (Pyrex glass) and an organic plastic (polyvinyl chloride) is illustrated in the figure.

1.4.3 Thermal conductivity and thermal diffusivity

Figures 1.10 and 1.11 illustrate the thermal conductivity and thermal diffusivity of a few technically important materials spanning the range from pure metals to

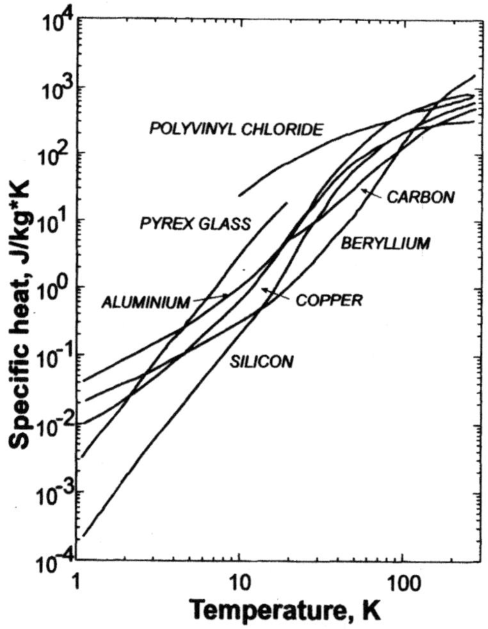

FIGURE 1.9 Specific heat as a function of temperature for several types of materials.

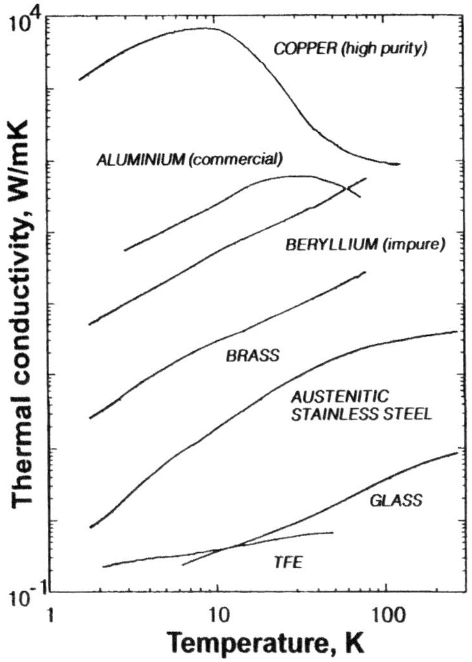

FIGURE 1.10 Temperature dependencies of thermal conductivity for some technically important materials.

structural insulating materials. These figures show only a single curve for each material, as is customarily done for simplicity's sake. Note that these are logarithmic graphs spanning about five orders of magnitude in thermal conductivity and eight in thermal diffusivity. If one also considers thermal insulation, the ranges are increased by several more orders of magnitude.

1.4.4 Thermal expansion

Most materials will contract when cooled, with a total length change of a few tenths of 1% when temperatures are reduced to near absolute zero. In general, this contraction is smooth, monotonic and similar in magnitude for similar materials. Broad bands of data are shown in Figure 1.12 for the general categories of glasses, metals, alloys, epoxies and various plastics, giving an indication of the size of the linear thermal contraction and the range of data possible for each category.

Most glasses, for example, have a small contraction, of the order of 0.1%, and the range is quite small. This is because the atomic bonds are fairly similar in their amorphous glassy structures, and are not strongly dependent on the kind of atoms.

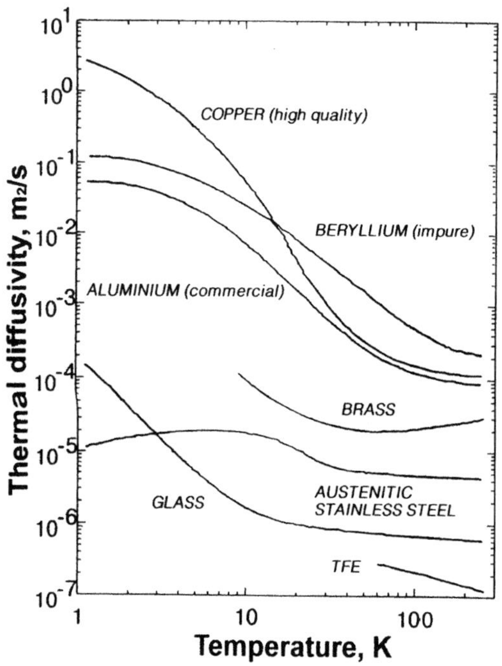

FIGURE 1.11 Temperature dependencies of thermal diffusivity for some technically important materials.

Some glasses and metals have been developed, however, to have expansions that match in order to make relatively stress-free glass-to-metal seals.

Metals, on the other hand, have a wide range of expansion, reflecting the wide variety of atomic structure. These can vary from tetragonal structure of indium, with a total contraction of about 0.7% upon reducing the temperature to near zero, to the BCC structure of chromium, with less than 0.1%. However, most alloy systems used as structural materials are based on iron, copper, aluminium, nickel or titanium, all of which have total contractions between 0.2 and 0.4%. Since alloying constituents do not alter the basic bond behaviour very much, the common alloys behave very much alike, particularly within an alloy class. One notable exception is the Fe–Ni system, which goes through the 'Invar' region near 36 wt.% Ni where the phase changes from BCC to FCC and the thermal expansion is drastically reduced.

There is a great variety of atomic bonding in the wide range of polymers in common use. Shown in Figure 1.12 are the epoxies or thermosetting resins that are an example of the strongly cross-linked polymers; they range in total contraction from 0.4 to 1.2%. This disparity in thermal expansion between components in fibre-reinforced composites, such as fibreglass/epoxy, can create large internal stresses upon cool-down.

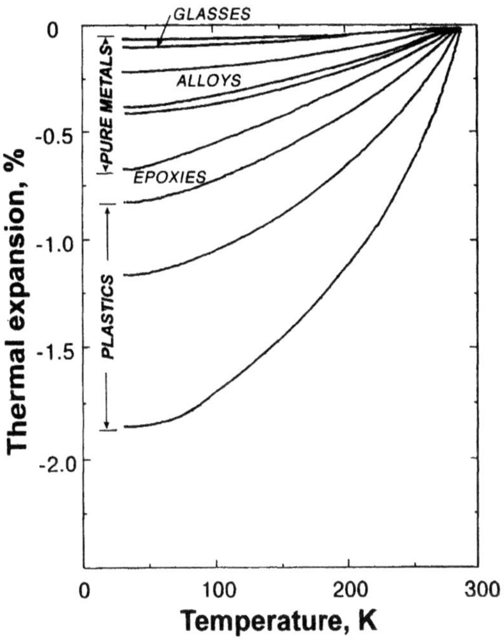

FIGURE 1.12 The range of thermal expansion values for various types of materials.

The group labelled plastics includes the more weakly cross-linked polymers, such as polytetrafluoroethylene, polyethylene terephthalate or polycarbonate. These are, in general, the more linear polymers which are frequently used as films or fibres. Their thermal expansion can be highly anisotropic. In addition, the properties of these materials frequently depend strongly on the degree of crystallinity and the value of their glass transition temperature. Total contraction to 4 K varies from 0.5 to almost 2.0%.

There are some materials with either very low or very high thermal expansions. On the low end are some special glasses and some of the 'Invar' alloys, with total contraction values of less than 0.05%. Some composites have been made utilizing the negative thermal expansion coefficient of graphite fibres to balance the positive contribution from an epoxy matrix, which results in a near-zero length change from room temperature to liquid-helium temperatures. On the high end of the expansion scale are the natural rubbers and elastomers, with total contractions to 4 K of 2 to 3%.

Thus, for most materials, the thermal expansion behaviour below room temperature is straightforward and quite small, a few tenths of a percent of contraction. In large structures, however, this can add up to a significant length change that must be accommodated. If not, internal strains on the order of the yield strains or, in some materials, even fracture strains result.

References

Barelko, V.V., Barkalov, I.M., Goldanskii, V.I. *et al.* (1988) Autowave conversions at low temperature solid state physics, *Adv. Chem. Phys.*, **74**, 339–84.

Benderskii, V.A., Makarov, D.E. and Wight, C.A. (1994) *Chemical Dynamics at Low Temperatures*, New York: John Wiley.

Calvet, E. (1963) *Recent Progress in Microcalorimetry.* New York: Macmillan.

A Handbook of Chemistry and Physics (1970) London–New York: Van Nostrand-Reinhold.

Kapitsa, P.L. (1989) Chemistry and physics of low temperatures, *Scientific Transactions*, Moscow; *Collected papers of P.L. Kapitsa,* New York: Macmillan, 1964–1986, vols 1–4.

Lomonosov, M.V. and Braun, I.A. (1959) *Experiments on Obtaining Solid Mercury and Investigating Its Properties.*

Malkov, M.I. (ed.) (1985) *Physico-chemical Fundamentals of Cryogenics: A Handbook*, Moscow.

Moskovits M. (ed.) (1989) *Chemistry and Physics of Matrix-isolated Species*, Amsterdam–New York: Elsevier Science Publ.

Reed, R.P. and Clark, A.F. (eds) (1983) *Materials at Low Temperatures*, Metals Park, Ohio: Am. Soc. Metals.

Rose-Innes, A.C. (1973) *Low Temperature Laboratory Techniques: The Use of Liquid Helium in the Laboratory*, English Universities Press.

Sergeev, G.B. (1981) *Cryochemistry,* Moscow: Mir Publ.

Sergeev, G.B. (1996) Cryochemistry of nanosize metal particles, *Chemical Physics at the Boundary of XXI Century. Dedicated to 100th Birthday of Acad. N.N. Semenov*, Moscow: Nauka.

Tretyakov, Yu.D., Olienikov, N.N. and Mozhaev, A.P. (1987) *Fundamentals of Cryochemical Technology*, Moscow: Vysshaya Shkola.

Veale, C.R. (1973) *Fine Powders: Preparation, Properties and Use*, New York: John Wiley, Halstead Press.

2

Main processes in cryochemical technology

Many key properties of prospective materials are structure sensitive. It means that the technology for their preparation should be aimed at attaining high chemical, phase and granulometric homogeneity. Relatively low efficiency of traditional 'heat and beat' technology was the reason for the development of alternative techniques based on preparation and treatment of more homogeneous precursors. These techniques may be divided into two groups based on equilibrium and non-equilibrium crystallization (precipitation) of salt precursors (Figure 2.1) (Tretyakov *et al.*, 1995).

The first group of techniques proved to be useful during the preparation of multicomponent materials (essentially ferrites) while the second group of techniques is more universal. Cryochemical technology belonging to the second group is based on the combination of low- and high-temperature processes used to avoid undesirable changes of the system (Figure 2.2).

Low-temperature processes are very different and include cryocrystallization, freeze-drying, cryoextraction, cryoprecipitation, cryoimpregnation and cryo-grinding. Besides them, cryochemical technology uses many traditional techniques carried out at room, moderate and high temperatures (Tretyakov *et al.*, 1987b).

Figure 2.3 demonstrates some possible processes of cryochemical technology. The choice of their combination depends on the chemical nature of materials, its desirable properties, available technological equipment and energetic, economic and ecological requirements for these processes. In most cases the process starts with the preparation of an initial aqueous solution where the ratio of salt concentrations corresponds to the stoichiometry of target material.

The principal question at this stage is the stability of the multicomponent solution based on compatibility of the salts used. In this case compatibility means

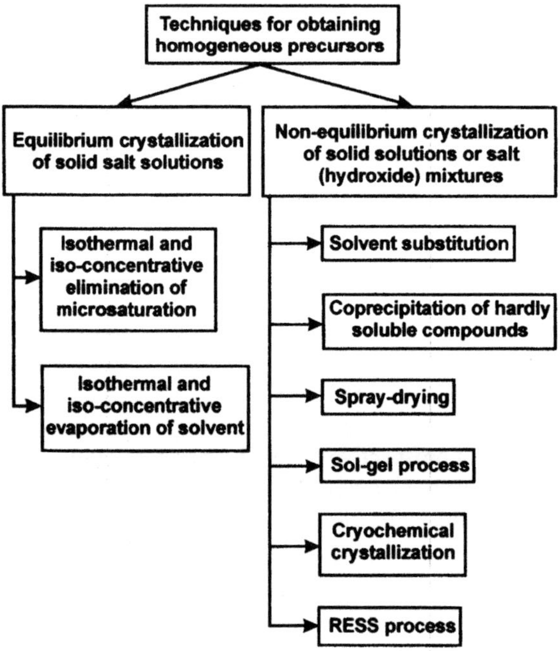

FIGURE 2.1 Classification of the methods of homogeneous precursor preparation.

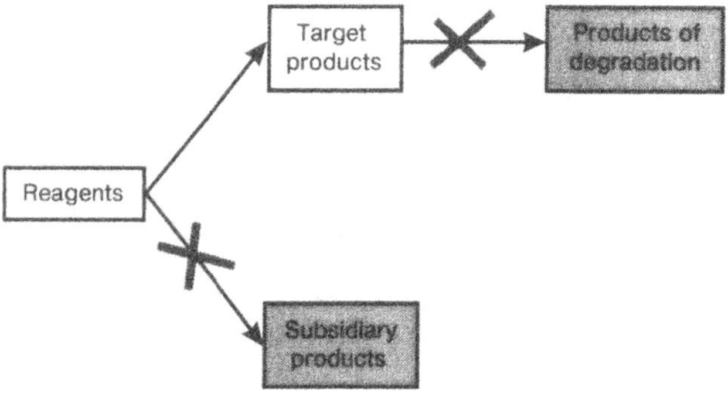

FIGURE 2.2 Combination of low and high temperatures as a way to prepare target product avoiding undesirable changes of the system.

that salts are able to form true solutions being present simultaneously. For example, $BaCl_2$ and Na_2SO_4 are incompatible when present in the same solution:

$$BaCl_2 + Na_2SO_4 = BaSO_4\downarrow + 2NaCl$$

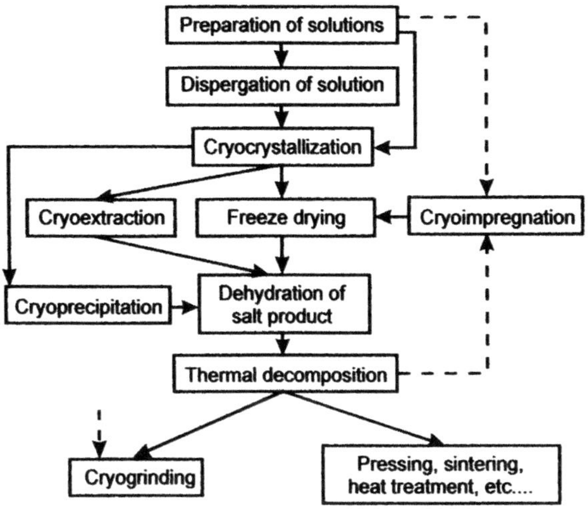

FIGURE 2.3 The main processes of cryochemical technology.

As another example, aqueous solutions of nickel, zinc and iron (II) sulphates used to synthesize nickel–zinc ferrite $Ni_{1-x}Zn_xFe_2O_4$ form true solutions at any required ratio of components. In this case the pH value should be kept in the range of 1.5–2.5. The lower pH value is accompanied by substantial deceleration of freeze-drying, the higher one – by the loss of solution stability due to intensive oxidation of Fe^{2+} and hydrolysis processes often leading to precipitation of ferrous hydroxosulphates. For these reasons solutions with $pH > 2$ cannot be kept for a long time and should be used immediately after preparation.

Another branch of cryochemical technology is based on application of colloidal solutions and suspensions. One such technique happens to be very effective in the synthesis of ferroelectric materials based on lead titanate–zirconate $Pb(Ti_{1-x}Zr_x)O_3$ (PZT). Aqueous solutions of titanyl- and zyrconyl-nitrates are known to be stable only at significant excess of nitric acid. However, such a presence of free HNO_3 in solution makes their freeze-drying practically impossible due to the formation of HNO_3–ice eutectics ($T_c = 207$ K) and the very low-temperature melting of frozen beads. This problem may be resolved by treatment of mixed solution of titanyl, zirconyl and lead nitrates with anion resin in OH^- form. The substitution of nitrate ions for hydroxyles results in the formation of stable colloidal solution which can be used for subsequent treatment.

When suspensions are used in the synthesis the level of components mixing is restricted by size of solid particles. However, this technique can also be effectively applied when one of the components is uniformly distributed not in bulk but on the surface of solid particles. The similar technique is used for synthesis of matrix-supported catalysts. For example, the substantial increase of catalytic activity of NiO catalysts utilized for oxidation of hydrocarbons is achieved by the technique of cryochemical deposition of nickel oxide layer on finely porous silicagel when suspension of silica particles in nickel sulphate solution is exploited.

Solutions and/or suspensions prepared at the first stage are subjected to cryocrystallization – fast freezing at high rates of cooling. There are several ways to carry out this process. The first and most effective one is the atomization technique yielding mono- or polydisperse microdroplets directed into a liquid cooling agent which is immiscible with water and non-reacting with salt solution. This is performed by means of different ejectors and nozzles. Fast freezing of the solution leads to the formation of frozen beads (cryogranules) sized from 0.01 to 1 mm for different devices, often having a spherical form and characterized by homogeneous distribution of salt components.

The cryocrystallization process may also be carried out by bulk freezing of the initial solution. In this case the tray or vessel with the initial solution is immersed in liquid nitrogen. The critical parameter in this technique is the small thickness of the solution layer to be frozen (5–10 mm) compared with the surface size ($0.25 \div 0.5\ m^2$). This condition is also valid for the subsequent process of freeze-drying. Among other techniques of cryocrystallization special attention should be paid to vacuum freezing the solution. In this case the solution to be frozen is sprayed directly into an evacuated vacuum chamber or the metal trays with thin layers of solution are placed in this chamber before pumping. The most complex syntheses need the technique of cryocrystallization by thin layer freezing on the metal surface precooled to liquid nitrogen temperature. This technique achieves the highest rates of freezing (10^4–10^6 K/s).

In most studies the removal of solvent (ice) from cryogranules is carried out by sublimation at sufficiently low temperatures and pressures to avoid the melting process in the macroscopic scale.

A time-saving alternative to freeze-drying is a cryoextraction technique based on treatment of cryogranules by dehydrating liquids at a sufficiently low temperature. In this case the ice is extracted by dissolution while the salt crystal framework remains practically unchanged. Another alternative often used for poorly compatible cations is the cryoprecipitation technique. Here the cryogranules are treated at low temperatures with solutions of acids, hydroxides or salts containing anions forming poorly soluble or completely insoluble products with cations of initial solution. In this case the extraction of ice is accompanied by the chemical interaction of cryogranules with extractant.

The consequent processes of dehydration and thermal decomposition lead to the preparation of finely dispersed, free-flowing and easily treated oxide powders.

The size of powder particles (crystallites) can be controlled from 0.01 to 0.5 μm by the proper choice of synthesis conditions.

Taking into account the unique role of aqueous solutions in most of the cryochemical syntheses, this chapter will be devoted to the consideration of thermodynamic and kinetic properties of water.

2.1 CRYOCRYSTALLIZATION PROCESSES

2.1.1 Some properties of water and ice

The phase diagram of water (Figure 2.4) is made up of the fields of vapour, liquid and nine fields corresponding to different polymorphs of ice (Table 2.1), though the majority of the latter ones are stable only at high pressures (Eisenberg, 1969; Fletcher, 1970).

The area concerned with the cryochemical synthesis is limited by temperatures from 77 to 373 K and pressures of up to 10^5 Pa. This part of the diagram consists of three fields of vapour, liquid and ice existence; the boundary lines (*Oa*, *Ob* and *Oc*) correspond to phase equilibria between them (Figure 2.5). The triple point *O* corresponding to the three-phase equilibrium vapour–water–ice is realized at the temperature $T = 273.15$ K and vapour pressure $p = 610.5$ Pa.

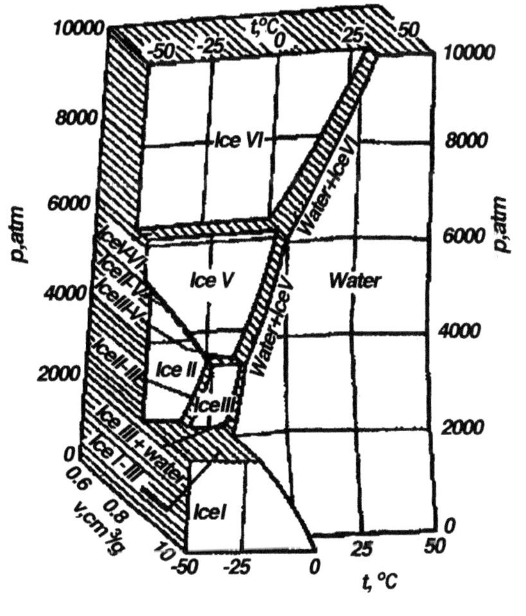

FIGURE 2.4 Three-dimensional diagram of water.

TABLE 2.1 Polymorphs of ice.

Type	*Density* (g/cm^3)	*Crystal system*	*Lattice parameters* (Å) *a*	*c*	*Number in cell*	*Proton order*
I_a	0.92	hexagonal	4.48	7.37	4	disordered
I_c	0.92	cubic	6.35	–	8	disordered
II	1.17	rhombohedral	7.78	$\alpha = 113.1°$	12	ordered
III	1.14	tetragonal	6.79	6.79	12	disordered
IV	1.28	–	–	–	–	disordered
V	1.23	monoclinic	9.22	10.35	28	disordered
			b7.54	$\beta = 109.2°$		
VI	1.31	tetragonal	6.27	5.79	10	disordered
VII	1.5	cubic	3.41	–	2	disordered
VIII	1.5	cubic	3.41	–	2	ordered
IX	1.14	tetragonal	6.79	6.79	12	ordered

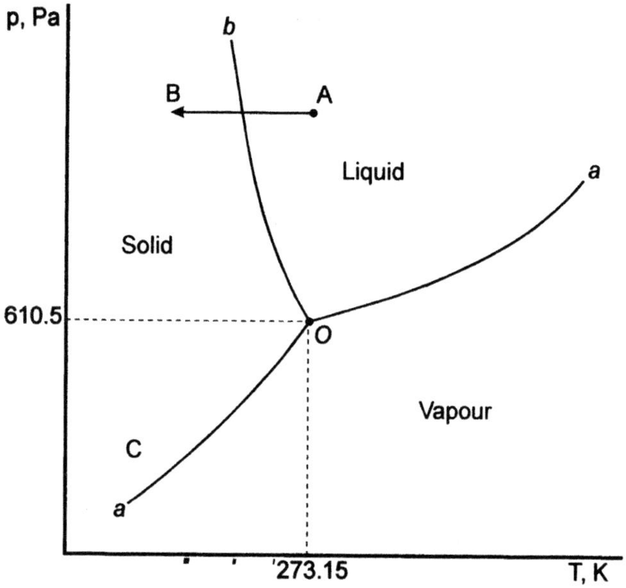

FIGURE 2.5 *p–T* phase diagram of water.

The following discussion will be devoted to the analysis of phenomena related to the phase transition water–ice (line *A–B* in Figure 2.5) which is the main process during the cooling of aqueous solutions. Processing at atmospheric pressure produced three ice polymorphs – hexagonal ice Ia, cubic ice Ic and amorphous ice.

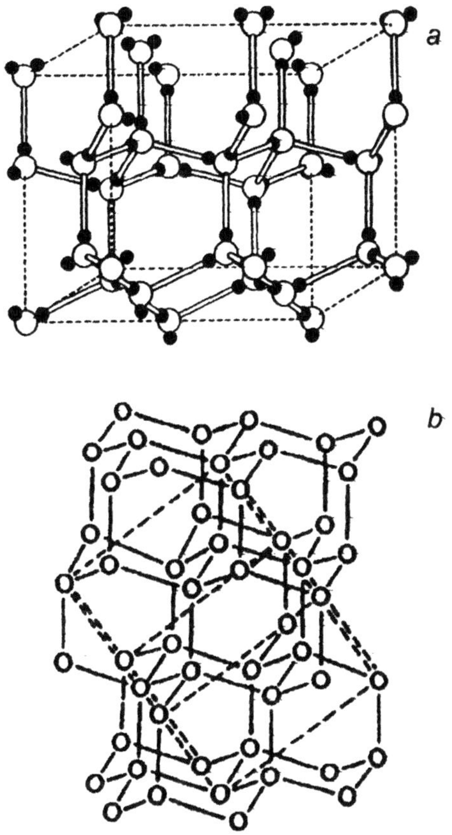

FIGURE 2.6 (a) Lattice structure of ice I_A ; (b) The arrangement of oxygen atoms in the structure of ice I_C (dashed lines are attributed to elementary cell).

The structure of hexagonal ice Ia (Figure 2.6a) is similar to the structure of wurtzite ZnS where Zn and S atoms are substituted for oxygen atoms. Each of these atoms is located in the centre of a tetrahedron formed by four identical neighbour oxygen atoms. This hexagonal structure has *D6h* symmetry. The general framework of Ia is constituted by layers perpendicular to the *c* axis and formed by hexagonal rings of water molecules. An important feature of this structure is a set of 'channels' oriented along and perpendicular to the *c* axis.

Cubic ice Ic (Figure 2.6b) can be obtained by condensation of water vapour on the cold surface at $T = 133 \div 153\,K$ or during the heating of amorphous ice or high-pressure ice polymorphs. Regardless of the preparation technique the cubic ice is converted into the hexagonal one at $T = 140 \div 210\,K$ (according to different authors, cubic ice passes into hexagonal polymorph at temperatures from 140 to 210 K; see also Skripov, 1984; Franks, 1986).

Cooling of the vapour trap surface below $T = 106$ K results in condensation of amorphous ice. The temperature of this devitrification of amorphous ice is estimated to be $131 \div 138$ K. (At real experimental conditions devitrification never proceeds at a single temperature but always within a more or less broad temperature range.) Transformation of amorphous to cubic modification is observed at $T = 140 \div 153$ K depending on its preparation history.

Indicated temperature limits of phase transitions are quite relative because of the important role played by kinetic factors. The structure of amorphous ice may be represented by the combination of disordered areas with ordered, closely packed layers similar to those of crystal modifications but having the statistical distribution of molecules within layers. The instability of disordered areas during heating leads to spontaneous crystallization of cubic phase. Crystals up to 400 nm in size fill the space between ordered areas.

Thus, fast freezing of water at conventional conditions leads to the single polymorph modification – hexagonal ice. The appearance of cubic or amorphous ice is observed only at special conditions like condensation of water vapour on cooled surfaces at very high cooling rates ($>10^6$ K/s).

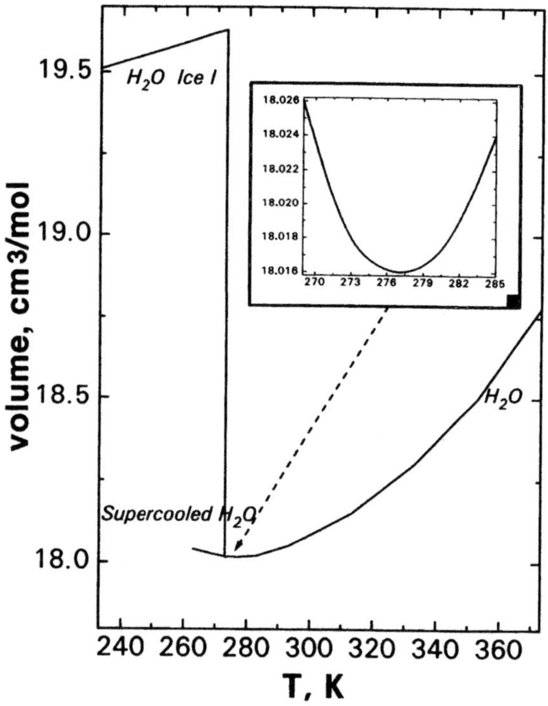

FIGURE 2.7 Specific volume of water and ice I_A *vs.* temperature.

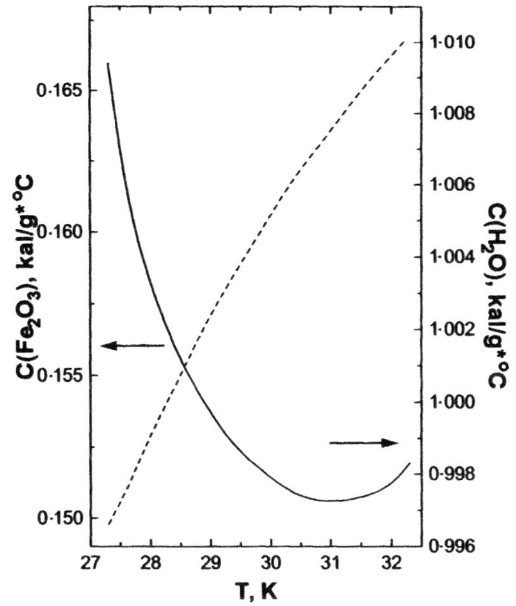

FIGURE 2.8 Heat capacity *vs.* temperature for water and ferric oxide.

Water is known to have some anomalous properties; the most familiar among them is non-linear dependence of density on the temperature (Figure 2.7). Water free of dissolved gases may be supercooled up to 200 K. Its density in this state is not decreased at $T < 273$ K but even increased monotonically with temperature.

Eleven per cent enhancement of specific volume of water during freezing results in the appearance of excess pressure reaching values up to 2500 kg/cm^2 in closed space. Water freezing temperature at normal pressure can increase above 273 K in the presence of gases forming clathrate hydrates (CH_4, chlorine, argon, xenon etc.). Water is also characterized by the abnormally high value of specific heat of melting which does not fall with temperature decrease. Another important feature that should be taken into account with respect to freeze-drying is the sevenfold excess in the evaporation heat of water over melting heat and the eightfold excess over the evaporation heat of mercury. The surface tension of water is much higher compared to that of the vast majority of liquids (alcohol, acetone, benzene etc.) with the sole exception of mercury.

Finally, the heat capacity of water is ten times higher than the values for the majority of substances except H_2 and NH_3. It also changes with temperature in the anomalous manner (Figure 2.8).

Until recent times it was customary to assume that ice can be transformed into the vapour phase, bypassing the liquid phase, i.e. via sublimation. Recent studies

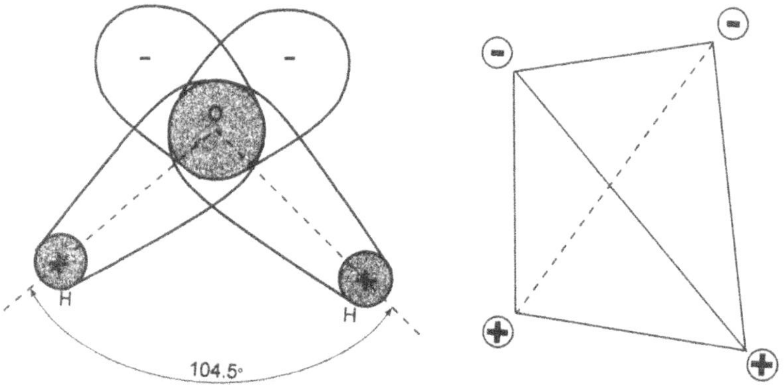

FIGURE 2.9 Geometry of charge distribution in water molecule.

demonstrated the presence of a supercooled quasiliquid film of water on the surface of ice at temperatures considerably lower than 273 K.

The angular shape of the water molecule H–O–H is the main reason for the anomalous behaviour of water and ice. Redistribution of electric charge results in the tetrahedral arrangement with two positive apices attributed to hydrogen ions and two negative ones appeared holding unshared electron pairs (Figure 2.9).

Having a permanent dipole moment of $1.83*10^{-18}$ e.s.u. cm, the water molecules tend to associate with each other up to the formation of the liquid phase at temperatures at which dipole-free molecules such as H_2S are still gaseous. The hydrogen bond O–H–O is the principal element of this condensed phase (de Quervain, 1975).

The existence of domains of ice-like structure in liquid water with increasing concentration towards the melting point is generally accepted though the nature of these domains remains underdetermined. The question is open as to whether they are comparable to molecular 'icebergs' or rather momentarily defined structural elements surrounded by distorted hydrogen bonds. Fletcher summarized the essential features of this model by postulating clusters of the size of some 10 molecules in a tetrahedral structure, surrounded by distorted hydrogen bonds as well as separated molecules, all being in a steady dynamic transformation (Fletcher, 1970). A mean lifetime of the clusters of 10^{-10}–10^{-11} s would correspond to the spectroscopic features of water. Such a model would explain the essential anomalies of water such as density, viscosity and specific heat.

2.1.2 Main features of structure and crystallization of aqueous solutions

Crystallization, i.e. formation and growth of crystals from melts, solutions or gaseous phase, is one of the well-known and thoroughly studied processes.

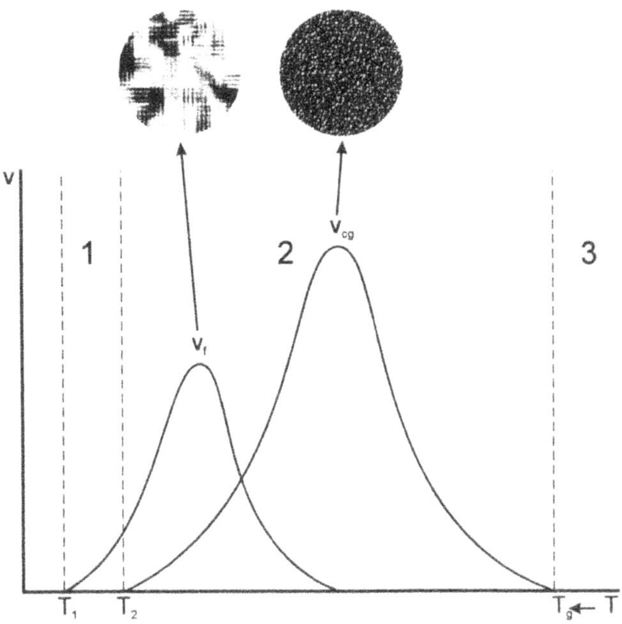

FIGURE 2.10 Temperature dependencies of nucleation rate (ν_f) and crystal growth rate (ν_{cg}): (1) liquidus temperature; (2) supercooling field; (3) field of amorphous phase.

Because extensive literature regarding this subject is available, it would be useful to specify the problems to be discussed in this section.

First, we shall consider non-equilibrium processes where diffusion in the liquid phase is rather important, while diffusion in the solid phase and on the interface between the solid and the liquid can be neglected.

Secondly, we shall pay attention to phase composition and to structural peculiarities of products formed during non-equilibrium crystallization.

The capability of liquids to supercooling was firstly noted by Fahrenheit. Then, Tamman proved experimentally and theoretically that two kinds of melt solidification – true crystallization and supercooling followed by vitrification – can occur (Tamman, 1935). He demonstrated that evolution of a system depends on the following three factors:

- nucleation rate (rate of crystallization centres generation, ν_{gf});
- rate of crystal growth (ν_{cg});
- viscosity of the cooling liquid (Figure 2.10).

Tamman regarded the main factor to be the nucleation rate because the nucleation is the necessary condition for the beginning of crystallization itself. The rate of crystal growth depends on the diffusion characteristics of the system, the surface energy of growing crystal and the temperature field around it. At very low cooling rates, crystallites of a large size are formed ($v_{cf} > v_{gf}$), due to formation of a small number of nuclei as shown in the left part of area 2 in Figure 2.10. An increase in cooling rate favours the appearance of a large number of relatively small crystallites as a result of intensive nucleation and low rate of crystal growth ($v_{gf} > v_{cg}$), as shown in the right part of area 2 in Figure 2.10. Further increase of the cooling rate and rapid increase in the viscosity of the system finally lead to the impossibility of spontaneous crystallization (area 3 in Figure 2.10) and to the formation of glass. It is obvious that at some intermediate ratio between nucleation and nuclei growth rates the system will contain both crystal and vitreous phases.

The possibility of forming vitreous phases from solution is largely determined by viscosity affecting the structural relaxation of the liquid. At high viscosity of the system existing molecular structures cannot follow changes of external conditions. Therefore, under cooling, a partial freezing of the structure occurs to corresponding higher temperature (Skripov and Koverda, 1984). Deviation from equilibrium state of the liquid increases under cooling and the relaxation rate decreases. Finally at the temperature close to some critical one, $T_g \pm \Delta T$ especially, great accumulation of non-equilibrium structural fragments occurs resulting in jump changes of temperature dependence of density, thermal expansion coefficient, heat capacity etc. At $T < T_g$ the substance behaves similarly to a non-crystalline solid. The regularities of liquid behaviour near their T_g were quite recently reconsidered by Angell (1995). As there is no sharp discrepancy between the liquid and amorphous states (short-range order for particle disposition exists in both states, but the site exchange between neighbour particles takes place only in liquids), glass is often considered as a supercooled liquid with a very high viscosity coefficient (10^{13}–10^{15} P) (Ubellode, 1982).

Let us examine processes occurring in vitreous phases under heating. At the temperature $T > T_g$ metastable liquid phase appears and its behaviour is determined by the heating conditions and the chemical nature of the substance under consideration. It was found (Beck *et al.*, 1983) that different transitions in crystal state may occur depending on the substance composition as a result of:

- **congruent crystallization**, i.e. crystallization without changes of chemical composition in metastable and stable crystal states;
- **preferential crystallization** of one phase (in this case the liquid phase loses its crystallizing component before the crystallization stops due to achievement of a new metastable equilibrium);
- **eutectic crystallization**, i.e. simultaneous crystallization of two crystal phases from solution or melt (it is characterized by the greatest moving force and can take place, for example in amorphous alloys, at any concentration; at both sides of the reaction front the components are separated between two phases).

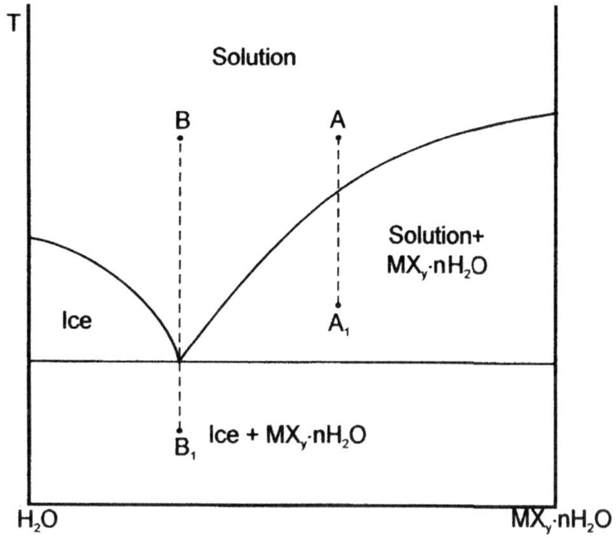

FIGURE 2.11 Schematic phase diagram of the 'water–salt' system and pathways for eutectic (BB_1 line) and preferential cryocrystallization (AA_1 line).

All kinds of transitions of amorphous substances into the crystal state have some peculiarities but at the same time they have some similarities with equilibrium crystallization. In the first place it concerns congruent and eutectic crystallization.

Preferential crystallization may be compared with extraction of one crystal phase of the binary system with a liquid phase composition which is initially very far from the eutectic one or which corresponds to a monotectic equilibrium (Kurz, 1980). Examples of different crystallization types are given in Figure 2.11.

At the same time some transitions observed in amorphous phases are not similar to equilibrium crystallization. In the first place it is the so-called explosive crystallization which occurs mainly at the centre of fluctuation origin even in the presence of impurity centres through the sample volume. (In this case the rate of homogeneous nucleation exceeds the rate of the heterogeneous one.) Secondly, it is the process of spinodal decomposition by microliquation of the amorphous phases (Beck *et al.*, 1983).

Consideration of the most general regularities of crystallization and vitrification of liquid phases would be incomplete without taking into account their connection with structural models of liquids. Low-angle X-ray scattering investigation of liquids proved the existence of short-range ordering. It was established that for any arbitrarily chosen molecule the neighbour molecules of the first and second coordination spheres are located at definite distances and the average number of neighbours can be characteristic for short-range ordering in liquids. The failure of non-structured liquid models strengthened the solid state (quasi-crystal) model (Frenkel, 1975). However, this model also has some drawbacks noted firstly by Bernall in the late 1950s. He stated that 'short-range ordering in liquids is considerably different from crystal state ordering' (Skripov and Koverda, 1984; Beck *et al.*, 1983). The simulation of tightly packed random structure has shown that there are such short-range ordering elements (for example, the presence of axial symmetry of fifth order) which exclude *ab initio* the possibility of a regular structure formation. At present this approach is intensively developed for simple liquids (Skripov and Koverda, 1984; Beck *et al.*, 1983), water (Franks, 1986), and aqueous solution (Dorosh, 1981). It should be noted that the consideration of liquid structure cannot be restricted by the configuration of the component alone. It is also necessary to take into account the dynamic properties of the liquid as manifested by its different real structure for different time intervals. Recently developed techniques of simulation (the Monte Carlo approach for calculation of equilibrium properties and the molecular dynamics approach for evaluation of kinetic characteristics) in some cases allowed investigation of the crystallization process depending on the character of the particular interaction. For the model of a supercooled rubidium particle made up of 500 atoms it was found that transitions in BCC and FCC lattices occur depending on the potential of interaction of the atoms (Beck *et al.*, 1983). It is possible that in simulation experiments one can observe not the crystallization itself but the preceding stage of the appearance of fluctuations in the supercooled system.

Accordingly, the behaviour of liquid phases under fast freezing is determined by dynamic and static changes in their structure. It is proved, firstly, by relations between the structure of glasses and corresponding liquid melts (Beck *et al.*, 1983; Dorosh, 1981). It was proposed that upon crystallization of oxide melts the first nuclei precipitated are those of the phase whose structural elements are most similar to structural elements of glasses (Bezborodov, 1981). Various static and dynamic phenomena displayed by glass-forming liquids, particularly those near the so-called 'fragile' limit, were considered as manifestations of the multidimensional complex topography of the collective potential energy function (Stillinger, 1995). These include non-Arrhenius viscosity and relaxation times, bifurcation between the α- and β-relaxation processes, and a breakdown of the Stokes–Einstein relation for self-diffusion. Primary formation of ice crystals on freezing of aqueous solutions is also connected with their structural peculiarities (Pushkar, 1977). Hence, the non-equilibrium crystallization and vitrification processes must be classed as structure-sensitive processes similar to structure-sensitive properties of solids. However, at the present time there are no experimental data to prove it.

In any case, before examining the processes of solution crystallization at high cooling rates it is necessary to consider the behaviour of the solvent (water) in the same conditions.

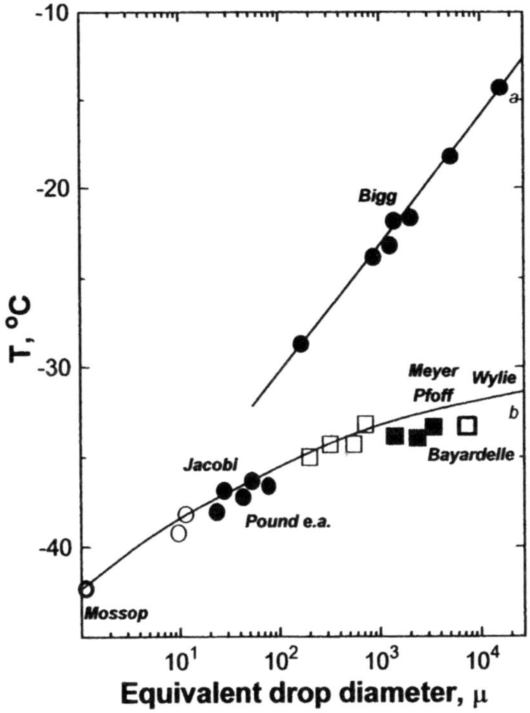

FIGURE 2.12 Dependence of cryocrystallization threshold temperature on the size of frozen water droplets: (a) heterogeneous nucleation; (b) homogeneous nucleation.

Vitrification of water is known to occur at the frequency of nucleation exceeding 10^{-2} cm^{-3} s^{-1} and the cooling rate 10^6–10^{10} K/s, kept up to the temperature $T = 0.6T_f$, where T_f is equal to 273 K. That is the reason why it was impossible to obtain amorphous ice by freezing water microdroplets in liquid nitrogen (cooling rate <10^5 K/s).

The maximum supercooling of water reached in microdroplets with a diameter of 10 μm was 233 ± 1 K (Franks, 1986). The degree of water supercooling (ΔT) relative to the equilibrium melting point (T_f) is small enough in comparison with other simple liquids having molecular and hydrogen bonding: for water $\Delta T/T_f = 0.15$, for the majority of simple liquids $\Delta T/T_f = 0.15$–0.35. This property of water is generally connected with its specific structure favouring considerably the process of spontaneous crystallization through the homogeneous nucleation (Franks, 1986; Skripov and Koverda, 1984). It should be noted that crystallization through the heterogeneous nucleation considerably decreases the value of $\Delta T/T_f =$ from 0.05 to 0.11 (see Figure 2.12). In the last case the heterogeneous 'nuclei' are, as a rule, solid impurity particles insoluble in water.

Another freezing behaviour is demonstrated by salt solutions. The presence of salt ions in water can cause several crystallization effects. Firstly, the linear crystallization rate decreases (Macklin, 1969; Lindenmeyer, 1966). Secondly, solutions can reach greater supercooling relative to water in similar conditions. Temperature decrease is connected with peculiarities of short-range hydration of ions (Rozenthal, 1972, 1973). Thirdly, ice crystals formed from aqueous solutions have different morphology depending on concentration, cooling rate and temperature, in contrast to ice crystallizing from pure water, and they always have the dendrite form (Pushkar, 1977; Franks, 1986).

The most part of dissolved substances proved to be squeezed out by the moving front of crystallizing solvent. In the restricted volume of the sample the solution is continually concentrated and remains liquid before the temperature is sufficient for its solidification. Upon rapid cooling, concentrated water solutions solidify in amorphous state (Angell and Sare, 1970; Sergeev, 1981). Studies carried out on processes of fast freezing of transition metals nitrate (Schmidt, 1976; Chaves and Garg, 1975; Vishnevskaya *et al.*, 1979; Karimova, 1981), chloride (Angell and Sare, 1970; Plachinda and Makarov, 1972; Cohen and West, 1972; Moerup *et al.*, 1976), perchlorate (Ruby *et al.*, 1971; Frohlich and Keszthelyi, 1973; Ohya and Ono, 1973) solutions confirm these conclusions. Upon freezing of diluted solutions all the authors note the formation of ice and amorphous areas containing ions of dissolved substance. The majority of investigators indicate that the concentration of dissolved salts in vitreous areas keeps a constant value independent of the initial solution concentration (Vishnevskaya *et al.*, 1979; Ruby *et al.*, 1971; Cameron *et al.*, 1973). Moreover, starting with a definite concentration the whole solution to be frozen is converted into an amorphous state (Rao *et al.*, 1973; Dupuy *et al.*, 1982; Angell and Tucker, 1980).

However, there is no definite opinion as to the quantity of salt in the amorphous phase. EPR studies of liquid nitrogen frozen solutions of $Mn(NO_3)_2$, $MnCl_2$, $MnSO_4$, $CuCl_2$, $Cr(NO_3)_3$ have demonstrated (Vishnevskaya *et al.*, 1978) that cations are concentrated in amorphous regions with constant local concentration close to the eutectic one.

Similar conclusions were made by Angell and Sare (1970), who investigated the crystallization of nitrates of Li^+, Mg^{+2}, Ca^{+2}, Cu^{+2}, Zn^{+2}, Mn^{+2}, Fe^{+2}, Co^{+2}, Ni^{+2}, Al^{+3}, Cr^{+3}, Th^{+3}, Ce^{+3}, La^{+3}. On the basis of these measurements an attempt was made (Hodge and Angell, 1978) to estimate total hydration numbers (numbers of water molecules strongly connected with cations and anions). The values of 6 to 8 for LiCl, of 17 to 20 for chlorides of bivalent cations and of 24 to 31 for chlorides of trivalent cations were obtained. Angell and Sare (1970) have established the correlation between vitrification temperature, charge and geometry of cations as well as the number of water molecules in hydrate complexes.

It was later shown (Naumov, 1982) that the composition of the amorphous phases differs considerably from the eutectic one. Characteristic DTA curves for

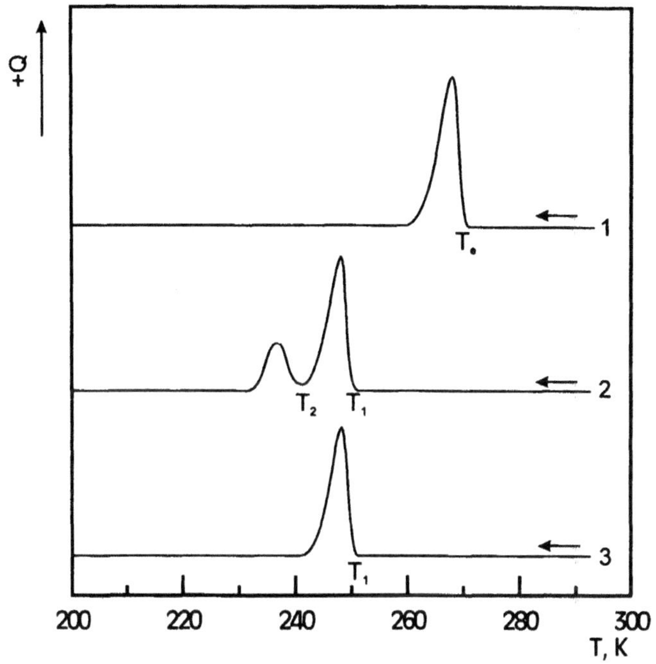

FIGURE 2.13 DTA curves for freezing of nickel sulphate solutions at different cooling rates.

solutions of nickel sulphate with eutectic concentration frozen at different cooling rates are shown in Figure 2.13.

At low cooling rates (1 K/min) an exothermic effect corresponding to the instant eutectic crystallization of the whole solution was observed (curve 1). At higher cooling rates (5 K/min) there are two exothermic processes (curve 2). The first exothermic effect is due to crystallization of a part of water with ice formation and the second one is due to crystallization of solution enriched by salt in comparison with the initial concentration. At further increase of the cooling rate (20 K/min) there is a single exothermic effect of ice crystallization whereas the solution enriched by salt does not truly crystallize but solidifies in a vitreous state.

At a very high cooling rate ($>10^4$ K/s) the whole sample is instantly converted into a vitreous state. Thus, one can separate three regions depending on the cooling rate:

1. at low cooling rates (1–5 K/min) the crystallization is realized according to quasi-equilibrium mechanism;
2. at relatively high cooling rates (5–10^4 K/min) a mixture of crystal (ice) and vitreous (salt and a part of solvent) phases is formed;

3. extra-high cooling rates (>10^4 K/min) result in transition of the whole solution in the vitreous state.

It is necessary to note that indicated intervals of cooling rate are quite conventional because of the essential dependence of cryocrystallization processes on solution concentration, the chemical nature of the salt and the presence of impurities and cryoprotectors in the solution. (A cryoprotector is a substance added to an individual liquid or solution to prevent or retard its crystallization during cooling. See also text near Figures 2.28 and 2.29 and references herein.)

It is important to clarify what occurs in vitreous phase later on (for example, under heating of frozen droplets) and to establish that the composition of the liquid phase is converted into a vitreous state.

The concentration of solution formed after quasi-equilibrium crystallization of water from the initial system was given the name 'critical concentration' which is denoted by C_c. As was expected, the value of C_c is practically independent of the initial solution concentration and the cooling rate. Therefore, the value of C_c is the characteristic parameter for each solution. In general, the behaviour of the majority of solutions on being cryocrystallized may be represented by the scheme in Figure 2.14.

Experimentally determined values of C_c are tabulated in Table 2.2, where the concentrations corresponding to eutectic solution compositions C_c are also given. It is clear that the critical concentrations differ from the eutectic ones. At the same time the values of C_c are close to calculated concentrations corresponding to the limit of the complete hydration (LCH) of solutions C_{LCH}. To facilitate comparison of C_{LCH}, C_{eut} and C_c values, the corresponding data for sulphates are given in Figure 2.15.

The term LCH (Mishchenko and Poltoratskiy, 1976) implies the electrolyte concentration at which the number of water molecules for one salt molecule is equal to the sum of coordination numbers of cation and anion. Such a solution in a structural sense may be likened to a system constituted of one-layered hydrated ions contacting directly by their hydrate shells. At the LCH the aqueous solutions of strong electrolytes are structurally uniform, being constituted by the structural elements of the single type. This model explains why the solutions with concentration close to the LCH are the most probable to vitrify: they contain no ice-like fragments of water structure and at the same time no crystalloid associations and clusters. This inhibits the formation of ice nuclei and corresponding crystallohydrates on solution freezing.

Using the technique of low-temperature optical microscopy it had been established (Naumov, 1982) that freezing of nitrate solutions of copper, nickel and cobalt with concentration below C_c results in segregation of ice and formation of vitreous phase. At concentrations higher than C_c salt crystals and

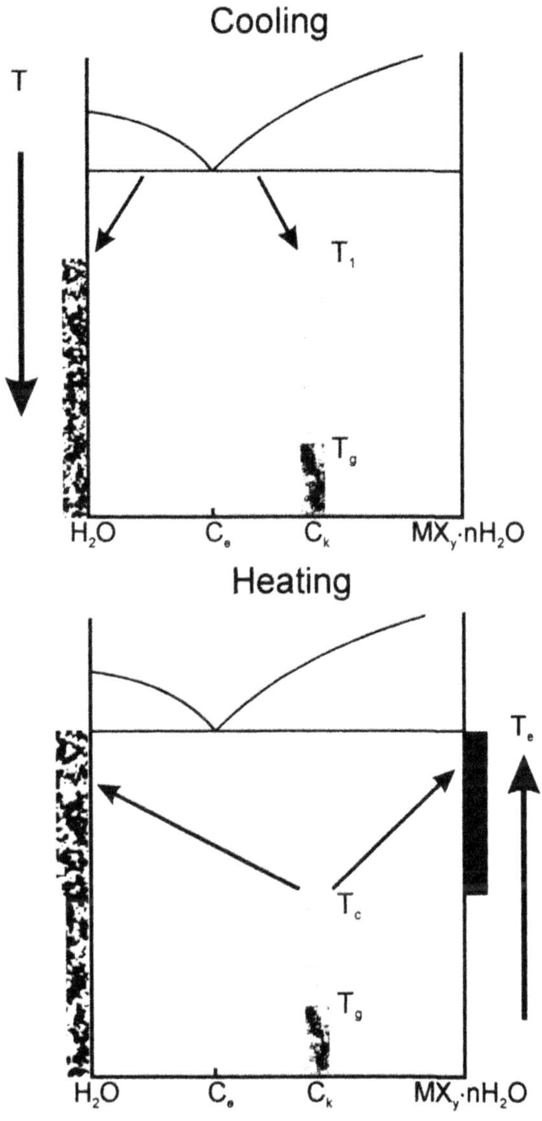

FIGURE 2.14 Scheme of solution behaviour during cryocrystallization.

vitreous phase are formed and the solution with concentration C_c is converted on cooling into a completely vitreous state.

Visual observations of cryogranulate on being heated as well as DSC and EPR data confirm that the transition of the vitreous phase into a metastable liquid occurs at devitrification temperature. Upon subsequent temperature increase the

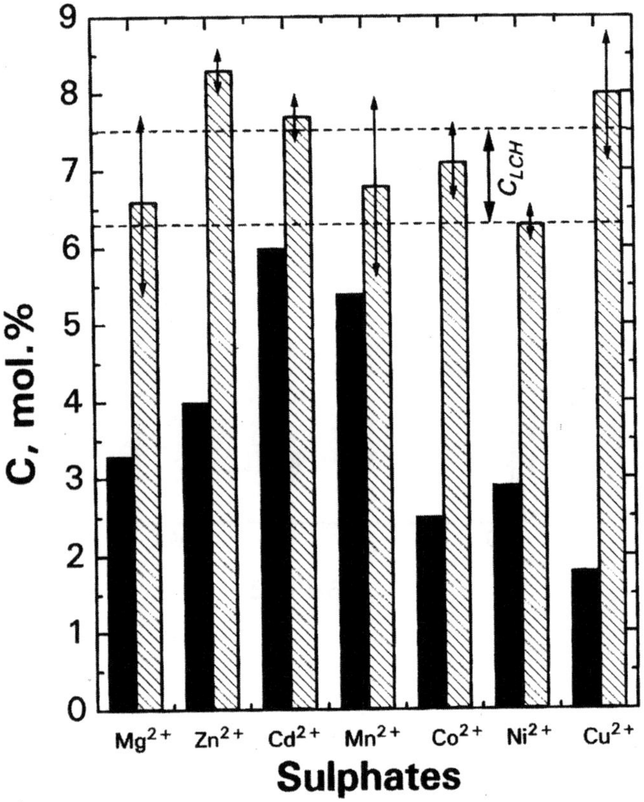

FIGURE 2.15 Characteristic concentrations for some sulphate solutions (dotted bars – C_{eut}; dashed bars – C_e).

crystallization of the liquid phase can take place with the formation of ice and a corresponding stable (or metastable) crystallohydrate. The specific feature of cryocrystallization consists of the fact that the crystallization of the salt itself takes place at temperatures considerably lower than the eutectic ones. For example, cryocrystallization takes place at temperature ranges from 230 to 250 K, 210 to 230 K and 150 to 190 K in sulphate, nitrate and chloride solutions correspondingly.

The behaviour of salt solutions varying in their chemical nature may be considerably different even under identical cooling conditions, for example, on being frozen in liquid nitrogen. The behaviour of vitreous phases is different on heating, too. Summary data pertaining to cryocrystallization processes in various water–salt systems are represented in Table 2.3.

Cooling rates of 10^3 K/min (the samples in size about $5*10^{-3}$ cm^3 were put in liquid nitrogen) do not make it possible to obtain vitreous state for many systems

TABLE 2.2 The values of characteristic concentrations of aqueous solutions of salts and coordination numbers of hydration of cations (h_K).

Salt	C_{eut}, %(mol)	C_e, %(mol)	C_{LCH}, %(mol)	h_K*
Li_2SO_4	5.77	6.9 ± 0.9	6.7–5.9	4
$MgSO_4$	3.31	6.8 ± 1.6	7.7–6.7	6
$Al_2(SO_4)_3$	1.93	2.0 ± 0.1	3.3–2.7	6
$ZnSO_4$	4.13	8.6 ± 0.3	7.7–6.7	6
$CdSO_4$	6.12	8.0 ± 0.6	7.7–6.7	6
$MnSO_4$	5.36	7.2 ± 1.6	7.7–6.7	6
$FeSO_4$	1.73	4.7 ± 1.0	7.7–6.7	6
$CoSO_4$	2.65	7.5 ± 0.6	7.7–6.7	6
$NiSO_4$	3.00	6.7 ± 0.1	7.7–6.7	6
$CuSO_4$	1.56	8.3 ± 1.4	7.7–6.7	6
$LiNO_3$	7.85	9.8 ± 0.3	11.1–9.1	4
$Mg(NO_3)_2$	5.58	6.4 ± 0.2	6.7–5.3	6
$Al(NO_3)_3$	3.58	4.1 ± 0.2	5.3–4.0	6
$Zn(NO_3)_2$	5.71	5.9 ± 0.2	6.7–5.3	6
$Cd(NO_3)_2$	4.26	6.5 ± 0.2	6.7–5.3	6
$Cr(NO_3)_3$	–	4.0 ± 0.2	5.3–4.0	6
$Co(NO_3)_2$	6.16	6.2 ± 0.2	6.7–5.3	6
$Ni(NO_3)_2$	5.25	5.7 ± 0.2	6.7–5.3	6
$Cu(NO_3)_2$	5.18	5.9 ± 0.2	6.7–5.3	6
LiCl	12.39	8.4 ± 0.6	11.1–9.1	4
$MgCl_2$	4.68	5.0 ± 0.5	6.7–5.3	6
$CoCl_2$	4.64	5.8 ± 0.2	6.7–5.3	6
$NiCl_2$	6.40	5.4 ± 0.1	6.7–5.3	6

Coordination numbers of hydration of anions are: $h_{SO_4^{2-}} = 6(+2)$, $h_{NO_3^-} = 4(+2)$, $h_{Cl^-} = 4(+2)$. Presented values show that some part of the water molecules are located far from the ion.

(solutions of salts of heavy alkaline-earth elements, potassium and ammonium chlorides, sulphates and nitrates, rare-earth sulphates). In contrast, solutions of lithium chloride and sulphate, transition-metal chlorides and rare-earth-element nitrates give a vitreous state with high probability. Vitreous phases formed in these solutions are converted into a metastable liquid which does not crystallize on heating, and prolonged isothermal heating times are necessary to obtain true crystalline products. An intermediate position is occupied by solutions of nitrates and sulphates of transition elements, solutions of salts of light alkaline-earth elements and chlorides of rare-earth elements. Therefore, the behaviour of solutions on being cryocrystallized correlates quite well with the physico-chemical nature of the cations resulting from their position in the Periodic Table. Figure 2.16 demonstrates various types of behaviour of nitrate solutions of different cations as an obvious verification of this fact.

In cryochemical synthesis of solid-state materials, as a rule, multicomponent water–salt systems are used. Are there any differences between cryocrystallization processes in binary and multicomponent systems? Is it possible to describe

TABLE 2.3 Characteristic behaviour of aqueous solutions of the salts upon rapid cooling (10^3 K/min) and following heating (5 K/min) of cryogranulate (the results characterized the behaviour of the solutions with critical concentration).

	NO_3^-		Cl^-		SO_4^{2-}	
Ion	*cooling*	*heating*	*cooling*	*heating*	*cooling*	*heating*
Li^+	G	C, M	G	–	G	–
Na^+	G	C_m, M_m	G	C, M	C	M
K^+	C	M	C	M	C	M
NH_4^+	C	M	C	M	C	M
Mg^{2+}	G	C, M	G	C, M	G	C, M
Ca^{2+}	–	–	G	C, M	–	–
Sr^{2+}	C_m	M_m	C	M	–	–
Ba^{2+}	C	M	C	M	–	–
Mn^{2+}	G	C, M	G	–	G	C, M
Fe^{2+}	G	–	–	–	G	C, M
Fe^{3+}	G	–	G	–	G	–
Co^{2+}	G	C, M	G	–	G	C, M
Ni^{2+}	G	C, M	G	–	G	C, M
Cu^{2+}	G	C, M	G	–	G	C, M
Zn^{2+}	G	C, M	G	–	G	C, M
Cd^{2+}	G	C, M	G	C, M	G	–
Al^{3+}	Gl	C, M	G	C, M	G	C, M
Pr^{3+}	G	–	G	C, M	C	M
Sm^{3+}	G	–	–		C	M
Gd^{3+}	G	–	–		C	M
Dy^{3+}	G	–	–		C	M
Ln^{3+}	G	–	–		C	M

Note: G – vitrification of the solutions, C – crystallization of the solutions upon cooling or heating, C_m – crystallization of metastable crystallohydrates, M – melting of the crystallized samples, M_m – melting of the crystallized samples containing the metastable forms of crystallohydrates.

the behaviour of the last ones on the basis of an equilibrium phase diagram and of data on cryocrystallization of binary solutions? The behaviour of multicomponent systems may be illustrated on the example of true solutions of copper and cobalt sulphates (Golovchansky, 1987). Isothermal section of the solubility diagram for the system $CoSO_4$–$CuSO_4$–H_2O and DTA curves for heating of solutions on being frozen with liquid nitrogen are given in Figures 2.17 and 2.18 respectively. By X-ray analysis it was shown that the crystallization products for metastable solutions with $x_n = 0.7$–0.9 (Figure 2.18 curve AA') are ice and salt solid solutions based on $CoSO_4$*$7H_2O$, and for a precursor with $x_n = 0.1$–0.5 the salt $CuSO_4$*$5H_2O$ crystallizes too (Figure 2.17). Experimentally founded values of C_c for this system differ considerably from additive values (Figure 2.18 curve BB'). It is clear that in cryocrystallization processes multicomponent aqueous solutions have their specific features, but it is possible to foresee their behaviour on the basis of the behaviour of binary solutions.

	I	II	III	IV	V	VI	VII	VIII			
I	^{1}H										^{2}He
II	**^{3}Li**	^{4}Be	^{5}B	^{6}C	^{7}N	^{8}O	^{9}F				^{10}Ne
III	**^{11}Na**	**^{12}Mg**	**^{13}Al**	^{14}Si	^{15}P	^{16}S	^{17}Cl				^{18}Ar
IV	^{19}K	^{20}Ca	^{21}Sc	^{22}Ti	^{23}V	^{24}Cr	**^{25}Mn**	^{26}Fe	**^{27}Co**	**^{28}Ni**	
	^{29}Cu	**^{30}Zn**	^{31}Ga	^{32}Ge	^{33}As	^{34}Se	^{35}Br				^{36}Kr
V	^{37}Rb	^{38}Sr	^{39}Y	^{40}Zr	^{41}Nb	^{42}Mo	^{43}Tc	^{44}Ru	^{45}Rh	^{46}Pd	
	^{47}Ag	**^{48}Cd**	^{49}In	^{50}Sn*	^{51}Sb	^{52}Te	^{53}I				^{54}Xe
VI	^{55}Cs	^{56}Ba	^{57}La*	^{72}Hf	^{73}Ta	^{74}W	^{75}Re	^{76}Os	^{77}Ir	^{78}Pt	
	^{79}Au	^{80}Hg	^{81}Tl	^{82}Pb	^{83}Bi	^{84}Po	^{85}At				^{86}Rn
VII	^{87}Fr	^{88}Ra	^{89}Ac**	^{104}Ku	^{105}Ns	106	107	108	109	110	

La:	^{58}Ce	^{59}Pr	^{60}Nd	^{61}Pm	^{62}Sm	^{63}Eu	^{64}Gd	^{65}Tb	^{66}Dy	^{67}Ho	^{68}Er	^{69}Tm	^{70}Yb	^{71}Lu
Ac:	^{90}Th	^{91}Pa	^{92}U	^{93}Np	^{94}Pu	^{95}Am	^{96}Cm	^{97}Bk	^{98}Cf	^{99}Es	^{100}Fm	^{101}Md	^{102}No	^{103}Lr

^{19}K - full crystallization of solution upon cooling;

^{29}Cu - crystallization of metastable liquid upon heating;

^{22}Ti - formation of glass from solution with the concentration c_k. Crystallization upon heating does not occur.

FIGURE 2.16 Crystallization behaviour of some nitrate solutions.

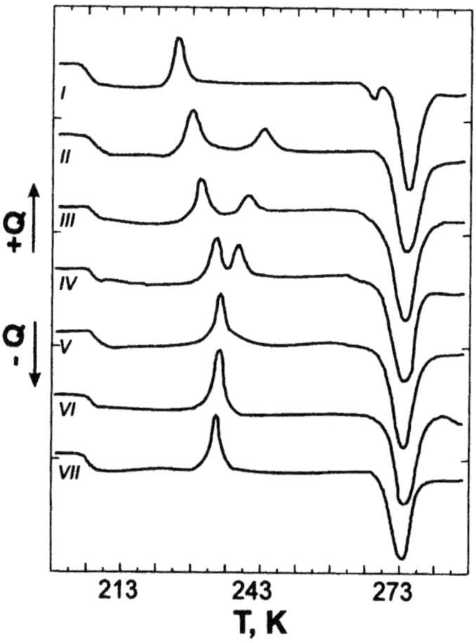

FIGURE 2.17 DTA curves for heating of $[nCoSO_4 + (1 - n)CuSO_4]$ frozen aqueous solutions: n is equal to (I) 0; (II) 0.1; (III) 0.3; (IV) 0.5; (V) 0.7; (VI) 0.9; (VII) 1.

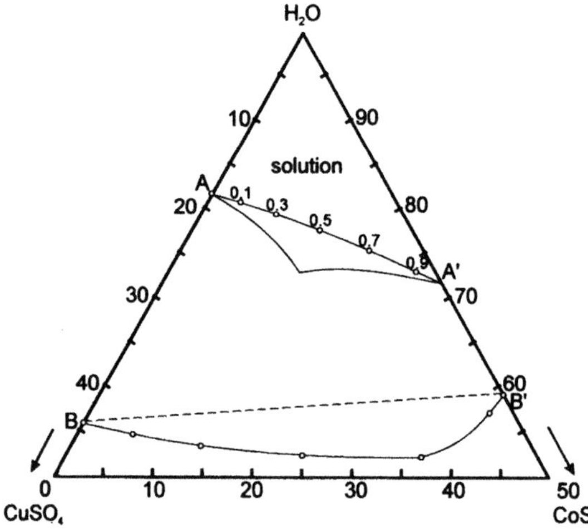

FIGURE 2.18 Solubility diagram for the system $CoSO_4$–$CuSO_4$–H_2O.

Another illustration could be the behaviour of yttrium, barium and copper nitrate solutions used in cryochemical synthesis of superconductive cuprates $YBa_2Cu_3O_{6+x}$ (Kulakov, 1989). Transfer from solution to crystal and amorphous products of freeze-drying is represented in Figure 2.19.

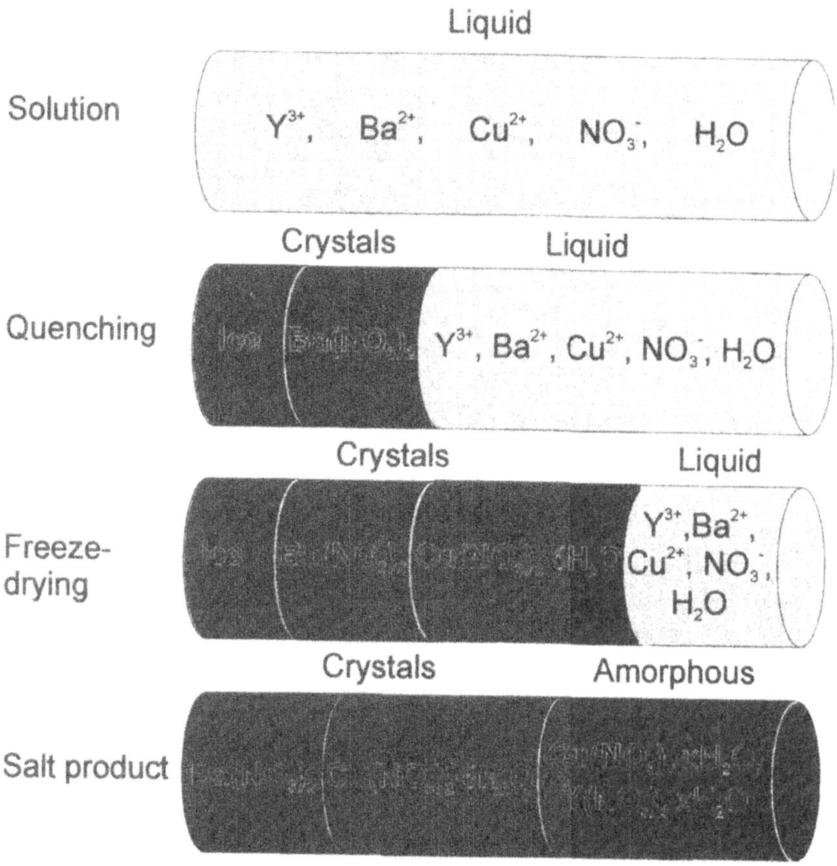

FIGURE 2.19 Crystallization behaviour of Y, Ba and Cu nitrate solution (Y/Ba/Cu = 1/2/3) during cryocrystallization and subsequent freeze-drying.

The elucidation of the nature of cryocrystallization processes is of primary importance in cryochemical technology because it helps to determine the correct method of subsequent cryogranulate treatment. If cryoproducts contain significant quantities of vitreous (or liquid) phase the performance of freeze-drying becomes quite impossible for the following reasons. Rather than simultaneous crystallization, consequent crystallization of the salt components of the system occurs upon solvent removal. This disturbs its chemical homogeneity considerably, while the initial cryogranule agglomeration during freeze-drying results in its losing its granulometric homogeneity. Therefore, an important requirement of the phase composition of cryogranules to be treated by freeze-drying is the presence of crystal phases only. To facilitate the crystallization process certain techniques are based on pH or composition of the solution changes as well as on the introduction of additives to the solution to facilitate the process of nucleation, i.e. formation of crystallization centres.

2.1.3 Preparation of initial solutions and freezing technique

The first stage of the cryochemical technology consists in the preparation of a more or less homogeneous precursor to be frozen and desolvated. An ideal precursor is a real solution of salts whose components should satisfy certain requirements. The most commonly used solvent is water and only in special cases is it replaced by non-aqueous organic or inorganic solvents (Hibbert and Tseung, 1979; Schaeffer *et al.*, 1991). Concerning the salt components forming the desired material, they are chosen taking into account the following requirements:

- high common solubility;
- easy transformation of precursor into the target product (salts, oxides, metals etc.) in the process of solvent removal, thermal decomposition or special treatment.

The last requirement is satisfied most easily by using salt with the same anion – in first turn halogenides, sulphates, nitrates and carbonic acid salts. If halides are not the end-products of cryochemical synthesis (as in the case of preparing solid electrolytes with anion or cation conductivity) they are rarely used because their transformation into oxide products necessitates high-temperature treatment with water vapour in an open system. This is not always convenient and it is accompanied by a change in the cationic composition of the multicomponent product because of the high volatility of some hydroxides and their selective chemical transportation.

$$(A_xB_yC_z)\tilde{A} + H_2O \rightarrow A_xOH + B_yOH + C_zOH + H\tilde{A}$$

where $\tilde{A} = \mathrm{Hal}$. If volatilities of A_xOH, B_yOH and C_zOH and their relative thermal stabilities are different the system behaviour is incongruent and the composition of the multicomponent product will be different from the composition of the halide precursor and, in addition, these changes will be irreproducible due to the open character of the system.

The use of sulphates, both simple and binary such as Mohr salt and alum, proved to be exclusively effective in the preparation of many ferrites (Levin *et al.*, 1978) with the exception of ferrites containing alkaline and some alkaline-earth metals (Ca, Sr, Ba). The sulphates of alkaline metals are well soluble in water but do not decompose practically into oxides and the last ones, in addition

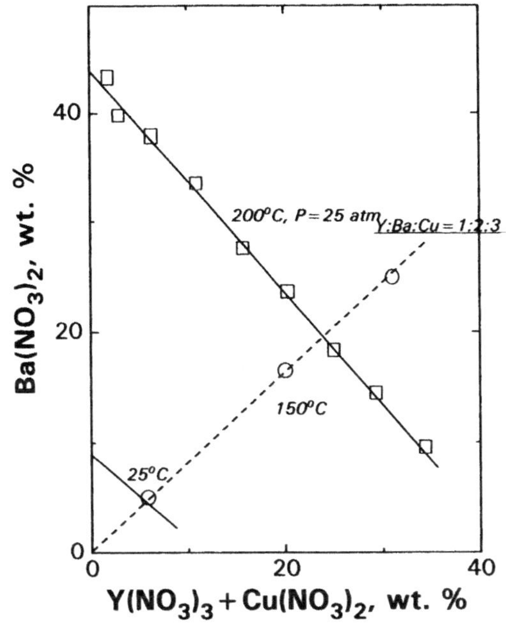

FIGURE 2.20 Solubility of $Ba(NO_3)_2$ in aqueous solutions of $[Y(NO_3)_3 + Cu(NO_3)_2]$.

to their thermal stability, are insoluble in water. For these reasons the use of sulphates in the cryochemical synthesis of an oxide superconductor proved to be impossible: the target products contain alkaline-earth metals and some of them include Tl and Pb which are unacceptable for the same reasons. Considerably more universal are nitrates (Tretyakov *et al.*, 1987b) possessing a high water solubility for practically all metals. Among the few exceptions is barium nitrate which should be replaced with barium nitrite (Mozhaev *et al.*, 1989; Zalishchansky *et al.*, 1988) or its solubility should be increased by increase of temperature and pressure (Petrenko, 1990a, 1990b) leading to a change in the cryochemical technology for so-called RESS (rapid expansion of supercritical solution) processes (Churagulov *et al.*, 1995). The solubility of $Ba(NO_3)_2$ in water solutions containing $Y(NO_3)_3$ and $Ca(NO_3)_2$ in ratios suitable for syntheses of $YBa_2Cu_3O_{7-x}$, at normal and elevated temperatures and pressures is represented in Figure 2.20.

The salts of carbonic acids – formates (Shabatin *et al.*, 1983), oxalates (Shishov, 1978), acetates (Thierauf *et al.*, 1992) etc. – proved to be similarly universal but not always appropriate precursors. All of them are highly soluble in water and decompose easily upon moderate heating, yielding metals (in the case of copper formate) or oxides with various metal oxidation degree (in the case of iron formate). A serious drawback of the salts of carbonic acids as precursors consists in the possible formation of elementary carbon or carbonates resulting in negative impact – for example, upon preparing high-frequency dielectric materials or oxide superconductors (Mozhaev *et al.*, 1989). In some cases salts with metal-containing anions (ammonium chromate, tungstate, molybdate, vanadate) are used as precursors for preparing phases containing Cr, W, Mo, V oxides. Sometimes the use of water-soluble salts as precursors creates the problem of their compatibility (Pankov *et al.*, 1982). For example upon preparation of hard magnetic materials such as barium, strontium and lead hexaferrites, the easiest way would be the use of aqueous solutions of $FeSO_4*7H_2O$ and $M(NO_3)_2*xH_2O$ (M = Ba, Sr, Pb) as the source of Fe_2O_3 and MO, respectively, but it was found to be impossible because of their irreversible interaction.

$$M(NO_3)_2 + FeSO_4 \rightarrow MSO_4 \downarrow + Fe(NO_3)_2$$

and segregation of the system as a result of complete precipitation of MSO_4 (M = Ba, Sr, Ca).

In some cases the incompatibility of salts as precursors can be overcome due to complex formation (for example, by using hexaniobic acid for cryochemical synthesis of alkaline metal niobates (Shishov, 1978)).

Turning back to nitrates, which are sometimes considered as unique universal precursors for cryochemical synthesis of advanced materials, it should be noted that this universality is quite conventional. The last remark concerns salts of multicharged ions of transition metals as precursors of many kinds of functional ceramics (Sharikov, 1991).

The main source of difficulty is the intensive hydrolysis of aqueous solutions of such nitrates resulting in the appearance of considerable amounts of free nitric acid which forms with water a series of low-temperature eutectics (−66.3 °C and lower). The latter do not allow effective sublimation of similar hydrolysed systems at technologically acceptable values of p and T. The attempts to neutralize an acid with a base lead to precipitation of the corresponding hydroxide. The solution of this problem was found (Fedotov, 1984) in cryochemical synthesis of ferroelectric materials based on lead titanates–zirconates as well as solid electrolytes with oxygen-ionic conductivity $ZrO_2(Y_2O_3)$. In both cases controlled hydrolytic polymerization was realized in aqueous solutions of $ZrO(NO_3)_2$ and $TiO(NO_3)_2$ through ion exchange with an anion exchange resin in OH^- form, which resulted in the formation of stable colloidal solutions at pH range of 2.5–4.0. These solutions are easily frozen and subjected to freeze-drying. In the process of treatment of the solution of hydrolysed salt with an anion exchange resin a partial neutralization of the solution takes place:

$$AnOH + NO_3 = AnNO_3{}^- + OH^-$$

$$OH^- + H^+ = H_2O$$

followed by formation of hydroxopolymers with high polymerization degree m

$$Me^{n+}(OH)(NO_3)_{n-1} + H_2O = Me(OH)_2(NO_3)_{n-2} + HNO_3$$
$$m[Me(OH)_2]^{(n-2)^+} = [Me_x(OH)_m]^{(n-x)m^+} + (2-x)mOH^-$$

Colloidal solutions prepared by anion exchange are considerably more stable as compared with the system formed by precipitation. Thus, the precipitation of $Fe(OH)_3$ from $Fe(NO_3)_3$ solution with ammonia takes place at pH = 2.2–2.3. At the same time a colloidal solution of $Fe(OH)_3$, obtained by an anion exchange technique, conserves its stability up to pH = 4.0. Such a difference arises from considerable decrease of the solution ionic force in the process of ionic exchange.

The anion treatment technique proved to be efficient for preparation of many advanced materials containing oxides of Fe, Al, Cr, Zr, Ti (Sharikov, 1991; Vertegel *et al*., 1995).

It was noted above that in some cryochemical syntheses non-aqueous organic and inorganic solvents are used instead of water (Hibbert and Tseung, 1979).

Three factors determine the optimum solvent for a given solute: (1) the rate of sublimation of the solvent at a given heat input; (2) the maximum depression of the freezing point, and (3) the solubility of the solute near-room temperature.

1. Solvents may be divided into two classes: those which under normal heating conditions (<1 W cm^{-2}) have a surface temperature below the melting point of the solution (class A); and those for which the heat input is limited by the melting temperature of the solution (class B).

In order to have an estimate of the so-called surface temperature, let's consider a dilute solution (e.g. an ionic salt in water) of concentration C_1, which is rapidly frozen and the solvent sublimed under vacuum. The temperature of the subliming surface (T_s) is determined by the heat input and characteristics of the solvent and remains approximately constant during sublimation. It is assumed that at the concentration C_1, $T_s < T_m$ where T_m is the melting point. As the drying process continues, the loss of solvent leaves the remaining solid solution notionally more concentrated, until at some new concentration C_2, melting temperature T_m', $T_s = T_m'$ and the frozen solution may melt. It is possible, however, that the concentration of the bulk frozen solution remains at C_1, with solid solute separating out. In this case, the melting point of the frozen solution will always remain at that corresponding to the initial concentration (T_m). This is shown in Figure 2.21.

The dashed line T_s represents the temperature of the surface during drying, and A–A' is the constant melting point of the solution discussed in the second case, in which the dried solute does not interact with the solid solution. The line A–B–C is the equilibrium freezing point curve with eutectic C. Thus if the solution becomes more concentrated during sublimation (i.e. it reaches equilibrium), melting will

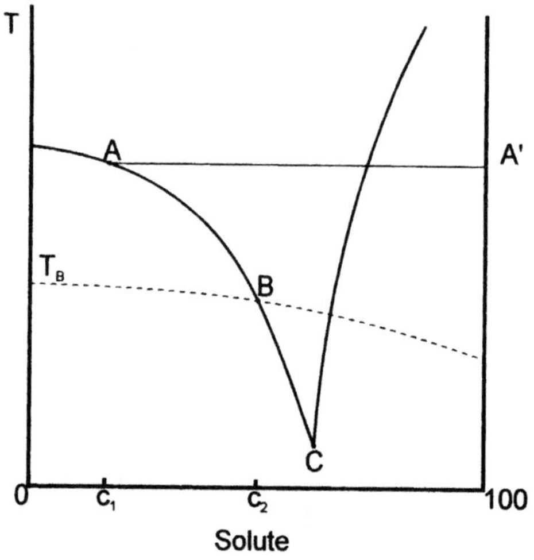

FIGURE 2.21 Surface temperature of sublimed solution during freeze-drying.

occur at *B*. It is seen that to avoid melting at any stage, the line of T_s should not cross the freezing-point curve.

Of the different solvents acetic acid is a class A solvent, and ethanol and glycerol are class B solvents. Water and DMSO are normally class A, but for solutes which cause a great depression of the freezing point (e.g. $CaCl_2$ in water) they may fall into class B. For solvents in class A, the rate of sublimation at a given heat input is solely determined by the heat of sublimation. Thus the relative rates of sublimation of acetic acid, DMSO and water seen in Figure 2.22 are explained. Class B solvents, however, are best compared by their equilibrium vapour pressures at the melting point of the frozen solutions, as discussed above.

2. The importance of the freezing-point depression is seen in the use of water as a solvent. For many solutes water acts as a class A solvent and the possible heat input is not limited. However, a solute which causes a great depression in the freezing point may effectively rule out water as a solvent. For example, the equilibrium vapour pressure of ice at −52 °C (the calcium chloride/water eutectic point) is 0.023 mm Hg, which may be compared with the value at 0 °C of 4.58 mm Hg, a difference of some 200 times.

3. The solvent with the greatest ability to dissolve a given solute is preferred when considering this factor. However, it should be noted that a high solubility may also lead to an increased lowering of the melting point.

Experience has shown that although acetic acid cannot dissolve as many ionic salts as water, the increase in sublimation rate outweighs this factor, and acetic acid has proved in some cases to be a most useful solvent.

It should be noted that solutions used as cryochemical precursors are as a rule (but not always) subjected to freezing in the form of small droplets in order to

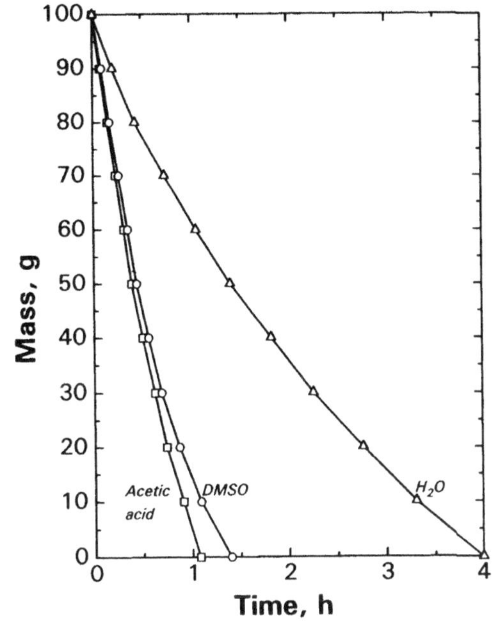

FIGURE 2.22 Calculated curves for drying of solvent (100 g), sublimed *in vacuo* at heat supply of 30 W.

increase the cooling rate and to prevent separation in multicomponent systems in the process of freezing. In practice, to carry out the process of dispersion and formation of cryodroplets, the different kinds of spray facilities are used. The simplest among them are nozzles acting on the basis of pneumatic dispersion of jet. The dispersion of the jet into droplets occurs at the jet output of the sprayer under the action of external and internal forces. External (aerodynamic) forces include those of interaction of the jet to be dispersed with the medium into which it is injected. They also include forces occurring by meeting of the jet with a liquid cooling agent. Internal forces include molecular forces and the stream

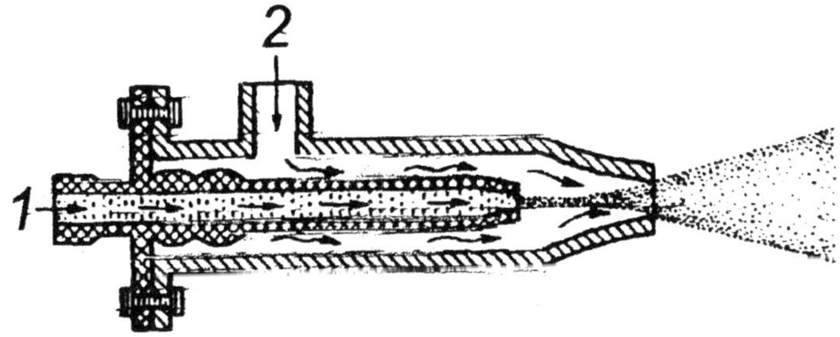

FIGURE 2.23 Principal scheme of pneumatic nozzle.

turbulence. A scheme of pneumatic nozzle is given in Figure 2.23. Dispersion of the liquid to be sprayed is characterized by average diameter of formed droplets: the smaller their diameter the better the dispersion. Homogeneity of dispersion is characterized by average square deviation of droplets diameter from the average value.

$$\sigma = \sqrt{\frac{\sum (d_i - d)^2}{n - 1}}$$

Besides pneumatic nozzles for dispersion of liquids the jet centrifuge and ultrasonic nozzles are used. The latter are very useful for the preparation of cryodroplets. In ultrasonic nozzles, the jet of a liquid is dispersed under action of ultrasonic oscillations produced by the generator. Such nozzles provide a good dispersion of liquids attaining in some designs 10–15 μm. At present, however, they are used in industry for processes with limited discharge of liquids to be dispersed (40–50 kg/h).

Of some interest is the technique based on electrochemical dispersion of solutions and melts of electrolytes which was proposed by Jolybois and developed for cryochemical applications by Barret (1970).

To obtain monodisperse solution droplets the following phenomenon was used: if an axial vibration with a frequency of $f = V/4.058d_0$ is applied to a laminar flow of liquid with diameter d_0 and rate V, the jet is broken into droplets of the same diameter $d = 1.89d_0$. A diagram of a system for obtaining monodisperse cryodroplets is shown in Figure 2.24.

The initial solution is placed into the vessel 4 from where under pressure of air or inert gas it is moved through the filter 5 to the calibrated draw plate 2 which is connected (6) with the vibrator 7 or with the piezocrystal. The flowing rate of the liquid is controlled by changing gas pressure with the regulator 3. Formed droplets are put into the reservoir filled with liquid cooling agent where their fast freezing occurs.

Initially, it was proposed to use hexane. The laboratory scheme originally developed by Schnettler (1968) and improved by Anastasyuk (Anastasyuk *et al.*, 1971; Anastasyuk, 1972) is represented in Figure 2.25.

In the vessel 3 with a fine net bottom 4 covered with filter paper hexane was placed and cooled with acetone–carbon dioxide mixture 5 down to a temperature of −65 to −70 K. Hexane was used because it is immiscible with water, does not dissolve salt components and ice, possesses sufficiently low melting point (95.3 K) and relatively high heat conductivity. Dispersion of aqueous salt solution was achieved using the pneumatic nozzle 2 at nitrogen pressure 1.3–1.5 atm. To facilitate the formation of cryodroplets, cooled hexane was intensively stirred with the mechanical stirrer 1. Under these conditions practically ‘instant’ freezing of microdroplets of sprayed solution took place with formation of finest ice grains

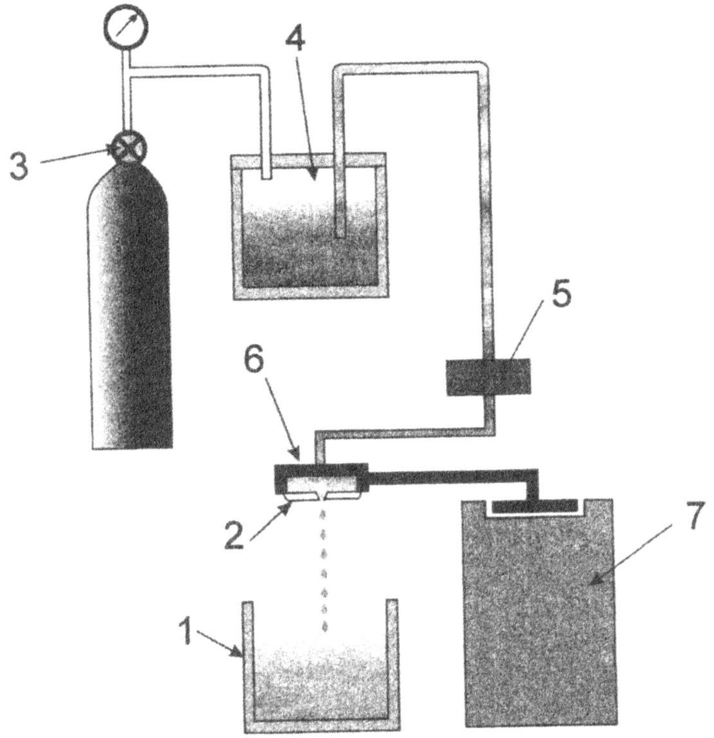

FIGURE 2.24 Device for preparation of monodispersed frozen beads.

with $d = 0.5$ mm. The product was separated from hexane and subjected to freeze-drying.

In a more recently developed freezing method (Figure 2.26) droplets of aqueous salt solution are introduced into the lower region of the freezer (Sauer and Lewis, 1972). They rise through an immiscible, non-flammable refrigerant of higher density than the solution, freeze and float as individually identifiable spherical droplets to the top of the liquid column. This technique can be successfully practised if the temperature of the refrigerant in the close environment of the injector is maintained slightly above the freezing point of the salt solution to prevent freezing in the nozzle of the injector, and if a sharp negative temperature gradient exists immediately above the injector in order to encourage rapid freezing and to prevent coalescence of the droplets.

This is accomplished in a two-phase liquid system in which the interface is a natural barrier to free convection and to uncontrolled heat transfer, phenomena which would occur in a single-phase refrigerant. The short injection zone is

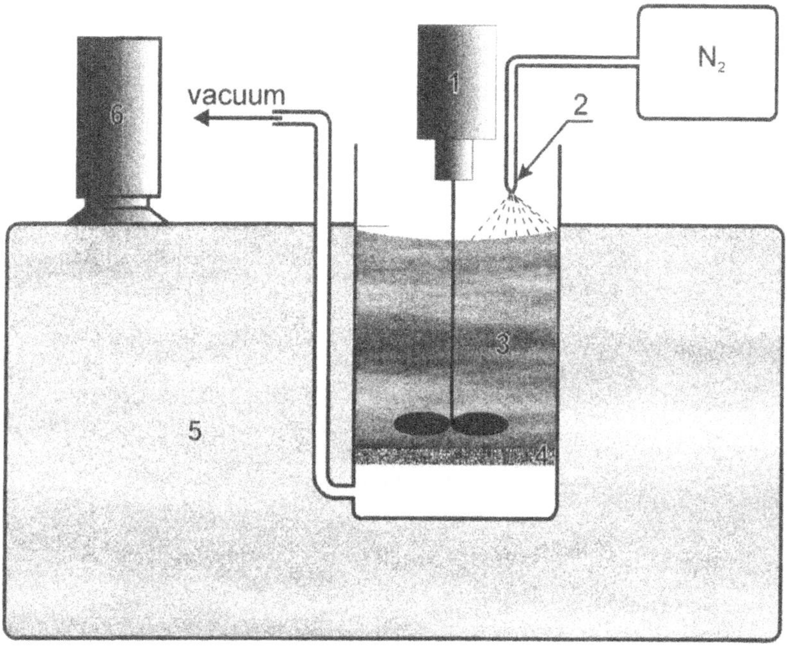

FIGURE 2.25 Laboratory apparatus for cryocrystallization in liquid organic solvents.

heated in the neighbourhood of the injector, while the much longer freezing zone is cooled over its entire length. This establishes a temperature difference between the zones, and a sharp negative temperature gradient in the injection zone. The gradient is maintained by a balance between heat transferred by conduction in narrow zones, or 'boundary layers', adjoining the vessel walls and the interface between the two fluids. The same qualitative description applies with even greater force to the single-phase case; but convection makes it impossible to maintain the bulk of the fluid at a suitably low temperature if the temperature at the injector is maintained above the freezing point. The appearance of striations, due to large temperature gradients in the flow, also confirms the presence of temperature boundary layers and makes the circulating flow in the lower fluid visible.

The most prominent feature of the flow is the pulsating motion of the interface between the two fluids, which plays an important role in the transfer of heat and momentum across the interface. As one would expect, when the densities of the two fluids are almost equal, the interface becomes unstable and the motion so violent that the interface may actually break up. At the opposite extreme, when the interface is almost stationary, the interfacial boundary layer is thickened and

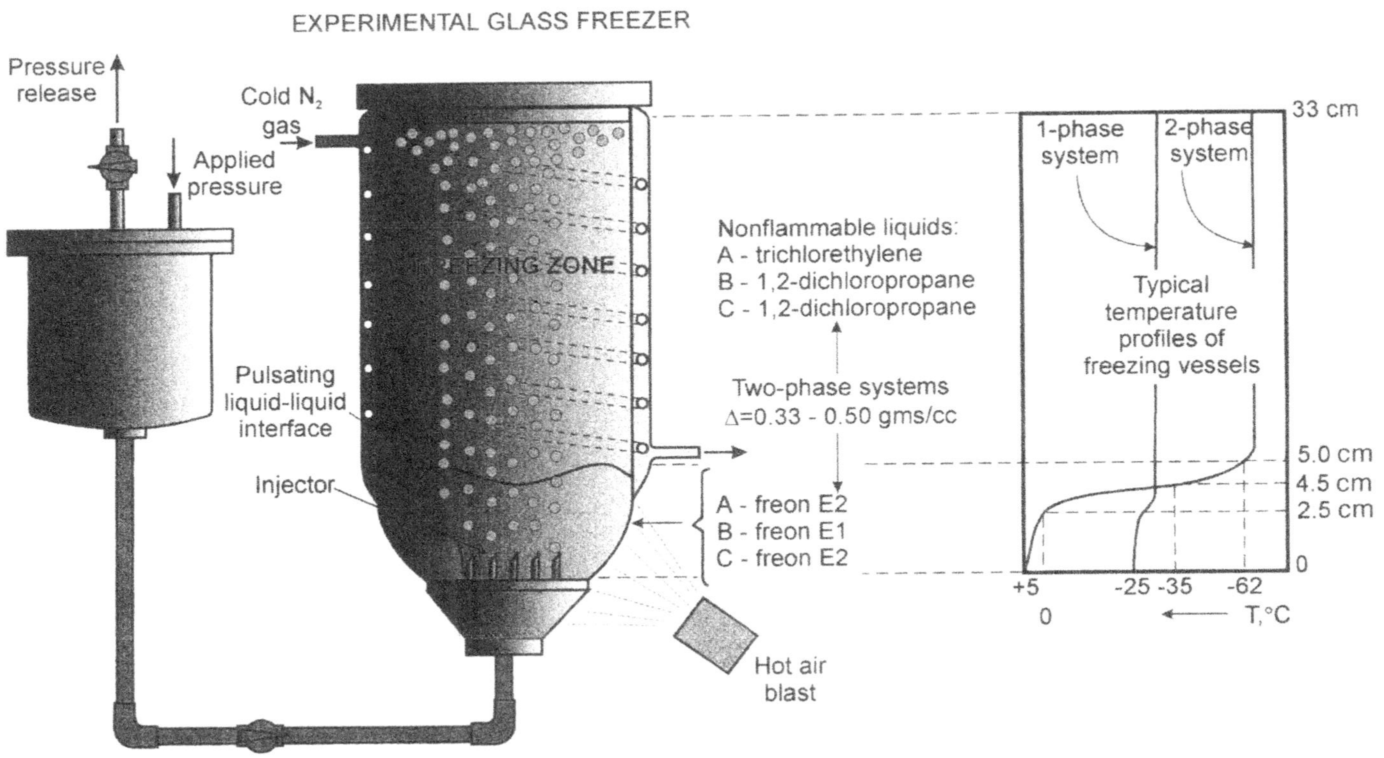

FIGURE 2.26 Scheme of installation for wet freezing of aqueous solutions.

cooled to the point where the droplet stream may have difficulty penetrating it. There is thus an optimum operating range where the interfacial disturbance keeps the boundary layer thin, but is not so pronounced that the interface breaks up.

Some combinations of non-flammable liquids, or liquids which do not sustain combustion at normal ambient temperatures, and which perform well are listed in Figure 2.26. For example, in the A combination, freon occupies the injection zone, and trichloroethylene, the freezing zone. For these liquids, the range of differential density is about 0.3–0.5 g/ml. Also in Figure 2.26 a typical, useful and controllable temperature profile is shown for a two-phase refrigerant in this experimental freezer. Thermocouples were located as indicated along the vertical axis of the two phases. The profile was determined by probing both phases axially along the entire length of the freezer. A temperature of +5 °C is maintained adjacent to the injector; about 1 in above it the temperature is 0 °C. From this point upward the temperature decreases rapidly to the freezing-zone temperature of −62 °C. In the operation of this facility it has been found that a freezing-zone temperature of −30 °C is usually sufficient to obtain complete freezing and to protect against agglomeration of the floating droplets. The temperature profile included in Figure 2.26 for a single-phase refrigerant in this glass system exemplifies the effect of uninhibited convection and uncontrolled heat transfer while attempting to attain an above-freezing temperature at the injector. The temperature of the frozen nozzle is −25 °C; in the freezing zone it is only −28 °C, a 3° temperature differential. A baffle for impeding free convection was installed between the injection zone and the freezing zone without appreciable benefit.

The freezing zone is cooled by passing cold nitrogen gas through an annulus of a double-walled zone. The injection region is maintained above the freezing point of the solution by a hot-air blast in the lower region of the injection zone or by an electrical heater attached to the injector. The rate of injection of droplets is monitored by varying the pressure of the liquid in a pressurized supply tank. A Neoprene 'O' ring supports a 5-orifice injector in the neck of the freezer. The dimensions of the extended stainless-steel orifices are 0.015 cm × 1.6 cm. An 18-orifice injector of the same over-all dimensions has been found useful only for low injection rates, at which the droplets are quite large and uniform in size, although shearing forces in the circulating refrigerant cause some variability in diameter. As the rate is increased, the droplets become smaller. At high tank pressures, the droplets are formed by the break-up of a stream of solution forced into refrigerant. These are much smaller and less uniform in size.

The most common freezing medium proved to be liquid nitrogen. In many cases droplets of aqueous solutions are immediately directed into the bath with liquid nitrogen but it is more convenient to use an apparatus like that shown in Figure 2.27.

The aqueous solution is injected onto a Teflon-coated stainless-steel block cooled to liquid nitrogen temperature. The resulting flakes of frozen material are

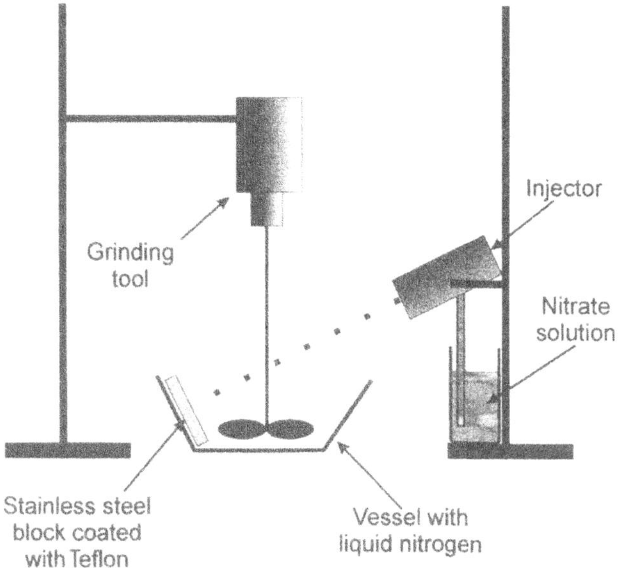

FIGURE 2.27 Device for ultrafast (>10^3 K/s) freezing.

ground using a high shear homogenizing tool to a particle size of less than 20 μm (Krishnaraj *et al.*, 1993).

The most effective way for large-scale freezing technology could be the so-called self-freezing processes based on decreasing the solution temperature down to the crystallization point and even lower under intense evaporation of water contained in it in a vacuum chamber. The solution is introduced into the vacuum chamber in a metal container or injected with nozzles. 'Self-freezing' techniques are perceived as a kind of energy-saving technology because the heat evolved in the process of crystallization is used for intensification of the liquid evaporation. In the last case the cooling agent is not necessary and the process efficiency is considerably enhanced. Use of cryochemical technology with the direct introduction of solutions into a vacuum chamber of continuously operated sublimators allows the preparation of a target product which is not inferior in its properties to the product obtained by freezing with hexane (liquid nitrogen) and superior in all characteristics compared with products produced by ordinary techniques (Gavrilova, 1978; Brazhnikov, 1983).

There are serious grounds to foresee that the industrial application of continuously operated sublimators with direct introduction of aqueous solutions into vacuum will give a considerable economical effect as a result of substantial costs reduction due to cutting down time-consuming operations.

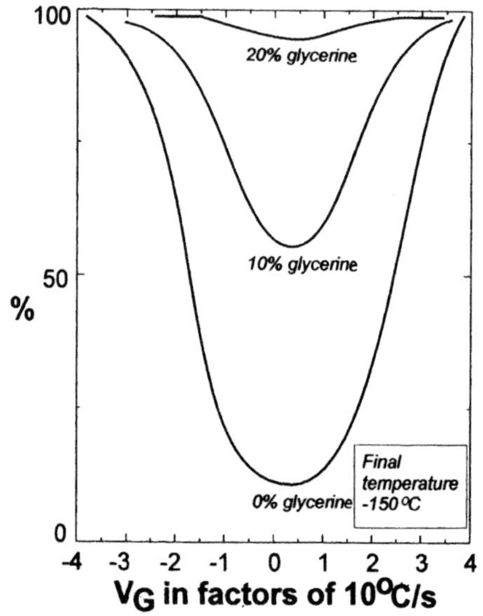

FIGURE 2.28 Influence of glycerine solution on the survival rate of yeast cells at different freezing speeds v_a.

In some cases freezing and self-freezing of solution is carried out not in dispersed state but just in bulk (Volfkovitch *et al.*, 1979; Zhiboedov *et al.*, 1977). It is clear that in the last case the chemical segregation of an incongruent multicomponent system seems to be possible because of the fractional crystallization of individual components (Naumov *et al.*, 1982). However, if the system behaves itself congruently in respect to the crystallization process or if it contains only one salt component it is not dangerous. Moreover, upon preparation of biological products crystallization should be avoided completely (for example, at blood cryoconservation) and so-called cryoprotectors (see p. 50) are used (Riehle, 1968; Mozhaev, 1988). Glycerin, dimethylsulfoxide, glucose, lactose, sucrose, albumen, polyvinylpyrrolidone and others are among them. Figure 2.28 shows the survival rate of yeast cells (Moor, 1964) as a function of the concentration of the protective solution (glycerine) and of the freezing rate.

In this case it is also the aim to achieve a glasslike state while freezing or to control the freezing process in such a way that the formation of ice crystals is restricted to very few crystals of extremely small size. A glycerine concentration of 20% will give satisfactory protection to these cells, even when the freezing process is slow. A similarly high concentration, however, may be inadmissible for other types of cells, due to its toxic effect, or its protective effect may still be insufficient. With a concentration of 20% glycerine the temperature will have to be lowered within a very short time – cool-down rate about 10^3 K/s – in order to reach the amorphous state and to avoid the formation of ice crystals.

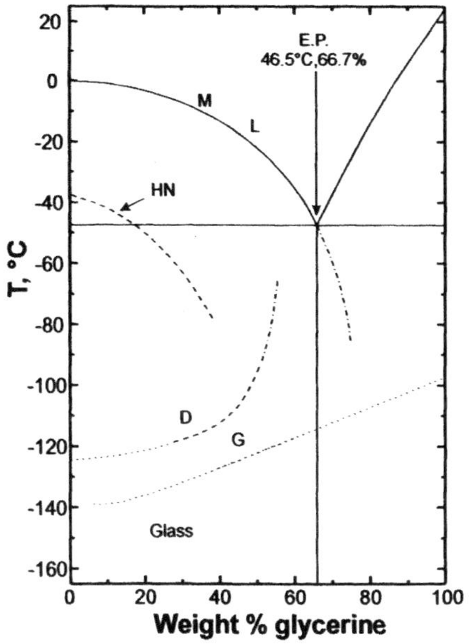

FIGURE 2.29 Phase diagram for water–glycerine system.

If freezing of a specimen is ended at too high a temperature, or if the specimen is stored at such a high temperature, it will not be possible to prevent crystal growth.

Figure 2.29 shows a phase diagram for a water/glycerine system. It illustrates the alteration in this model substance if it is transformed by sufficiently rapid cooling to the amorphous phase and then warmed up.

The line *G* indicates the limit below which the amorphous phase remains stable. The values for small concentrations are extrapolated, as it is known from other measurements that for pure water this limit is between −137° and −144 °C. The individual changes appear from the temperature ranges and designations (Luyet, 1960; Luyet *et al.*, 1968; Dowell and Rinfert, 1960; Rasmussen and McKenzie, 1971). For the purposes of the present chapter it will suffice to establish that with admissible concentrations (i.e. concentrations that are reasonable and not too high for practical purposes), temperatures between −140° and −130 °C will be required in order to exclude transformations in the product.

For many viruses and cells the freezing speed is not fast enough as yet – since this has too many technical problems or is not yet possible – nor are they freeze-dried at temperatures which could theoretically be maintained. But acceptable results are nevertheless obtained.

Voss *et al.* (1970) reported the successful production scale freeze-drying of yellow fever vaccine. He used a freezing technique allowing a cool-down rate of

$20\,°C\ s^{-1}$ and a freeze-dryer having a shelf area that could be cooled down to −60 °C (sublimation pressure of 10^{-2}–10^{-3} torr). The condenser of the unit is cooled by liquid nitrogen to −190 °C. McKenzie and Luyet (1964) describe a small unit for the temperature range between −50 °C and −100 °C, also provided for the use of liquid nitrogen. Possibly, for such systems a condenser temperature of −120 °C might be sufficient. In most cases, however, it is more difficult and more expensive to produce such temperatures by methods other than with liquid nitrogen.

For freeze-drying viruses and cells, it should be examined from case to case in which way satisfactory results can be obtained. Generally speaking, the following rules should be taken into account:

1. the concentration of the protective solution should be as high as admissible in view of its compatibility (i.e. too high concentration of cryoprotector may have a negative influence on room-temperature solubility);
2. cooling-down should be made as fast as possible; and
3. the product temperature during freeze-drying should be as low as possible though for economic reasons the drying time might be decisive here. (The drying time depends on the temperature and increases with lower temperatures.) To improve the freeze-drying of cells and viruses, protective solutions and other freezing methods will have to be tested from the point of view of their efficiency.

2.2 FREEZE-DRYING PROCESSES

2.2.1 General information on freeze-drying processes

The principal method of solvent removal from frozen products of cryocrystallization is freeze-drying – transition of ice from solid state into vapour without the formation of liquid phase. Formation of the latter even in a small amount deteriorates, to a considerable degree, the chemical homogeneity of the product achieved at the freezing stage. To perform freeze-drying, partial vapour pressure in environment should be constantly kept below the pressure of saturated vapours corresponding to the frozen solution.

Let us consider the phase diagram of water (Figure 2.30). It is known from the phase rule that in a single-component system the number of degrees of freedom is equal to zero if three phases coexist. In the phase diagram the triple point O corresponding to the existence of ice, water and vapour occurs at the temperature 273.15 K and partial pressure of water vapour of 610.5 Pa. The sublimation process is shown by the curve aO. It means that heat supply from the outside results in sublimation of the frozen water if the external pressure is below 610.5 Pa (Table 2.4).

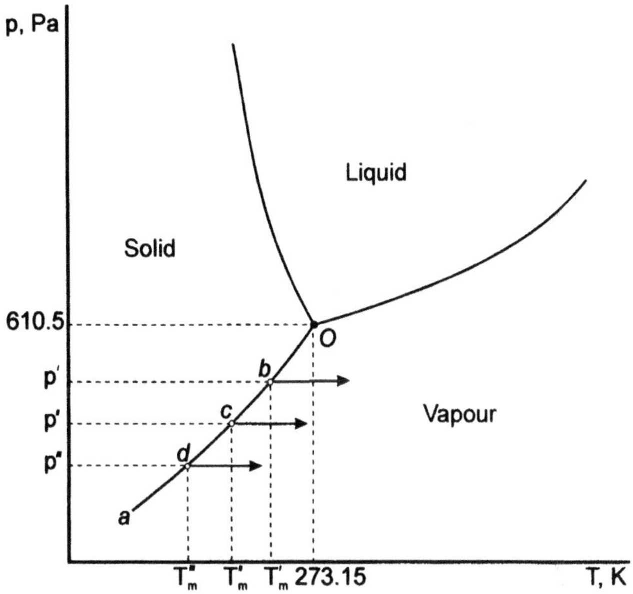

FIGURE 2.30 Phase diagram of water.

Therefore, ice sublimation process may be carried out at a wide range of temperatures and pressures (regimes *b*, *c*, *d*).

On the other hand, according to the phase rule, a single-component two-phase equilibrium system possesses only one degree of freedom and, therefore, the ice temperature should be equal to the temperature determined by the pressure value in the sublimation chamber accordingly to the curve *aO* for the ice–vapour phase transition. In other words, sublimation pressures p', p'', p''' in the chamber (regimes *b*, *c*, *d*) ensure the establishment of corresponding ice temperatures $T_m{}'$, $T_m{}''$, $T_m{}'''$ (Figure 2.30). In this case the equilibrium temperature T_m may be observed only in ice volume where the pressure of water vapours differs

TABLE 2.4 The water vapour pressure over the pure ice versus temperature.

p, Pa	*T*, K	*T*, *p-regions of sublimation and condensation*
610.5	273.15	
437.2	269.15	The most characteristic temperatures
260.0	263.15	of the product during freeze drying
104.0	253.15	
46.6	246.15	
13.3	233.15	The most characteristic temperatures
3.1	221.15	on the surface of the condenser
0.6	209.15	
0.13	185.15	

insignificantly from the equilibrium pressure. The developed ensemble of capillaries and pores within the ice bulk is completely filled with water vapours ensuring conservation of the quasistationary state: the volume is in local equilibrium state, and at internal surfaces the removal of small ice quantities occurs continuously.

The temperature on the ice surface is determined by the rate of freeze-drying. Because the ice sublimation is an endothermic process ($\Delta H_s = 51.83$ kJ/mole), the temperature in the sublimation zone decreases with an increase in sublimation rate (its maximum value tends to quasi-equilibrium temperature $T_m{}^0$). In such conditions the rate of ice sublimation is insignificant. For this reason it is necessary to equalize heat losses for ice sublimation with external heat supply. To carry out the freeze-drying process the following ways of heat supply are the most frequently used:

1. conductive (or contact) supply;
2. convective supply;
3. radiative supply.

Conductive heat supply is ensured by a contact of the solid to be dried with the heat source. Commonly its role is played by heating plates of sublimators on which containers with materials to be dried are disposed.

Convective heat supply for freeze-drying is realized by collisions of gas phase molecules with a solid. The role of convective heat input at lowered pressure is decreased (external heat exchange), but it can be considerably increased by heat exchange processes in capillary porous solids (internal heat exchange).

Radiative heat supply takes place continuously as a result of complex intra-atomic processes whose rate is determined by the temperature of the solid. Radiative energy represents the energy of electromagnetic oscillations with various wavelengths. To intensify the processes occurring in sublimators the heat supply is carried out using special generators producing radiation in the IR and UHF ranges. The most powerful energy sources (up to 10^{13} W/cm^2) are lasers.

It is interesting that the penetrating abilities of different kinds of heat supply are not the same. By conductive or convective input of heat to subliming ice the whole energy is absorbed by its surface (absorption coefficient α equals infinity). In this case temperature gradient in the ice bulk is absent ($\Delta T \propto 1/\alpha$) and the sublimation process takes place at its surface.

On the other hand, at radiative heat supply the α coefficient has finite values not equal to zero and depending on the wavelength. For example, the α values for ice depend on radiation energy in the following way:

α, 1/cm	0	290	29	2, 9
Wavelength, µm	0–2	2–3	3–4	4–10

In this context a considerable gradient of temperature and pressure arise in sublimated material by radiative heat supply. It means that the sublimation zone is not restricted at geometrical limits of the solid submitted to freeze-drying but is spreading into its volume, i.e. the process takes place on ice–vapour interface located on the surface of internal capillaries and pores. Therefore, the appearance of external as well as internal interfaces results from considerable difference between drying processes by the radiative heat input and by the conductive (or convective) heat supply.

Efficiency of any kind of heat supply is best demonstrated at the given regimes of gas flow unambiguously determined by the residual pressure value in the system.

Three kinds of gas-flowing regimes are commonly considered. This classification is based on Knudsen numbers determined by the relation:

$$\mathrm{Kn} = L/d \tag{2.1}$$

where L is a mean free path of water molecule calculated by the equation:

$$L = k\frac{\eta}{p}\sqrt{\frac{T}{M}} \tag{2.2}$$

where η, p, T and M are the medium viscosity, pressure, temperature and molecular weight of water, respectively, k is the proportionality coefficient, d is the diameter of capillary.

The possible gas-flow regimes and corresponding pressure ranges (Lebedev, 1973) are as follows.

Viscous regime	$p \geq 133.3$ Pa; Kn ≤ 0.001
Molecular-viscous regime	$1.3 < p \leq 133.3$ Pa; $0.001 <$ Kn ≤ 10
Molecular regime	$p < 1.3$ Pa; Kn > 10

Note: Kn = Knudsen number, one of the standard complex parameters used in gas and liquid dynamics.

In Figure 2.31 the dependencies of flow regime for gas (water vapour) on heat transfer characteristics is represented. It is clear that in the molecular viscous gas-flow regime all three kinds of heat transfer coexist: convective, conductive and radiative.

At pressures $p \geq 130$ Pa the convective process is predominant in the system whereas the conductive and radiative ones are practically absent. Actually, at viscous gas-flow regime the pressure and heat transfer are large and for this reason the process looks like an intermediate between evaporation and freeze-drying. In the molecular regime the convective heat transfer is absent (Samokhvalov, 1975).

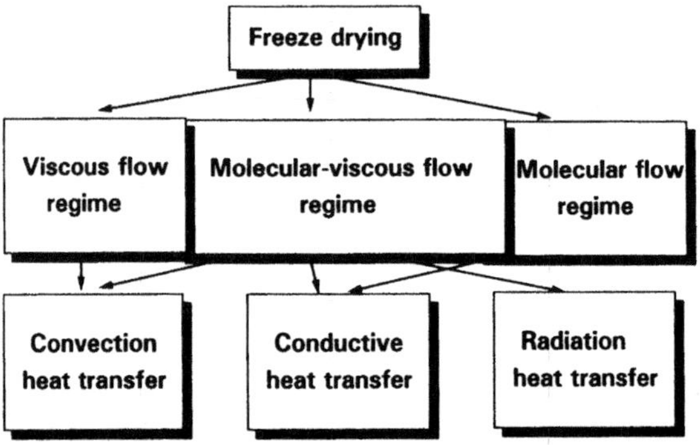

FIGURE 2.31 Possible regimes of flow and character of heat transfer during freeze-drying.

As for the regularities of freeze-drying independent on the kind of heat supply, the following concepts should be taken into account:

1. At a given pressure in sublimation chamber ($p_s = \text{const}$) there is the limit heat supply value, below which the correlation between the pressure and temperature of a phase transition is no longer observed and the ice temperature goes below the equilibrium temperature. Hence a limited heat input from the outside of the sublimation process occurs to a significant degree at the expense of the ice heat content.
2. On the curves of the dependence of the rate of ice sublimation j versus the total pressure in sublimation chamber p_s a maximum connected with the change in limiting stages of the process in the range 13 to 130–250 Pa is observed (Figure 2.32).

   The increase of j with a decrease of total pressure in the sublimation chamber (at $p_s > 130$–250 Pa) is connected with the increase of the water vapour diffusion coefficient from the sublimation zone to the condenser. The further decrease of j at $p < 1.3$–13 Pa is determined by the lowering of the temperature T_m of quasi-equilibrium ice volume and therefore by the decrease of the rate of transfer process at the solid–vapour interface.
3. The efficiency of the condenser's work depends mainly on the difference in temperatures of the material T_m and condenser's surface T_c determined by the difference in partial vapour pressures of water at the surfaces of the material p_m and condenser p_c, i.e. by ($p_m - p_c$) value proportional to the water vapour diffusion coefficient from the sublimation zone. In general, the process rate j increases rapidly by increase of the difference $\Delta T = T_m - T_c$ up to 30–40 K. At $\Delta T > 40$ K the dependence of j on ΔT is slightly expressed.

Possible ways for improvement in the freeze-drying process were analysed (Kelly, 1978). The slowest step of the whole procedure is the sublimation of ice from the frozen salt solution. However, to increase the rate, heat may not be

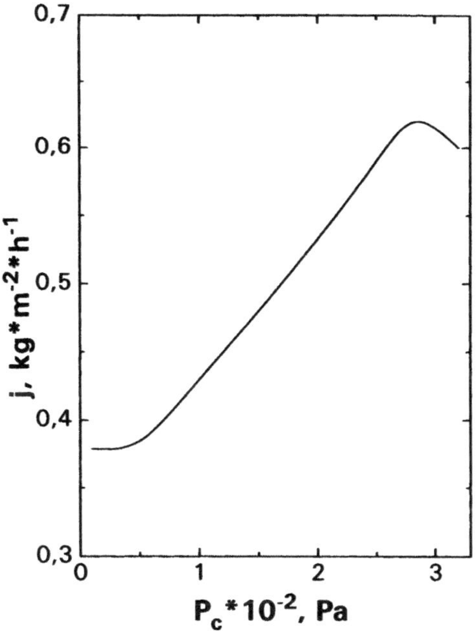

FIGURE 2.32 Evaporation intensity *vs.* total pressure p_c in vacuum chamber.

supplied indiscriminately because of the possibility of melting and fractionation of the solution.

Certain factors involved in the drying of various biological materials have been noted (Kramers, 1958), and the application of heat to increase the rate of sublimation has been used. However, the optimum rate of heat supply is usually determined empirically. The analysis was carried out via an examination of the steady-state heat balance and mass balance of the system illustrated in Figure 2.33.

The supply of heat

In the absence of any external heating source the heat will reach the ice from the outside by convection and radiation to the surface of the vessel (surface 1) followed by conduction through an ice layer to the subliming surface (surface 2). Radiation from the surroundings to surface 2 may also occur.

Convective heat supply to surface 1 If surface 1 is at temperature T_1 and of area S while the room temperature is T_r, then (McAdams, 1954)

$$H_{conv} = h*S*(T_r - T_1) \tag{2.3}$$

where H_{conv} is the rate of convective heat supply to surface 1, and h is the heat

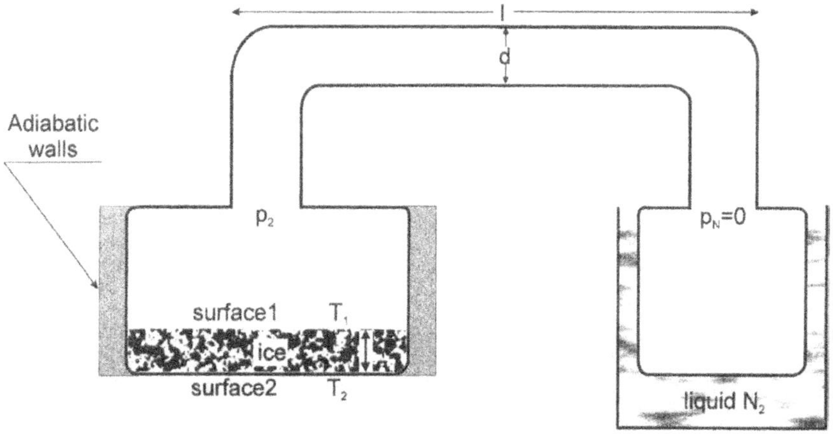

FIGURE 2.33 Principal scheme of the model freeze-drying device.

transfer coefficient. For free convection in air within the normal freeze-drying temperature range and with a downward-facing, square, cold plate of length l (m),

$$h = 0.021\,39\left(\frac{T_r - T_1}{l}\right)^{1/4} \quad (\mathrm{WK^{-1}\,m^{-2}}) \tag{2.4}$$

Radiative supply of heat to surface 1 The rate of radiative heat supply to surface 1 (R_1) is given by Stefan's Law:

$$R_1 = S\sigma(\alpha T_r^{\,4} - T_1^{\,4}) \tag{2.5}$$

where S is Stefan's constant ($5.67*10^{-8}\ \mathrm{Wm^{-2}\,K^{-4}}$), and α is the absorptivity of surface 1 (numerically, for wet ice at 273 K, α is 0.97).

Conduction of heat from surface 1 to surface 2 If the thermal conductivity of the frozen solution is q, the cross-sectional area is S and the thickness t, then

$$H_{\mathrm{cond}} = \frac{qS}{t}(T_1 - T_2) \tag{2.6}$$

where H_{cond} is the rate of heat flow from surface 1 to surface 2.

Radiative heat supply to surface 2 This may also be evaluated from Stefan's Law

$$R_2 = S\sigma(\alpha T_r^4 - T_2^4) \tag{2.7}$$

The steady-state flow of heat At the steady state, the flow of heat to surface 1 should be equal to the rate of conduction to surface 2, thus:

$$H_{\text{cond}} = H_{\text{conv}} + R_1 \tag{2.8}$$

The total heat supply to surface 2, H_2 is therefore:

$$H_2 = H_{\text{cond}} + R_2 = H_{\text{conv}} + R_1 + R_2 \tag{2.9}$$

Loss of heat from surface 2 Under steady-state conditions, all the heat supplied to surface 2 should be lost, and the mechanism for the loss of heat is sublimation. The rate of heat loss, H_1, and the heat of sublimation, ΔH_s, of the solvent:

$$W_s = \frac{H_1}{\Delta H_s} \tag{2.10}$$

Loss of solvent from surface 2 At a given partial pressure of solvent over surface 2, p_2, the rate of sublimation may be calculated from kinetic theory. The rate of condensation of solvent in mole per 1 m^2 of surface per second, W_{C_2}, is:

$$W_{C_2} = \frac{p_2}{(2\pi RMT_g)^{1/2}} \tag{2.11}$$

where M is the molecular weight of the solvent and T_g is the temperature of the gas phase solvent molecules. It is assumed that every molecule hitting surface 2 condenses.

The rate of sublimation, W_{S2}, from surface 2 at temperature T_2 is derived from the equilibrium vapour pressure of the frozen solvent (p_0) at that temperature:

$$W_{S_2} = \frac{p_0}{(2\pi RMT_2)^{1/2}} \tag{2.12}$$

The net sublimation rate is, therefore:

$$W_S = W_{S_2} - W_{C_2} \tag{2.13}$$

Pumping away solvent vapour The solvent vapour is pumped exclusively by liquid nitrogen traps whose speed for condensable gases is 15 l sec^{-1} (Guthrie and Wakerling, 1949) for 1 cm^3 of trap surface. Rotary oil pumps serve to reduce the background pressure of air. The pumping system is thus represented (see Figure

2.33) by a pressure gradient between the surface of the ice (partial pressure p_2) and the cold trap at a partial pressure of solvent vapour p_n. For most solvents p_n is very small (e.g. ice at 77 K has a vapour pressure of $<10^{-7}$ torr) and may be taken as zero. The flow of vapour, W_f, is fixed by the dimensions of the apparatus and the pressure gradient (p_2).

Assuming mixed molecular and viscous flow, which cover the range of pressures encountered in this work, the vapour flow rate W_f for water (g sec^{-1}) is

$$W_f = \frac{p_2}{RT_r} * \left[0.188 \frac{d^4}{l} p_2 + 15.554 \frac{d^3}{l} \frac{(1 + 2.067*10^{-4}\, dp_2)}{(1 + 2.542*10^{-4}\, dp_2)}\right] \tag{2.14}$$

(source: Kelly *et al.*, 1978).

Here d is the diameter in cm and l the length of the tube in cm through which the flow is taking place, at a pressure difference of p_2 torr.

Calculation of model parameters p_2, T_1 and T_2, and hence the rate of sublimation and heat flow may be determined from the above equations and the constraints, as follows:

- for heat flow through surface 1:

$$H_{cond} = H_{conv} + R_1 \tag{2.15}$$

- for heat flow through surface 2:

$$H_2 = H_1 \tag{2.16}$$

where H_1 is the heat lost from surface 2 by sublimation;

- and for the mass flow in the system:

$$W_f = W_s \tag{2.17}$$

There being no analytical solution, a numerical method has been devised and implemented on the computer.

Two binary search routines are used to determine T_1 and T_2. At a given T_2 a binary search routine for T_1 (between the limits $T_1 = T_2$ and $T_1 = T_r$) is used to bring $H_{conv} + R_1 - H_{cond} = 0$. The value of the heat supply H_2 from Equation 2.9 is equated to H_1 (Equation 2.16) and thus W_S (Equation 2.10) can be calculated. For water as solvent the equilibrium vapour pressure of ice at temperature T_2 is given by the following equation:

$$\log_{10} p_0 = -2445.6/T_2 + 8.23\,\text{antilog}_{10}T_2 - 0.0168T_2 + 1.205*10^{-2}T_2^{\,2} - 6.757 \quad (2.18)$$

From Equations 2.11–2.13 and 2.18 p_2 may be calculated from W_S, and used to determine the flow W_f from Equation 2.14. The function W_f *vs.* W_S is calculated. T_2 is varied by the binary search routine between the melting point of the frozen solution and an arbitrary low value (taken as 150 K) until $W_f - W_S = 0$. A system which tries to converge on a value for T_2 which is greater than the melting point of the solution is deemed to have melted.

To take into account the decreasing size of the ice block the procedure is operated for a number of short time intervals, Δt. At the steady-state flow, the amount of solvent evaporated in time Δt is $W_S\Delta t$. The thickness of the ice block is recalculated and the entire procedure repeated to find new values of T_1, T_2, p_2 etc. In this way, the time to sublime a given quantity of frozen solvent may be determined.

The computer program was written to accommodate a spherical flask containing the frozen solution (as was used in practice), and a supply of heat to surface 2 from an external source. In the latter case, the heat supply H_2 is fixed and no longer depends on the conduction–convection–radiation mechanism described above.

Results of the computer model Table 2.5 contains the parameters used in calculating the time taken to sublime 100 g of ice. The dimensions of the flask and pumping tube were taken from the experimental apparatus. Figure 2.34 shows the predicted rate of sublimation and the temperature of the surface T_2, during the process.

For the maximum rate of sublimation, T_2 should be near the melting point of the frozen solution in order to maximize p_0 in Equations 2.12 and 2.13. Figure 2.34 shows that for heat supplied from the room alone by convection, conduction and radiation, the rate of sublimation is much less than could be possible. The

TABLE 2.5 Dimensions of the freeze-drying apparatus and input parameters for the computer model.

Pumping tube	*diameter* = 2.0 cm *length* = 15.0 cm
Ice block	diameter of spherical flask = 6.8 cm thickness of ice block = 2.12 cm initial area of ice = 40 cm^2
Latent sublimation heat of ice	46 643 J mol^{-1}
Thermal conductivity of ice	0.021 W cm^{-1} K^{-1}

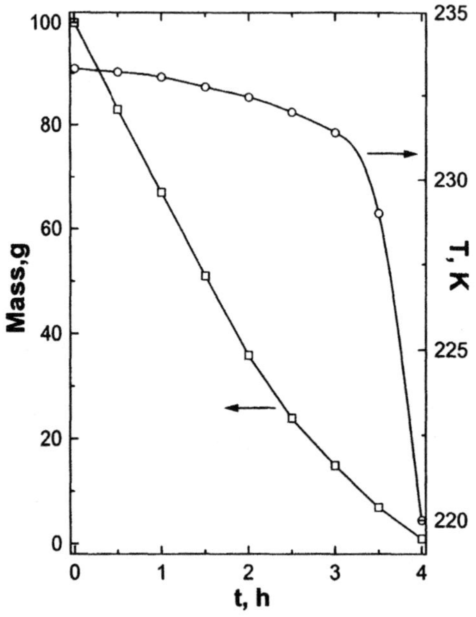

FIGURE 2.34 Calculated curves for sublimation of ice (mass of ice remained *vs*. temperature of the sublimation surface). No external heat supply.

surface of the ice was calculated to be receiving no more than 0.05 W cm^{-2}, whereas the computer model showed that up to 2.4 W cm^{-2} could be supplied without melting. This was calculated by increasing the heat supplied for a given pumping speed until T_2, the temperature of the ice surface, reached 273 K. We found for the arrangement in Table 2.5 a heat supply of 98 W on a 40 cm^2 surface resulted in a value of T_2 of 272.7 K. In calculating the limiting heat supply, the dimensions of the pumping tube exert a profound influence. If the diameter of the tube is reduced from 2.0 to 1.5 cm, the heat required to give $T_2 = 272.6$ K is now 40.3 W.

Figure 2.35 is calculated for the same system but with a constant heat supply of 0.7 W cm^{-2}. The estimated time required to sublime 100 g of ice falls from 40.3 to 4.2 h.

Increasing the pumping speed by changing the dimensions of the apparatus was found to change the time of sublimation by only 5% for low levels of heating. Thus the model of the drying process suggests that a major improvement would be brought about by supplying up to 2.4 W cm^{-2} to the surface of the frozen solution.

Validity of the computer model To test the model given above, 100 g of ice were sublimed in a freeze-drying apparatus, first with no external heat and then

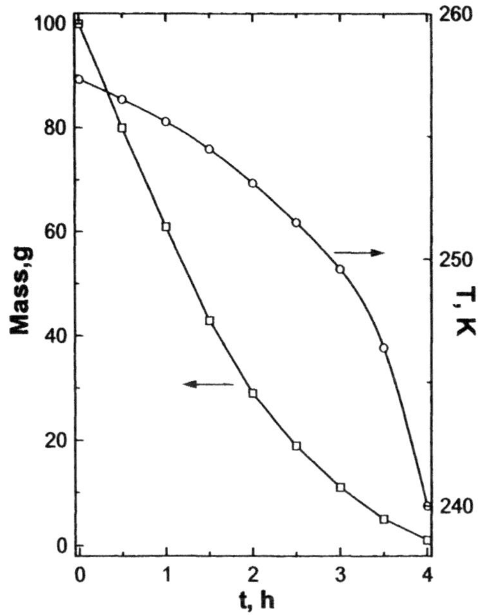

FIGURE 2.35 Calculated curves for sublimation of ice (mass of ice remained *vs.* temperature of the sublimation surface). 0.7 W*cm^{-2} heat supplied to the ice surface.

with a 100 W infra-red lamp suspended above the ice surface giving, from the geometry of the arrangement, an irradiation of 0.7 W cm^{-2}. The ice was held in a 250 ml round-bottomed flask. The dimensions of the tubes connecting the flask and liquid nitrogen trap are given in Table 2.5. The background air pressure was kept below 5*10^{-3} torr by 50 l sec^{-1} two-stage rotary oil pump.

The time taken to sublime 100 g of ice with and without external heating is given in Table 2.6. Also shown are the computer model predictions and the time taken to dry the lithiated nickel sulphate samples.

Table 2.7 contains surface area and conductivity data of 10% lithiated nickel sulphate samples prepared by boiling to dryness, and freeze-drying with and

TABLE 2.6 Drying times for different heating rates and systems.

System	*Weight of solvent* (g)	*Heat supply*	*Drying time* (h)	*Computer prediction* (h)
Ice	100	None	42	40.3
Ice	100	0.7 W cm^{-2} from IR-lamp	3.8	4.2
10% $Li^+/NiSO_4$	200	None	80	–
10% $Li^+/NiSO_4$	200	0.7 W cm^{-2} from IR-lamp	5	–

TABLE 2.7 Specific conductivity and surface area measurements of lithiated nickel sulphate, and nickel sulphate prepared by different methods.

System	*Preparation*	*Specific conductivity* (W^{-1} cm^{-1})	*Surface area* (m^2 g^{-1})
$NiSO_4$	Boiled to dryness	8.9×10^{-9}	–
10 at % $Li^+/NiSO_4$	Boiled to dryness	4.8×10^{-8}	5
10 at % $Li^+/NiSO_4$	Freeze-dried, no external heat	3.5×10^{-7}	14
10 at % $Li^+/NiSO_4$	Freeze-dried, external heat 0.7 W cm^{-2}	4.0×10^{-7}	13

without an external heat source. It can be seen that the use of an external heat source does not affect the drying process or the quality of the dry powder.

Freeze-drying of non-aqueous solutions brings a number of new problems if, for instance, alcohols, benzene, acetone, dioxane, chloroform and also ammonia and carbon dioxide are involved. Table 2.8 shows the temperatures to be attained by freezing for some of these solvents. Compared with ice, such temperatures

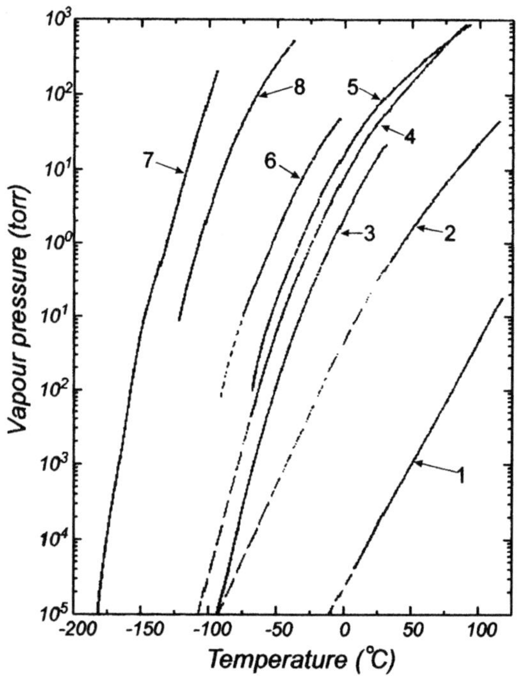

FIGURE 2.36 Vapour pressures of different solvents: (1) glycerin; (2) DMSO; (3) water; (4) ethanol; (5) benzene; (6) acetone; (7) CO_2; (8) NH_3 (Willemer, 1975).

TABLE 2.8 Thermophysical parameters of some organic solvents.

Name	*Formula*	*Molecular weight*	*Evaporation* T (°C)	*Evaporation* ΔH (cal/g)	*Melting* T (°C)	*Melting* ΔH (cal/g)
Water	H_2O	18.02	100	539	0	79.5
DMSO	C_2H_6OS	80.14	192	132	18.4	20
Methanol	CH_4O	32.04	64.7	264	−97.9	23.6
Ethanol	C_2H_6O	46.07	78.4	202	−114.7	25.8
Acetone	C_3H_6O	58.08	56.1	120	−96.5	23.6
Benzene	C_6H_6	78.11	80.1	94	5.6	30.5
Carbon dioxide	CO_2	44.01	−78.11	138	−78.3	–
Ammonia	NH_3	17.03	−33.3	328	−77.9	79.4
Dimethylformamide	C_3H_7ON	73.09	70.5	138	−61	
Dioxane	$C_4H_8O_2$	88.1	101.6	97.2	11.1	34.8
Carbon tetrachloride	CCl_4	153.84	77	46.4	−22.9	3.5
Chloroform	$CHCl_3$	119.39	61.44	59.7	−6.5	20.1
Diethylamine	$C_4H_{11}N$	73.13	55.5	92.5	−48	

may be much lower, but the vapour pressure may be higher or lower, as shown by some examples in Figure 2.36.

We are not dealing here with solvents the vapour pressures of which at room temperature are higher than atmospheric pressure. For the freeze-drying of such substances entirely different concepts of apparatus will have to be considered. But even when excluding this type of problem, there remain quite a number of other factors to be taken into account for the design of a freeze-drying plant for non-aqueous solutions. To maintain for instance a product dissolved in acetone in its frozen state, a shelf temperature of about −120 °C will be required, at least at the beginning of freeze-drying. It will not suffice to cool the shelf area, but its temperature, and thus the energy supply, should be very carefully controlled. Furthermore, the specimen and the shelf surface should be protected from uncontrolled external irradiation in order to avoid temperature fluctuations and non-uniform drying. Such precautions are very important as the heat of sublimation of these products is normally of the order of 20–30% of that for ice.

The model of vacuum sublimation (Kelly *et al.*, 1978) was extended to solvents other than water (Hibbert and Tseung, 1979). The time taken to sublime 100 g of five solvents: water, acetic acid, dimethyl sulphoxide (DMSO), ethanol and glycerol, was calculated for two heating rates, first at a constant rate of 30 W (a typical experimental value), and secondly at the maximum heating rate. The parameters used in the calculation are given in Table 2.9. Those for the vacuum system are those used in experimental arrangement used in this work. Physical data were obtained from the literature where possible. The heat of sublimation (ΔH_s) at the melting point was calculated from the heat of vaporization at the

TABLE 2.9 Parameters for computer model for drying process. Apparatus: Radius of flask = 6.34 cm, pumping tube radius = 1.5 cm, length = 2.5 cm, vapour temperature = 293 K, and trap temperature = 77 K.

Solvent	*Latent heat of sublimation* (J mol^{-1})	*Density of solid* (g cm^{-3})	*Viscosity of vapour* (μP)	*Molecular weight*	*A*	*B*
					lg(*P*, mmHg) = $A + B/T$	
Water	50 900	0.917	96.0	18		
Acetic acid	40 780	1.1	80	60	8.269	−2130.2
Glycerol	110 758	1.3	100	92	15.0177	−5785.6
Ethanol	49 173	0.80	60	46	12.266	−2568.6
DMSO	69.125	1.1	80	78	12.2313	−3690.76

boiling point (ΔH_v), the heat of melting (ΔH_m) and specific heat data, using the Laws of Hess and Kirchhoff:

$$\Delta H_s = \Delta H_m + \Delta H_v + (C_p{}^v - C_p{}^s)(T_m - T_b) \qquad (2.19)$$

where $C_p{}^v$ and $C_p{}^s$ are the specific heats at constant pressure of vapour and solid respectively, and are taken to be independent of temperature in the range considered. T_m and T_b are the melting and boiling temperatures of the solvent at a temperature T. p_T was derived from ΔH_s and the tabulated triple point pressure and temperature (p^*, T^*) using the Clausius–Clapeyron equation:

$$\log_{10}(p_T, \text{mm Hg}) = A + B/T \qquad (2.20)$$

where $A = \log_{10}(p^*, \text{mm Hg}) + \Delta H_s/2.30RT^*$; $B = -\Delta H_s/2.303R$.

Calculated drying curves for 100 g water, DMSO and acetic acid, for an applied heat of 30 W are shown in Figure 2.37. The model predicted that ethanol and glycerol should melt at this rate of heating.

A second calculation was performed to determine the maximum heating rates possible with the different solvents. The results are given in Table 2.10.

The effect of an added solute on the drying rate is twofold. First, the melting temperature is lowered as discussed above. The computer model was modified to take the melting point of a solution as the maximum possible temperature. Secondly the addition of solute causes a depression of the freezing point of the solvent. For low concentrations Raoult's Law is obeyed and the vapour pressure lowering is proportional to the mole fraction of solute. The calculated solvent vapour pressure was modified by multiplication by $1 - x_s$, where x_s is the mole fraction of solute, up to a maximum value of x_s corresponding to the mole fraction of solute at the eutectic. However, the most concentrated salt solutions in water have mole fractions less than 0.1, which affect the model only slightly. For

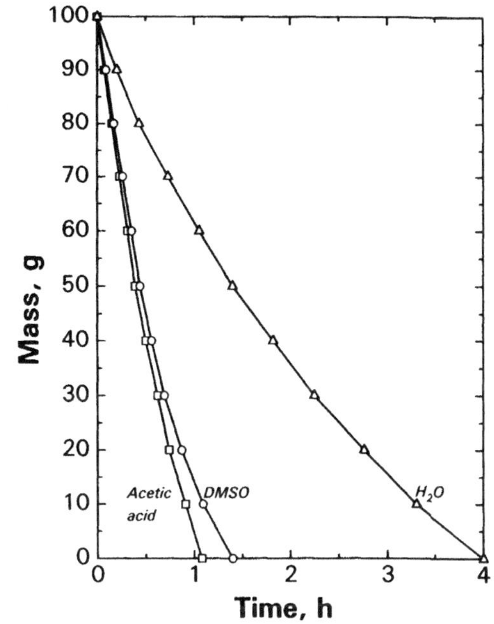

FIGURE 2.37 Calculated curves for drying of solvent (100 g), sublimed *in vacuo* at heat supply of 30 W.

TABLE 2.10 Calculated maximum drying rates for different solvents in the experimental apparatus.

Solvent	*Melting point* (K)	*Maximum heat input* (W)	*Maximum sublimation rate* (g sec^{-1})
Acetic acid	289	4000	0.575
Water	273	4800	0.166
DMSO	291.5	40	4.4×10^{-3}
Ethanol	156	0.0003	3×10^{-8}
Glycerol	293	0.002	1.6×10^{-8}

example, a calculated drying curve of 10% sodium chloride solution in water leads to an increase in surface temperature, compared to pure ice, of 0.5 °C. The effect was more marked with other solvents but was still relatively small.

The results of the experiments with frozen sodium chloride and calcium chloride solutions show conclusively that vacuum sublimation of frozen aqueous solutions leads to a concentration of the solutions as postulated above. The effect is akin to that produced by spreading salt on ice to cause melting except that here, sublimation of ice from a solution creates its own salt.

As the solvent sublimes the local surface, concentration increases to the maximum solubility of the solute. If the temperature of the surface, which is determined by the rates of heating and sublimation, is above the melting point of the frozen solution, local melting will occur. The resulting solution will be made more dilute by the dissolution of the less concentrated bulk solution, and when its melting point reaches the temperature of the surface, the solution will re-freeze. This process continues until the entire solid solution has reached that concentration, the melting point of which corresponds to the surface temperature. At this point the frozen solution melts. It is clear from the above discussion that any solution will eventually melt if the temperature of the surface is above that of the eutectic. Thus in the experiments conducted above, a surface temperature of −30 °C was above that of the calcium chloride/water eutectic (−52 °C) but below that of the sodium chloride/water eutectic (−21 °C).

The rate of removal of solvent vapour is determined, for a given pumping system, by the presence of vapour above the subliming surface. This in turn may be shown to be related to the equilibrium vapour pressure at the temperature of the surface, and the heat input.

If the pressure above the surface is p_g, and the temperature of the vapour T_g, the number of moles striking 1 m^2 of surface is $p_g(2\pi MRT)^{-1/2}$ sec^{-1} where M is the gram molecular weight of the solvent. The number of moles evaporating from the surface is given by a similar expression containing the equilibrium vapour pressure p^0 and the temperature of the surface T_m, viz: $p^0(2\pi MRT)^{-1/2}$ sec^{-1}. The net evaporation rate is the difference between these expressions and is equal to the number of moles subliming, given by $H/DH_s{*}S$, where H is the heat input, ΔH_s the latent heat of sublimation and S is the area of the surface. Therefore:

$$p_g = p^0\left(\frac{T_g}{T_m}\right)^{1/2} - H(2\pi MRT)^{-1/2}{*}(\Delta H_s{*}S)^{-1} \tag{2.21}$$

For a system at constant heat input, the rates of sublimation of different solvents are dependent on the heat of sublimation, assuming the melting point has not been exceeded. If the maximum heat input is attained, when the subliming surface is at its melting point, the rate is better calculated (in the absence of the

TABLE 2.11 Vapour pressure data for different solvents.

Solvent	*Melting point* (K)	*Vapour pressure of solid at m.p.* (mmHg)
Water	273	4.58
Acetic acid	289	7.91
Glycerol	293	1.9×10^{-5}
Ethanol	156	6.3×10^{-5}
DMSO	291.5	0.37

value for the heat input) using Equation 2.21 for p_g, from the pumping characteristics of the system. In this case the analysis of Equation 2.21 reveals that the dominating factor is the equilibrium vapour pressure at the melting point. Table 2.11 contains vapour data for five solvents at their melting points (Willemer, 1975).

Using this criterion for the rate of sublimation at the maximum heating rate, acetic acid is the optimum solvent with ethanol and glycerol having very low vapour pressures. A more accurate test is given by the computer model of maximum sublimation rates (Table 2.10) which takes account of different heats of sublimation and of the variation of vapour pressures with temperature. It is seen, however, that the results agree with the order of solvents (acetic acid > water > DMSO > ethanol, glycerol) suggested by the values of equilibrium vapour pressures at the melting point. It should be noted though that the maximum heating rates calculated for acetic acid and water could not be realized in practice. Of interest is that the melting temperature of a solvent is not a useful guide to the rate of sublimation.

2.2.2 Mechanism of freeze-drying

The macroscopic description of sublimation process given in section 2.2.1 does not concern its mechanism. It assumes that the process occurs uniformly at the whole surface. Actually, the energy supplied to the surface is not distributed uniformly. In this context two models for sublimation of water molecules from ice surface have been proposed.

The first model assumes the existence of various surface irregularities (grain interfaces, dislocations) characterized by an excess of free energy. The process of energy absorption by surface does not occur uniformly, and at the imperfect solid surface there are some molecules which are able to be converted into an adsorbed state and then into a gaseous one. For the model (Verkin *et al.*, 1980), assuming direct transfer of separate molecules into the gaseous phase and ignoring partial recovery of these molecules, the mass evaporation rate is:

$$\mathrm{d}m/\mathrm{d}\tau = \alpha p_s{}^0/\sqrt{2\pi RT_s} \tag{2.22}$$

where α is the evaporation or condensation coefficient; and $p_s{}^0$ is the saturated vapour pressure at the surface temperature T_s (where T is the temperature of vapour near the surface).

Taking into account that a part of vapour molecules returns to the surface, Equation 2.22 transforms to

$$\mathrm{d}m/\mathrm{d}\tau = \alpha[p_s{}^0/\sqrt{2\pi RT_s} - p_s/\sqrt{2\pi RT}] \tag{2.23}$$

where p_s is the vapour pressure near the surface S, and p_s^0 is the saturated vapour pressure at $T = T_s$.

Hence, according to the model of direct transfer of separate molecules, the variation in intensity of energy supply should lead to changes in the surface temperature T_s.

In the second model (Savchenko, 1982) it is assumed that due to external energy supply some heat fluctuations occur leading in turn to formation of liquid micro-areas from which a part of the liquid evaporates (microphase evaporation) and another part freezes again, i.e. is converted into a solid state.

This mechanism may be described using the following relations:

$$1H_2O_{(s)} \rightarrow 1H_2O^*_{(s)} \quad (2.24)$$

$$1H_2O^*_{(s)} \rightarrow mH_2O_{(l)} + (1 - m)H_2O_{(s)} \quad (2.25)$$

$$mH_2O_{(l)} \rightarrow nH_2O_{(g)} + (m - n)H_2O_{(s)} \quad (2.26)$$

where $H_2O_{(s)}$, $H_2O_{(l)}$, $H_2O_{(g)}$ and $H_2O^*_{(s)}$ are water molecules in solid, liquid, gaseous and excited (at the imperfect solid surface) states, respectively.

In this case the rate of mass change may be represented by the equation

$$dm/d\tau = kS_p(p_s^0 - p_s) \quad (2.27)$$

where S_p is the interface limiting the total process rate, k is the constant rate of the process and τ is the time coordinate of reaction.

As a result of studies of water removal in model systems (for example, water–sand) it was established that while freeze-drying:

1. considerable temperature and pressure gradients appear independently of the type of energy supply (conductive, convective or radiational);
2. non-uniform distribution of water is observed demonstrating that besides the regular progress of evaporation zone the sublimation takes place from the whole volume of the solid.

Products of cryocrystallization in the majority of 'water–salt' systems have cellular structure (Figure 2.38). The walls of the structure are constituted by the low-temperature crystallized solution with critical concentration (three-dimensional structure) and its pores are filled with separate ice crystals (one-dimensional structure). Contacts between ice crystals are quite absent, i.e. each ice microgranule represents a localized sublimation zone. In similar systems the energy supply to the external surface leads mainly to radial distribution of heat front inside the salt framework. (Heat front is a zone of high local temperature gradient in the drying product.) The effect is intensified because of the sufficiently fast rupture of the salt framework – an ice granule structure due to

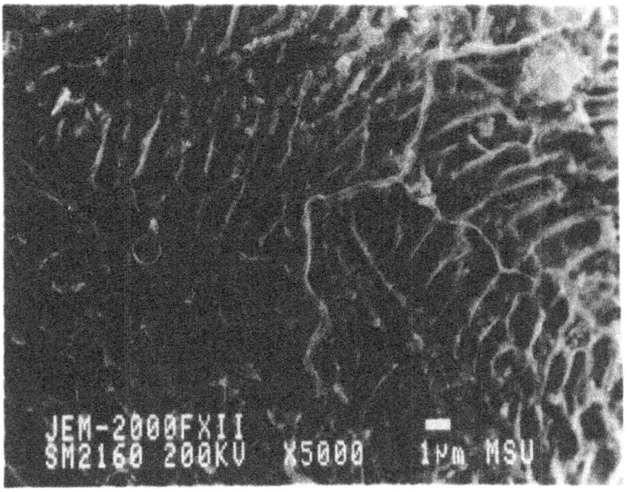

FIGURE 2.38 SEM image of porous salt framework formed during freeze-drying.

sublimation process at the 'ice–salt' interface. (Sometimes ice crystals can form a two- or three-dimensional structure. The existence of the last may facilitate considerably the process of removing water vapours from cryocrystallization products.)

The vapour formed due to sublimation is initially accumulated in micropores. The role of convective heat exchange increases. Then, pressure increase leads to the local destructions of the walls of salt framework and the formation of perforations in it.

As a result of this process the system of separate micropores transforms into an interconnected system. This process, however, occurs behind the developing heat front.

In addition, it is assumed that in the area of contacts of liquid microzones with solid phase a local three-phase equilibrium is obtained (gaseous phase is always present in the system) and it causes maintenance of a temperature near 273 K in sublimation microzones. In this case $p_s^0 \gg p_s$, and Equation 2.27 may be simplified:

$$dm/d\tau = kS_p p_s^0 \qquad (2.28)$$

i.e. for ice sublimation:

$$dm/d\tau = \beta S_p = \beta S_p^0 \exp(-E_a/RT) \qquad (2.29)$$

where $\beta = kS_p^0$ is the proportional coefficient.

The experimental value for activation energy $E_a = 8.4 \pm 2.1$ kJ/mole is close to the heat of melting of ice ($\Delta H_f = 6.69$ kJ/mole). This favours the microphase melting model. According to this model, the changes in heat supply do not lead to changes in surface temperature which is maintained constant in sublimation microzones and equal to $T_s = 273$ K. As to supplied energy, it is consumed for the formation of liquid phase or more precisely of the reaction interface.

It should be pointed out that considered models are not contrary to each other because they describe the process in different temperature ranges. Actually, the coincidence of experimental curves calculated using Equation 2.22 with condensation coefficient $\alpha = 1$ is observed in the range of 190–200 K only. At $T > 200$ K it is necessary to introduce a value of $\alpha(T) < 1$, and this means that the mechanism of direct transition of separate molecules into vapour is substituted for another one with temperature increase. Finally, at the temperature $T = 273$ K direct transition is completely absent ($\alpha = 0.15$) and in this temperature range the mechanism of microphase melting prevails.

At the same time, the model based on the concept of jet evaporation from microzones of liquid phase and describing the satisfactory kinetic behaviour of systems under sublimation (Guhman and Ermakova, 1953) explains the peculiarities of heat and mass exchange in the process of ice sublimation very well, in particular the increase of heat exchange coefficient in the range of pressures of 10–10^3 Pa. 'Jet evaporation' of ice in a vacuum at these pressures multiplies the volume of substance by 10^4–10^6. Jets formed at discrete surface areas create very complex and disordered motion in the working chamber of the sublimator due to the appearance of flow with a great multiplicity (multiplicity of circulation for some substances can reach 30–130). It results in the local increase of vapour pressure at the sublimate surface by a factor of 2.5–3.3.

An alternative explanation for a sharp increase of heat exchange coefficient in the pressure range of 13.3–1333 Pa is given by the hypothesis of 'volume evaporation' developed by A.V. Lyikov (Lyikov, 1968). It is based on the experimentally observed phenomenon of the carrying away of fine ice crystals from the surface without heat losses for evaporation due to fluctuations in surface development.

Processes of ice sublimation are considerably more complicated in multi-component systems. It was noted above that, as a result of the freezing of water–salt solutions followed by their crystallization upon heating, there occurs a specific salt structure (Figure 2.38) which determines some peculiarities in the processes of freeze-drying.

First, this structure considerably decreases the probability of development of sublimation with participation of local fluctuation mechanisms such as 'jet evaporation' (Guhman, 1953, 1968) and 'volume evaporation' (Lyikov, 1968).

Secondly, the presence of dry salt framework at the surface of cryogranules creates additional difficulties for vapour transport from the sublimation zone and enhances the contribution of internal and external diffusion in the whole sublimation process. It is to be noted that in this case the internal diffusion means the diffusion occurring inside granules and the external diffusion means the diffusion through the layer of granules.

Vapour transfer from the boundary of granules layer to the desublimator (condenser) is usually not considered. Such a transfer limits the whole process only in cases where it is carried out under conditions of high energy supply to sublimating materials (pure ice or solutions with low concentration of solute) or in some kinds of (laboratory) sublimators with separated sublimation and condensation zones (see, for example, section 2.2.1).

Let us consider the process of freeze-drying of frozen aqueous salt solutions for the simplest case of reaction zone: flat interface (see Figure 2.39). The rates of different stages of the process at an instant of time $\tau = 0$ may be represented by the following equations (Lyikov and Vasil'ev, 1980):

1. rate of ice sublimation at the *AB* interface within a cryogranule:

$$w_{sp} = 1/S^* \, dm/d\tau = K_{sub}(P_0 - P_v) \tag{2.30}$$

2. rate of internal diffusion (vapour transfer from sublimation zone *AB* through salt framework to the surface of cryogranule *CD*):

$$w_{sp}' = 1/S^* \, dm/d\tau = K_D'(P_v - P_s) \tag{2.31}$$

3. rate of external diffusion (from the geometrical boundary of separate cryogranules *CD* to the boundary of substance under sublimation *EF*):

$$w_{sp}'' = 1/S^* \, dm/d\tau = K_D''(P_s - P_k) \tag{2.32}$$

P_0 is the equilibrium water vapour pressure over ice in the centre of each separate cryogranule; for capillary-porous solids (e.g. pure ice) the limit value of P_0 is the value of total pressure in the sublimator P_c;

P_v is the pressure in sublimation zone (*AB* boundary);

P_s is the pressure near the geometrical surface of separate cryogranules (*CD* boundary);

P_k is the pressure near the desublimator (condenser) surface existing in practice near the surface of the whole ensemble of cryogranules under dehydration (*EF* boundary);

K_{sub}, K_D', K_D'' are constants describing process stages – sublimation, internal and external diffusion, respectively.

Taking into account that in the stationary regime $w_{sp} = w_{sp}' = w_{sp}''$, it is possible to obtain the following solutions for intermediate pressures P_s and P_v:

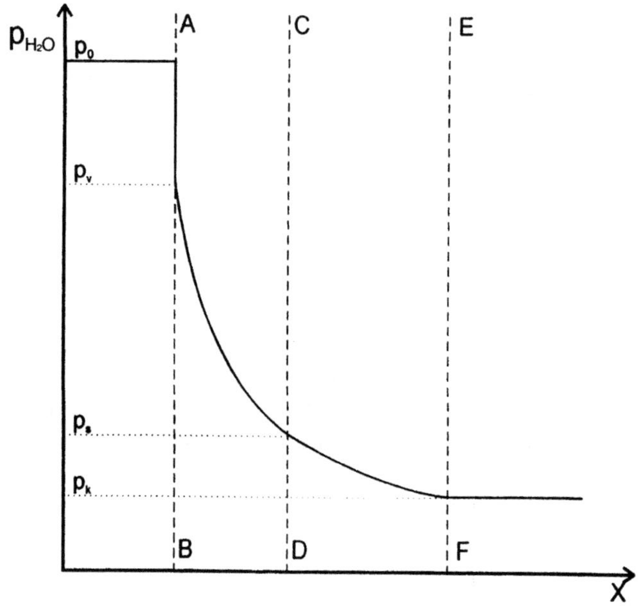

FIGURE 2.39 Scheme of vapour pressure distribution upon freeze-drying. *AB* – front of ice sublimation; *CD* – geometrical boundary of cryogranule; *EF* – boundary of substance layer.

$$P_s = (K_D{}'P_v + K_D{}''P_k)/(K_D{}' + K_D{}'') \tag{2.33}$$

$$P_v = (K_{sub}K_D{}'P_0 + K_{sub}K_D{}''P_0 + K_D{}'K_D{}''P_k)/(K_D{}'K_D{}'' + K_{sub}K_D{}' + K_{sub}K_D{}'') \tag{2.34}$$

Simultaneous analysis of these equations allows one to make the following conclusions:

1. if the limiting stage of the freeze-drying process is the sublimation at the *AB* boundary ($K_{sub} \to 0$), then $P_v = P_s = P_k$ (Figure 2.40, curve 1);
2. if the process is limited by the external diffusion ($K'' \to 0$), then $P_v = P_s = P_0$ (Figure 2.40, curve 2).
3. if the limiting stage is the internal diffusion ($K' \to 0$), then $P_v = P_0$, $P_s = P_k$ (Figure 2.40, curve 3).

This analysis indicates that the intermediate pressures P_v and P_s (at boundaries *AB* and *CD*, respectively) may be changed over a wide range from P_k to P_0. In these conditions the maximum value P_v within a separate cryogranule under freeze-drying is attained in the case when:

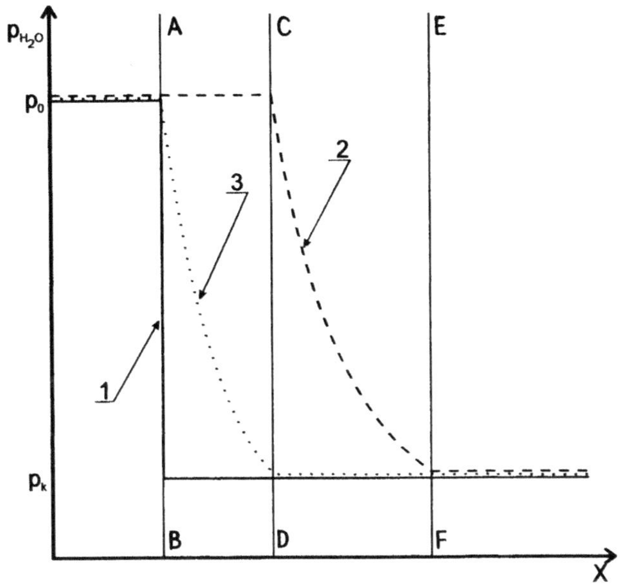

FIGURE 2.40 Profiles of various types of freeze-drying process. The limiting stage is: (1) ice sublimation; (2) external diffusion; (3) internal diffusion.

1. the process is limited by diffusional transfer;
2. the substance conserves its properties of capillary-porous solidity, first of all permeability for gases, during freeze-drying.

If the last condition is not met the water vapour's value in the centre of a separate cryogranule P_0' can overcome considerably the P_0 value characteristic for capillary-porous solids. A similar situation commonly occurs if a thick salt layer is formed at the surface of cryogranules under sublimation near a *CD* boundary as the result of intensive restructuration of dry salt framework (see Figure 2.40). The thick salt layer complicates gasodynamic conditions within a separate cryogranule: the role of diffusional impedance increases as well as the absolute value of pressure P_0' and value of temperature T_0', corresponding to the ice–vapour equilibrium at the pressure P_0'.

Similar processes can explicate the phenomenon observed by so-called 'drop technique' (see Figure 2.41). This technique for the measurement of temperature within capillary-porous solids (ice) under freeze-drying allows a plateau to be observed on the curve $t = f(\tau)$. Deviation from parameters characteristic for capillary-porous solids leads to monotonically decreasing function $t = f(\tau)$ (see Figure 2.42), manifesting the increase of water vapour pressure inside separate granules.

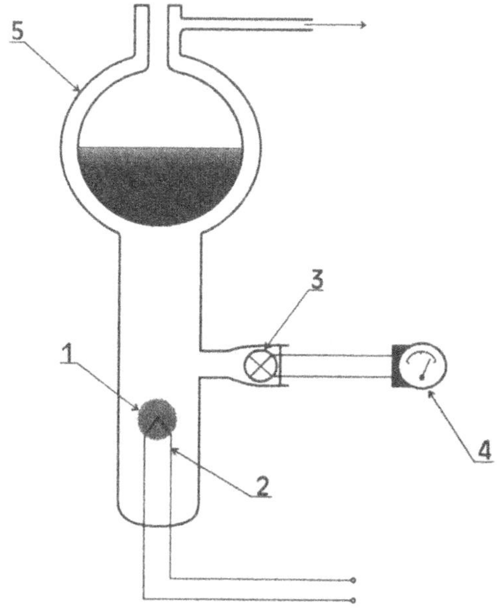

FIGURE 2.41 Device for temperature measurements in the object under freeze-drying by 'droplet method'. (1) droplet of frozen solution; (2) thermocouple; (3) pressure gauge; (4) vacuummeter; (5) cold trap.

The increase of pressure in water vapours and the simultaneous increase of temperature within cryogranules leads to the increasing influence of these vapours on the substances which are in a metastable state, followed by considerable changes in the microstructure of the salt framework formed at the first stage of freeze-drying.

The change in the microstructure under such conditions occurs through the well-known mechanism of 'low-temperature sintering' (Ivensen, 1971) and results in shrinkage of the salt product layer. This, in turn, can lead to further increase of pressure in water vapours within a salt framework and to further restructuring of the last one.

Thus, the considered kinetic model allows one to conclude that the principal factor affecting the structure of freeze-drying products is the existence of water vapours within separate cryogranules. Their pressure value is greatly determined by internal diffusion caused by vapour transport through the salt framework.

On the basis of many experiments performed by the 'frozen drop' technique (Figures 2.41 and 2.42) the following conclusions can be made:

1. for a restricted number of salt systems with low eutectic temperatures T_{eut}, including some nitrates, the water vapour pressure inside cryogranules can exceed pressure value P_{eut}, corresponding to T_{eut} at the phase diagram, which results in product melting;

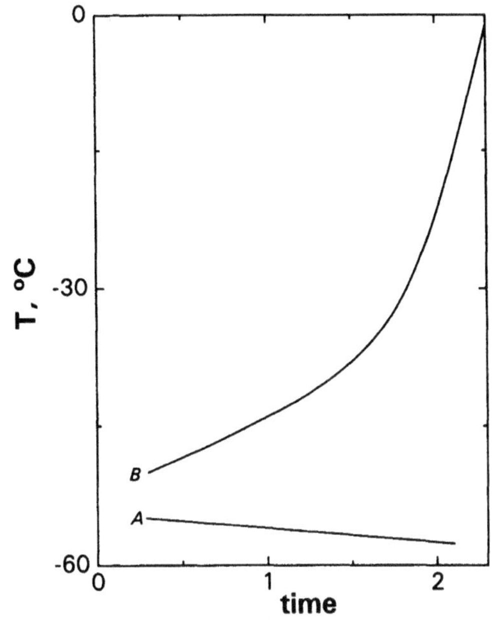

FIGURE 2.42 Temperature *vs.* time upon freeze-drying: (a) porous solid; (b) frozen solution of salts.

2. for some products like frozen solutions of many ammonia salts the internal diffusion is practically absent;
3. finally, for the great majority of salt products the internal diffusion proved to be limiting stage of reaction: the decrease of salt concentration in solution results in the possibility of freeze-drying without appreciable changes in salt framework structure (Figure 2.43).

The model gives a chance to explain results of measuring the specific surface of K_2SO_4, obtained by freeze-drying of frozen solutions with eutectic concentration (6.7 mass %, $T_{eut} = -1.51\,°C$). After drying of initially neutral K_2SO_4 solution (pH = 7) the product with value of $S_{sp} = 10\,m^2/g$ was obtained. Addition of small amounts of KOH (up to pH = 12.5) or H_2SO_4 (up to pH = 1.0) decreases the value of specific surface down to $3\,m^2/g$. In both cases the negative role of liquid phase is obvious because the temperature of eutectics in binary systems KOH–H_2O and H_2SO_4–H_2O is considerably below the similar value in the system K_2SO_4–H_2O.

The phenomenon of volume shrinkage of the system relative to the volume of frozen product and microstructural changes during the freeze-drying process was observed by different authors (Savchenko, 1982; Lloyd *et al.*, 1989) and received the name 'collapse'.

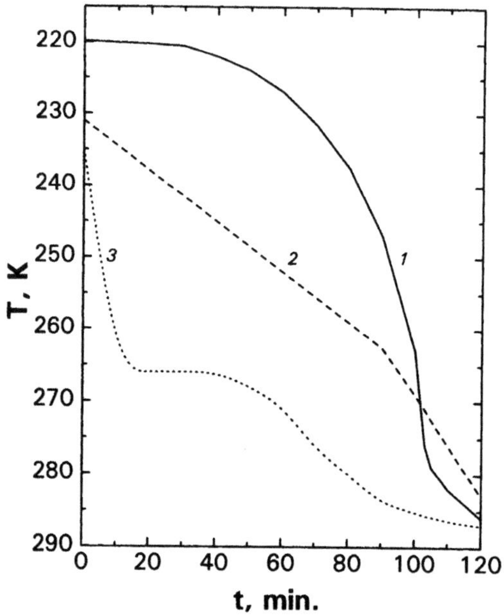

FIGURE 2.43 Temperature *vs.* time dependence for frozen $ZnSO_4$ solutions of the following concentrations: (1) $c = 7.0$ wt.%; (2) $c = 27.8$ wt.%; (3) $c = 37.0$ wt.%.

The explanation of collapse is based on dependence of freezing temperature on the curvature of water surface. Thus, the freezing temperature of micro-droplets with radius $r > 100$ nm, corresponds to 273 K. However, further decrease of this radius leads to considerable depression of water-freezing temperature (Blachere and Younge, 1972). A decisive role is played by multiple channels in the frozen salt framework (Figure 2.44).

Thus, to explain the irreversible changes in the salt framework structure this model assumes permanent presence of some quantities of 'unfrozen' liquid water in the system cooled down to $T < 273$ K.

Irreversible changes in salt framework have an outstanding influence on the synthesis of materials with desired properties. The experience of applying cryochemical technology to materials engineering shows that the stage of freeze-drying is one of the most important in controlling structure-sensitive properties of materials.

This can be illustrated by correlation between coercive force H_c of ferrites and heat supply value at the stage of freeze-drying of a frozen solution used as a ferrite precursor.

As is shown in Figure 2.45, the high values of coercive force H_c for ferrite $Ni_{0.5}Zn_{0.5}Fe_2O_4$, obtained by cryochemical technique from sulphate solutions are observed at minimum temperature (−20 °C) of heating plates maintained during the freeze-drying of cryogranules. With the increase of heat input intensity by

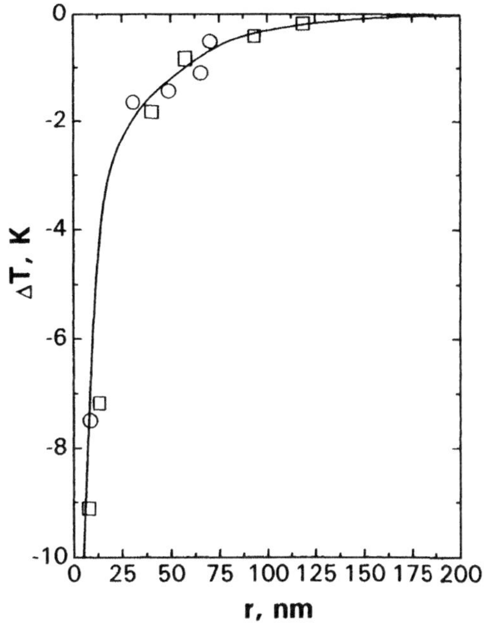

FIGURE 2.44 Dependence of melting temperature decrease ΔT on capillary radius (circles and triangles – experimental results, solid line – theoretically calculated by Laplace–Kelvin equation).

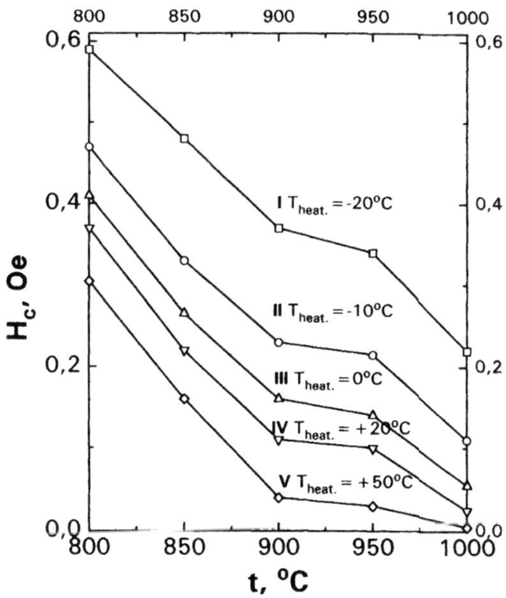

FIGURE 2.45 Coercive force H_c vs. temperature of heat treatment for $Ni_{0.5}Zn_{0.5}Fe_2O_4$ ferrite powders synthesized at different heat supply during freeze-drying.

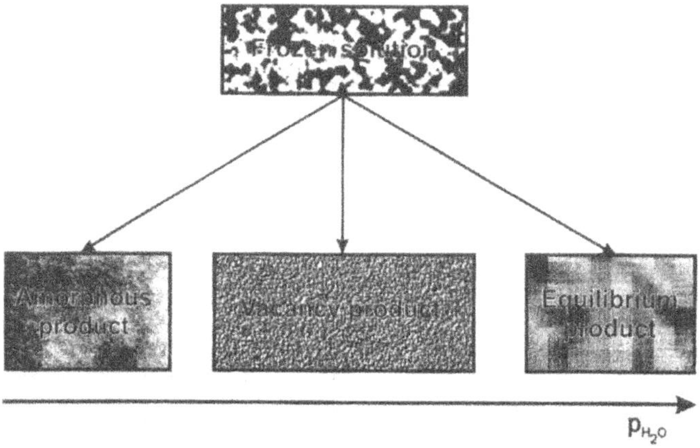

FIGURE 2.46 The main types of freeze-drying products.

rising heating plate temperature to +50 °C, the H_c value decreases considerably. A similar effect is observed also with increasing temperature of heat treatment of salt products (from 800 to 1000 °C) (Shlyakhtin, 1985).

Taking into account that the value of H_c is inversely proportional to the average size of crystallites in ceramics it may be supposed that the total suppression of water vapour influence on recrystallization processes in salt products under freeze-drying is impossible. It is clear that these effects should be taken into account during freeze-drying of frozen water–salt systems with relatively low eutectic temperatures. Alternative techniques of solvent removal, such as cryo-extraction and cryoprecipitation, also cannot resolve these problems completely.

In conclusion let us consider another aspect of freeze-drying connected with the fact that the change in water vapour pressure may lead to obtaining different kinds of salt products. For instance, freeze-drying of frozen aqueous solutions of bivalent metal sulphates $MeSO_4$ (where Me = Mg, Mn, Fe, Co, Ni, Cu, Zn) results in the formation of three products with different structure and properties (Figure 2.46), namely equilibrium lower hydrates, lacunary phases (vacancy structures) and X-ray amorphous hydrates.

The possibility of obtaining one or another product is determined by the different factors, including:

1. the chemical nature of the salt whose solution is frozen and submitted to freeze-drying;
2. the physical and chemical properties of the solution (pH, concentration etc.) used as a precursor (Figure 2.47);
3. the parameters of freeze-drying – consecutive decrease of the water vapour pressure P_{H_2O} causes the transition from the equilibrium product to the vacancy product and further to the X-ray amorphous one.

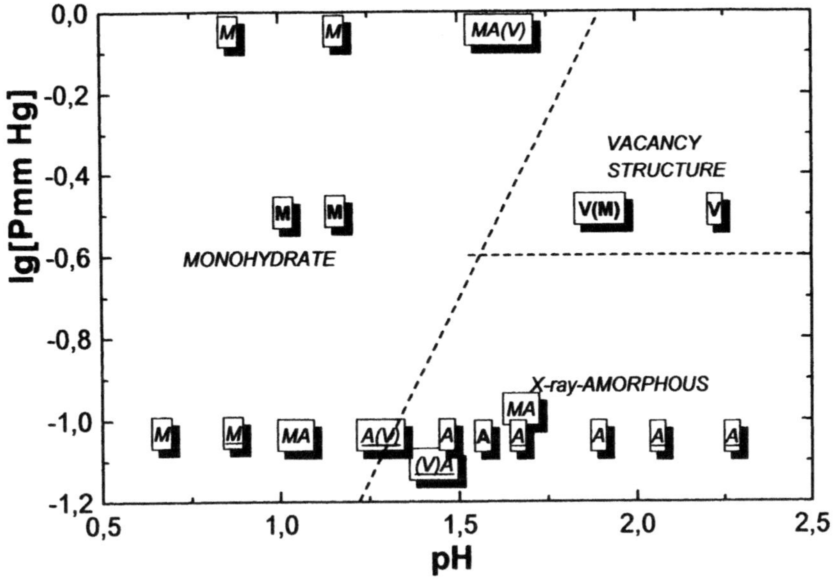

FIGURE 2.47 Dependence of the most probable phase composition of freeze-dried products on pH of initial solution and p_{total} in the vacuum chamber.

Some feasible recommendations to exclude permanent presence of liquid phase while freeze-drying include the following actions.

1. It is necessary to optimize the heat supply: increase of heat supply intensity, especially at first stages, above some critical values leads to the permanent presence of the liquid phase.
2. It is necessary to optimize the size of the object under drying. Reducing the size of cryogranules to a certain limit decreases the role of internal diffusion while further reducing this size favours the external diffusion as a limiting stage. In this case the most characteristic feature is a tendency of liquid-phase sintering of the product during freeze-drying. (Liquid-phase sintering is the intensive shrinkage of a product, promoted by the presence of small amounts of liquid phase at grain boundaries. See also 'low-temperature sintering', above, p. 96.)
3. It is necessary to control the microstructure of the salt framework at crystallization stage: the framework structure should be more permeable for water vapours. A specific interest is represented by the self-freezing process which only takes place in the sublimation chamber at the first stage of freeze-drying due to the intensive evaporation of a part of solvent. In this case it is possible

to carry out uniform and rapid freezing of large volumes of solutions. In addition, self-freezing facilitates the process technologically: the solution is introduced at once into a vacuum sublimation chamber and then the pumping of vapour–gas mixture begins.

Self-freezing is also advantageous from the energy-saving point of view because with the freezing of liquid the evolution of latent heat of ice melting occurs in the material which is used for sublimation. This phenomenon decreases power consumption.

Finally, during self-freezing the crystallization occurs gradually from the surface to the volume of the solid, ensuring the creation of a permeable microstructure in frozen layers due to high vapour pressure in internal unfrozen layers.

One of the disadvantages of self-freezing is its layer-by-layer character which can lead to the segregation of salt components and the appearance of inhomogeneities. Optimizing the thickness of the frozen solution layer gave good results in some cases and served as the basis for the application of the self-freezing technique in multicomponent inorganic materials technology.

The microstructure of cryogranules may be controlled not only by using various cooling rates but also by the variation of final temperatures of freezing in the range $170\,K < T < 240\,K$. At optimum cooling rate and temperature it is possible to obtain a product consisting of two (ice and salt) interpenetrating one-dimensional structures similar to a eutectic one characterized by minimum resistance to internal diffusion.

2.2.3 Kinetics of ice sublimation from frozen aqueous salt solutions

To characterize the kinetics of the process the following relations are usually considered (Figure 2.48):

1. the variations of dehydration degree α with the time (curve 1); in this case the dehydration degree is the ratio of water mass m_i removed from the frozen solution during time τ_i to the mass of water m_0, contained in it at initial time τ_0;
2. the variations of dehydration rate $d\alpha/d\tau$ with time (curve 2);
3. the variations of product temperature with time (curve 3).

For ordinary water–salt systems, freeze-drying kinetic curves $\alpha = f(\tau)$ (Figure 2.49, a) contain a short initial section where the stabilization of vacuum and temperature occurs and two main sections with a constant ($d\alpha/d\tau$ = const) and decreasing ($d\alpha/d\tau \neq$ const) dehydration rate (Kelly *et al.*, 1978).

For the systems where the formation of crystallohydrates occurs, freeze-drying curves are more complicated (Figure 2.49, b) and include at least three main

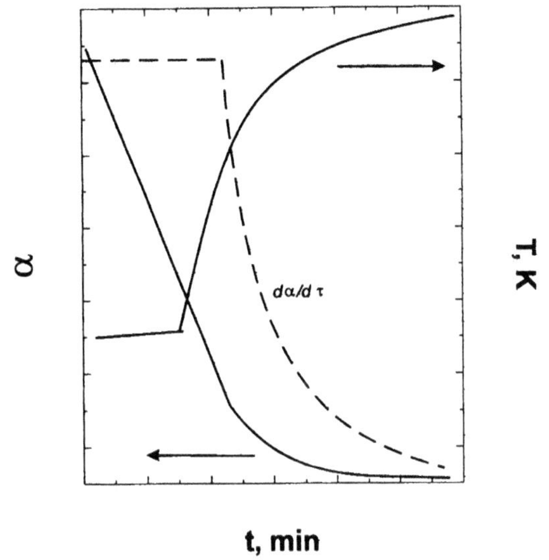

FIGURE 2.48 Dependencies of dehydration degree α (1), rate of freeze-drying dα/dτ (2) and material's temperature on process duration.

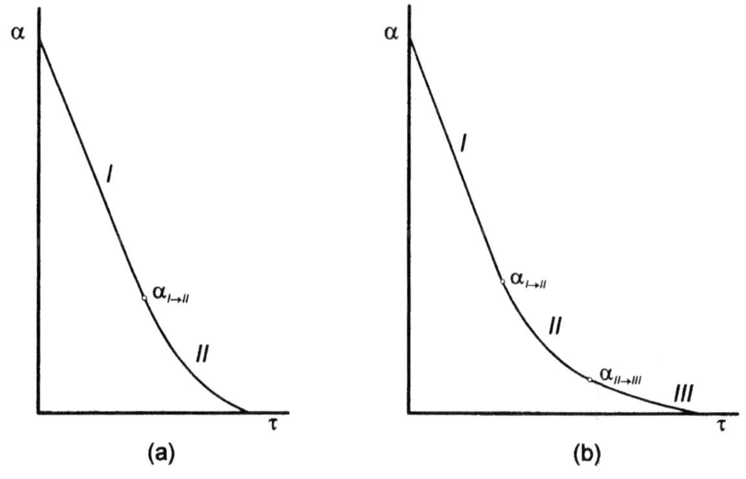

FIGURE 2.49 Kinetic curves of freeze-drying process for products obtained by cryocrystallization of water–salt systems: (a) without formation of crystallohydrates; (b) with formation of crystallohydrates.

sections: one with a constant rate and two with decreasing rate of freeze-drying, which are described by the following formal kinetic equations (Savchenko, 1982):

$$\text{I section}: \; d\alpha/d\tau = \text{const} = k_1 \quad (2.35)$$

$$\text{II section}: \; d\alpha/d\tau = k_2(1-\alpha)^{2/3} \quad (2.36)$$

$$\text{III section}: \; d\alpha/d\tau = k_3(1-\alpha) \quad (2.37)$$

It is reasonable to fix some points in the curve $\alpha = f(\tau)$ corresponding to:

$\alpha_{\text{I–II}}$ – transition from the stage with a constant rate of freeze-drying to the stage with decreasing rate;
$\alpha_{\text{II–III}}$ – change in mechanism of dehydration at the stage with decreasing rate.

Special attention should be given to the fact that kinetic equations describing the process at all three stages can be obtained assuming that process is limited by reactions at phase boundary which results in the equation:

$$d\alpha/d\tau = k(1-\alpha)^n \quad (2.38)$$

where $n = 0$, $\frac{2}{3}$ and 1 for stages I, II and III respectively.

Particular emphasis in this model is placed upon influence of internal diffusion in local areas on the process mechanism. At the same time internal diffusion is not the limiting stage of the whole process. It occurs by a series of reasons, including:

1. space separation of reaction microzones;
2. self-regulating character of heat transfer in the reaction microzones.

By kinetic analysis of ice sublimation in frozen water–salt systems it was established that $\alpha_{\text{I–II}}$ and $\alpha_{\text{II–III}}$ are key characteristics of the process: at a given state of initial solutions the values of $\alpha_{\text{I–II}}$ and $\alpha_{\text{II–III}}$ are independent of the initial mass of the substance submitted for freeze-drying.

From values of $\alpha_{\text{I–II}}$ and $\alpha_{\text{II–III}}$ parameters $h_{\text{I–II}}$ and $h_{\text{II–III}}$ equal to the number of water moles per one salt mole were estimated assuming a constant quantity of water in the system:

$$h_{i-i+1} = (1 + \alpha_{i-i+1})(100M_c - M_cC - 18nC)/18C + n \quad (2.39)$$

where n is the number of water moles per one mole of anhydrous salt in the final product; M_c is the relative molecular mass of anhydrous product; and C is the concentration of initial solution, mass %.

It was established that for many investigated water–salt systems including Me_2X–H_2O and MeY–H_2O ($Me = K^+$, Na^+, NH_4^+; $X = SO_4^{2-}$; $Y = NO_3^-$, HCO_3^-, as well for $MeSO_4$–H_2O ($Me = Cu^{2+}$, Fe^{2+}, Co^{2+}, Ni^{2+}, Mg^{2+}, Zn^{2+}, Mn^{2+}) and $MeSO_4$–$(NH_4)_2SO_4$–H_2O ($Me = Fe^{2+}$, Co^{2+}, Mn^{2+}), parameters $h_{\text{I–II}}$

TABLE 2.12 Parameters of the equation $h_{i\to i+1} = AM_c/C + B$.

System	$h_{i\to i+1}$	T_n, K	*A*	B
Me_2X–H_2O MeY–H_2O ($Me = K^+, Na^+, NH_4^+$; $X = SO_4^{2-}, Y = NO_3^-, HCO_3^-$)	$h_{I\to II}$	273	2.16 ± 0.12	-0.50 ± 1.10
		296	1.46 ± 0.10	-0.30 ± 0.80
		313	0.65 ± 0.28	-0.80 ± 2.30
$MeSO_4$–H_2O ($Me = Fe^{2+}, Co^{2+}, Ni^{2+}, Mg^{2+}, Zn^{2+}, Mn^{2+}, Cu^{2+}$)	$h_{I\to II}$	273	2.18 ± 0.02	2.40 ± 1.20
		296	2.03 ± 0.08	2.50 ± 3.20
		313	1.98 ± 0.07	1.80 ± 0.70
	$h_{II\to III}$	273	0.84 ± 0.24	4.00 ± 0.70
		296	1.19 ± 0.16	0.70 ± 0.50
		313	1.34 ± 0.24	-1.80 ± 0.70
$(NH_4)_2SO_4$–$MeSO_4$–H_2O ($Me = Fe^{2+}, Co^{2+}, Mn^{2+}$)	$h_{I\to II}$	273	2.79 ± 0.05	-6.90 ± 0.50
		296	2.67 ± 0.07	-6.30 ± 0.50
		313	2.58 ± 0.12	-6.50 ± 0.50
	$h_{II\to III}$	273		
		296	1.01 ± 0.07	-0.80 ± 0.20
		313		

and $h_{II\text{–}III}$ have a linear dependence on the inverse concentration of initial solutions C (Savchenko, 1982), i.e.

$$h_{i-i+1} = AM_c/C + B \tag{2.40}$$

where A and B are constants whose values are listed in Table 2.12, and kinetic curves for freeze-drying may be calculated from experimental value $d\alpha/d\tau = \text{const}$ only.

It is evident that in the points $\alpha_{I\text{–}II}$ and $\alpha_{II\text{–}III}$ the condition of equality of rates of two subsequent stages of process is observed:

for α_{I-II}: $$(d\alpha/d\tau)_I = (d\alpha/d\tau)_{II} \tag{2.41}$$

for α_{II-III}: $$(d\alpha/d\tau)_{II} = (d\alpha/d\tau)_{III} \tag{2.42}$$

The set of relations (2.41)–(2.42) and formal kinetic equations (2.35)–(2.37) lead to equations

$$(d\alpha/d\tau)_I = k_1 = k_2(1 - \alpha_{I-II})^{2/3} \tag{2.43}$$

$$k_2 = (1 - \alpha_{II-III})^{2/3} = k_3(1 - \alpha_{II-III}) \tag{2.44}$$

which allow the empirical constants, necessary to calculate the whole curve, to be estimated.

Rules found for two-component systems allow the characteristics of kinetic curves for freeze-drying in more complex systems to be forecast.

2.2.4 Peculiarities of the freeze-drying of colloidal solutions

The analysis of various salt systems (section 2.1) has shown that nitrates are the most suitable compounds for cryochemical synthesis. In fact, most of them are well-soluble in water, compatible with each other in solution and readily decompose forming oxides at relatively low temperatures. At the same time, the nitrates of multicharged ions (+2 and, especially, +3 and +4) are remarkably susceptible to hydrolysis in aqueous solutions:

$$Me(NO_3)_n + H_2O = Me(OH)(NO_3)_{n-1} + HNO_3 \tag{2.45}$$

Such a process can be regarded as transformation of a rather simple binary system system '$Me(NO_3)_n$–H_2O' into a more complex ternary '$Me(NO_3)_n$–HNO_3–H_2O'. The presence of free HNO_3 will complicate subsequent cryocrystallization and freeze-drying to a considerable extent. The melting points of eutectics in the binary system HNO_3–H_2O (−60 to −70 °C) are much lower compared to the temperature of ice surface (−40 to −45 °C) upon freeze-drying under standard conditions (1–10^2 Pa). As a result, in the course of freeze-drying there appear certain amounts of liquid phase, leading to the so-called collapse of the salt framework and to the loss of product homogeneity.

However, the problem may be resolved if true solutions are transformed into colloidal ones, for example, by anion exchange treatment (Fedotov, 1984). The soaking of OH-form anionite resin in the solutions results in a controlled shift of ionic hydrolytic equilibrium:

$$\text{An-OH} + NO_3^- = AnNO_3 + OH^- \tag{2.46}$$

Partial neutralization of salt and a further shift of hydrolytic equilibrium leads to the formation of colloidal particles of hydroxopolymers with high values of polymerization degree *m*

$$OH^- + H^+ = H_2O \tag{2.47}$$

$$Me(OH)(NO_3)_{n-1} + H_2O = Me(OH)_2(NO_3)_{n-2} + HNO_3 \tag{2.48}$$

$$m[Me(OH)_2]^{(n-2)+} = [Me(OH)_x]_m{}^{(n-x)m+} + (2-x)mOH^- \tag{2.49}$$

Colloidal solutions obtained in this way are more stable than the systems formed simply by the addition of the neutralizing agent to the salt solution. Thus, the precipitation of $Fe(OH)_3$ from $Fe(NO_3)_3$ solutions by aqueous ammonia takes place at pH = 2.2–2.3. At the same time, the colloidal solution of $Fe(OH)_3$ obtained by anion exchange remains stable up to pH = 4.0 (Sharikov, 1991). The reason for such phenomena is a considerable decrease of ionic strength of the solution in the course of ion exchange.

The stability of colloidal solutions of Fe(III) hydroxopolymers can be still enhanced by their treatment in periodic regime, when short-time soakings of anionite resin are followed by stabilization of the solution for some time without anionite, to reach an equilibrium between the solution and the forming of colloidal particles.

Taking into account that anion exchange allows the preparation of colloidal particles of practically pure hydroxides, the aqueous solutions of chlorides, instead of nitrates, can also be used in these syntheses.

Let us follow the process of iron oxide formation from $Fe(NO_3)_3$ and $FeCl_3$ solutions ($c = 0.5$ mole/l). The composition and properties of iron hydroxonitrate and hydroxochloride $Fe(OH)_{3-x}(Y)_x$ (where $Y = NO_3^-$, Cl^-) solutions prepared by anion exchange are listed in Table 2.13.

The results of observations on pH and pNO$_3$ (or pCl) changes after single short-time treatment of the solution by anionite can be summarized as follows:

1. The first (fast) stage of the process is the reaction of neutralization due to the replacement of NO_3^- or Cl^- ions for OH^- ions, accompanied by the decrease of total ionic strength of the system (Livage, 1985).

TABLE 2.13 Composition and properties of the $Fe(Y)_x(OH)_{3-x}$ ($Y = NO_3^-$, Cl^-) colloidal solutions at $C = 0.5$ mol/l. Composition is determined using ionometric analysis of NO_3^- and Cl^- ions in the solution.

Y	*Sample N*	pH ±1%	pNO$_3^-$(Cl^-) ±5%	*Range of melting temperatures* (°C) ± 1°C	*Composition*
NO_3^-	1	0.80	−0.15	−60 to −45	$x = 3$
	2	1.23	0.45	−45 to −35	$x = 1.7; (3 - x) = 1.3$
	3	1.45	0.75	−40 to −30	$x = 0.8; (3 - x) = 2.2$
	4	1.65	1.00	−35 to −30	$x = 0.5; (3 - x) = 2.5$
	5	2.00	1.45	−18 to −16	$x = 0.15; (3 - x) = 2.85$
	6	2.50	1.70	−10 to −8	$x = 0.10; (3 - x) = 2.90$
	7	3.00	2.05	−7 to −5	$x = 0.05; (3 - x) = 2.95$
	8	4.00	2.50	−5 to −4	$x = 0.02; (3 - x) = 2.98$
Cl^-	9	2.50	1.75	−10 to −9	$x = 0.25; (3 - x) = 2.75$
	10	3.50	2.15	−5 to −4	$x = 0.04; (3 - x) = 2.96$
	11	5.00	3.2	−3 to −2	$x = 0.01; (3 - x) = 2.99$

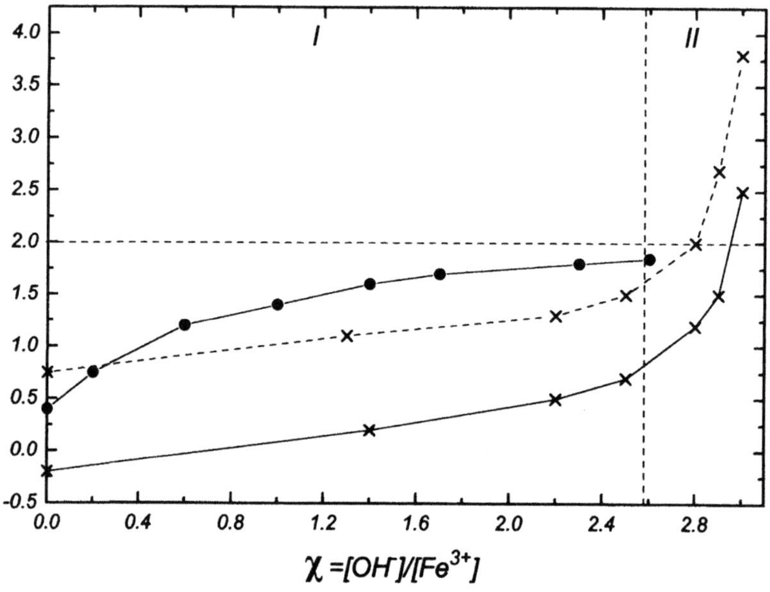

FIGURE 2.50 Dependencies pH(pNO$_3$)$=f(\chi)$ for the solutions of Fe(III) hydroxopolymers: I: area of colloidal solutions obtained by addition of base; II: extended area of stable colloidal solutions prepared by ion exchange.

2. The second (slow) stage can be considered as system ageing, when diffusion and, then, adsorption layers of colloidal particles slowly adapt to new equilibrium conditions. When ion exchange is carried out in strongly non-equilibrium conditions, the coagulation of the colloidal system occurs due to a decrease in its ζ potential.

The comparison of pH-*vs.*-χ curves ($\chi = [OH^-]/[Fe^{3+}]$) for anion exchange treatment of ferric nitrate solution (Sharikov, 1991) and alkaline titration of the latter (Mylay and Selwood, 1954; Spiro *et al.*, 1966) shows that anion exchange allows a substantial extension of the pH–χ area of stability for colloidal solutions of Fe(III) hydroxopolymers, including area II, the most appropriate for freeze-drying synthesis (Figure 2.50). A remarkable increase of pH, pNO$_3$ (pCl) values, as well as higher melting points of frozen solutions, is observed in this area.

The possibility for different colloidal solutions to be freeze-dried can be estimated by low-temperature differential scanning calorimetry DSC (Figure 2.51). Sharikov (1991) observed the direct correlation between the increase of solutions' pH, systematic increase of their melting points and reducing of melting range.

For most of the solutions with pH value below 2.0, freeze-drying seems to be practically impossible: the experimentally observed endothermic peaks actually

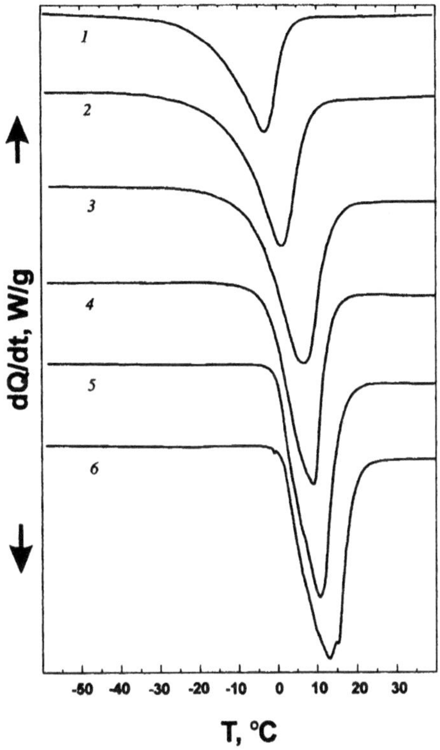

FIGURE 2.51 DSC curves for the melting process of frozen solutions: of $Fe(OH)_x(NO_3)_{3-x}$ at different pH of frozen solution: (1) pH = 0.8; (2) pH = 1.23; (3) pH = 1.45; (4) pH = 2.00; (5) pH = 2.50; (6) pH = 3.00.

represent a superposition of two effects corresponding to two simultaneous processes – freeze-drying itself and melting of the solutions. The freeze-drying of solutions with pH $\geq$ 2.0 shows no sign of melting, as it follows from the smooth continuous DSC curve for this process. The behaviour of solutions with pH = 2.5–3.0 upon freeze-drying is just indistinguishable from that of pure ice. The total absence of chemically bound water, belonging to hydration shells of separate ions, indicates the completion of hydroxopolymer particle formation.

The analysis of $dQ/d\tau = f(\tau)$ curves for the freeze-drying process of colloidal solutions allows three different regions to be distinguished. In the first region the rate of freeze-drying is constant and can be described by a formal kinetic equation of zero order $d\alpha/d\tau = k_1$. This region is clearly distinguished on thermogravimetric curves and corresponds to the removal of chemically unbound water, i.e. ice drying. The second region, with rate decreasing, corresponds to the removal of water chemisorbed in hydroxopolymeric framework. The rate of water

TABLE 2.14 Chemical composition of the product after freeze-drying of the colloidal solutions with different prehistory.

	Composition (wt% ± 0.5%)			
Solution prehistory	Fe_2O_3	H_2O	$NO_3^-(Cl^-)$	*Formula of the product*
N5 (pH = 2.0)	49.2	42.6	8.2	$Fe(OH)_{2.85}(NO_3)_{0.15}$
N6 (pH = 2.5)	50.2	44.6	5.6	$Fe(OH)_{2.90}(NO_3)_{0.10}$
N7 (pH = 3.0)	51.3	45.6	2.8	$Fe(OH)_{2.95}(NO_3)_{0.05}$
N8 (pH = 4.0)	51.9	46.9	1.2	$Fe(OH)_{2.98}(NO_3)_{0.02}$
N9 (pH = 2.5)	51.3	44.8	3.9	$Fe(OH)_{2.75}(NO_3)_{0.25}$
N10 (pH = 3.5)	51.9	45.4	2.7	$Fe(OH)_{2.85}(NO_3)_{0.15}$
N11 (pH = 2.5)	52.2	47.4	0.6	$Fe(OH)_{2.99}(NO_3)_{0.01}$

removal during the third stage, when the destruction of the hydrate shell of hydroxopolymers occurs, is the lowest and often defines the rate of the whole process. The duration of this stage can be considerably reduced by using solutions with pH = 3.0–4.0.

To study the influence of chemical prehistory on the properties of freeze-dried products, the colloidal solutions obtained from hydroxonitrates (Table 2.13, Nos. 5–8) and hydroxochlorides (Nos. 9–11) were chosen.

Chemical (ionometry) and thermal (TG and DSC) analyses of the freeze-drying process showed the identity of chemical composition for the products and corresponding hydroxopolymer solutions (Table 2.14). All these products were XRD-amorphous and non-hygroscopic; their low bulk density and properties related to it remained practically unchanged for a long time (up to three years of observations). This feature gives these products an essential advantage over highly hygroscopic freeze-dried products prepared from conventional solutions and undergoing irreversible changes (shrinkage, recrystallization etc.) during their storage.

The high stability of freeze-dried colloidal solutions was proved by Mössbauer studies, when the nearest surrounding of Fe^{3+} ions was observed during the transition from the initial solution to the freeze-dried product (Figure 2.52).

The spectrum of the initial solution (pH = 0.80) at $T = 77$ K have the only minimum without its splitting in the central part despite slight broadening caused by extra-fine relaxation effects. This corresponds to the presence of separate $Fe(H_2O)_6^{3+}$ ions, which do not interact with each other and are octahedrally surrounded. The spectrum of the colloidal solution (pH = 3.00) characterized by quadrupole splitting $\Delta = 6.67$ mm/s and isomeric shift $\delta = 0.22$ m/s corresponds to the formation of linear chain polymeric structures. The spectrum of freeze-dried products does not alter upon dehydration at 100 °C and is indistinguishable

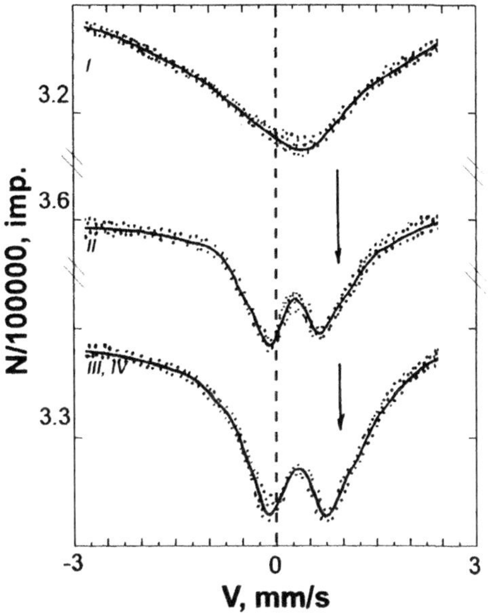

FIGURE 2.52 Mössbauer spectra of frozen solutions of Fe(III) hydroxopolymers and freeze-dried products. $T_{exposition}$ = 77 K. (I) initial solution, pH = 0.80; (II) colloidal solution pH = 3.0; (III) salt product of freeze-drying (prehistory 7, Table 2.13); (IV) the same salt product after dehydration at 100 °C.

with the spectrum of colloidal solution. Some increase of peak amplitude in the course of freeze-drying may be ascribed to the increase of Lamb–Mössbauer factor *f*, due to the higher rigidity of the polymeric structure of freeze-dried product compared to colloidal solution (Goldansky, 1968). These observations prove the high stability of the nearest surrounding of Fe^{3+} ion, formed at the stage of hydrolytic polymerization, against such powerful influences as freeze-drying and thermal dehydration.

Mössbauer studies of thermal decomposition of the freeze-dried product showed that at $T = 450\,°C$ the process of α-Fe_2O_3 crystallization was accomplished after treatment for 10 hours. The product of crystallization was formed in the so-called 'superparamagnetic state', when particle size does not exceed 120Å (Wertheim, 1964) (Figure 2.53). The similar processes are observed upon thermal decomposition when initial colloidal solution is obtained via hydroxochlorides.

The DSC analysis of thermal decomposition revealed some other important features. As the composition of colloidal particles, due to an increase of solution pH, approaches 'pure' ferric hydroxide, the temperature range of oxide crystallization extends and shifts to higher temperatures (Table 2.15). The

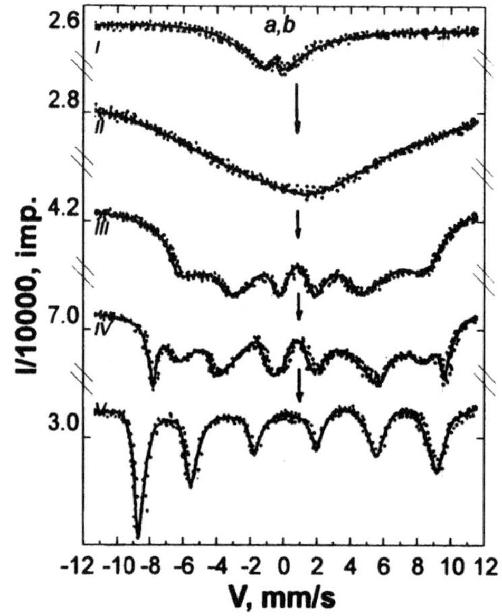

FIGURE 2.53 Mössbauer spectra at $T = 77$ K, corresponding to the formation of α-Fe_2O_3 fine powders during thermal treatment of amorphous $Fe(OH)_{3-x}$ $(NO_3)_x$ at 450°C. $T_{exposition}$ = 77 K, synthesis performed according to prehistory 8 (see Table 2.14).

crystallization behaviour of ferric oxide obtained from hydroxochloride differs considerably from the behaviour of oxides with hydroxonitrate prehistory due to much lower crystallization temperatures of the first and sharper peaks corresponding to its crystallization. The difference may be attributed to the influence of a residual chloride ion promoting the crystallization by local melting of ferric chloride (380 °C).

TABLE 2.15 The influence of hydoxonitrate and hydroxochloride compositions on the temperature range of crystallization of the products of thermolysis.

x in formula $Fe(OH)_{3-x}(NO_3)_x$	*Temperature interval of crystallization of products of thermolysis,* °C	*x in formula* $Fe(OH)_{3-x}Cl_x$	*Temperature interval of crystallization of products of thermolysis,* °C
0.15	350–420	0.25	290–330
0.10	360–460	0.04	300–350
0.05	400–550	0.01	340–420
0.02	420–600		

Therefore, to prepare stable dispersed oxide systems from colloidal solutions we should comply with the following requirements:

1. Anion exchange treatment should be periodically changed by ageing, to reach ionic equilibrium between colloidal particles and solution. High uniformity of colloidal particles achieved in this way determines retarded mass transfer during the subsequent heat treatment.
2. The freeze-drying of colloidal solutions seems to be the only way to preserve homogeneity of colloidal particles formed upon anion exchange treatment. Use of evaporation at 100 °C or dehydration at higher temperatures leads to agglomeration of separate colloidal particles into continuous 3D-framework, as it often occurs upon dehydration of precipitated (or co-precipitated) metal hydroxides. The role of water as a powerful surfactant in this process is assumed.
3. Treatment of colloidal solutions should be carried out until the highest possible pH values are reached (see Table 2.15). This is especially important, since in this case the better removal of residual alien ions, strongly affecting the system behaviour, can be achieved. The value of $pH = 7$ is the optimal if it can be reached in accord with the chemical nature of the system.

Syntheses of oxide materials conforming to these requirements enable the highly dispersed state of the system to be maintained up to rather high temperatures. This can be clearly illustrated by X-ray data for the samples prepared by thermal decomposition of freeze-dried products obtained from a colloidal solution of $Al(NO_3)_3$ with $pH = 7$ (Figure 2.54).

It is interesting to note that crystallization in such a system occurs only at temperatures $T > 1000\,°C$, while for hydroxides prepared by traditional co-precipitation technique the temperature does not exceed 700 °C.

Created technique of reproducible preparation of iron(III) hydroxopolymers colloidal solutions with high values of pH and concentration enabled them to be widely used for cryochemical synthesis of ferrites. In the last case, two different methods of chemical system formation were used:

1. The solution for cryocrystallization was prepared by mixing of salt solutions of ferrite-forming compounds ($LiNO_3$, $Ba(NO_3)_2$ etc.) with a previously obtained colloidal solution of iron(III) hydroxopolymers.
2. The solution for cryocrystallization was prepared by anionite treatment of solution-containing nitrates of all ferrite-forming elements.

In the first case, the introduction of other salts into already formed colloidal solution causes a sharp increase of ionic strength of the solution leading to loss of the stability and subsequent coagulation. For this reason, the freezing should be carried out immediately after homogenization. In this case colloidal solutions consist of monodisperse particles with a size of 100–120 Å. Such a degree of

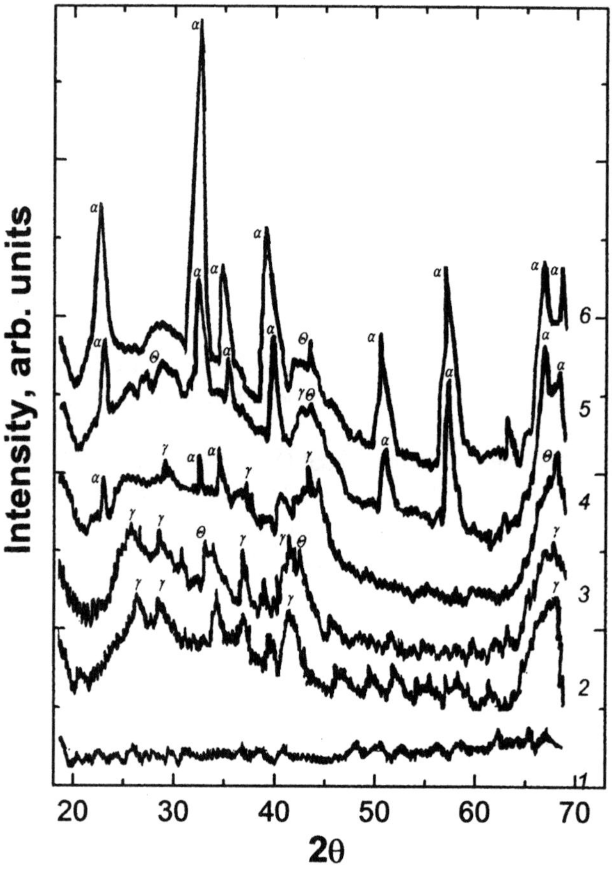

FIGURE 2.54 XRD patterns of freeze-dried aqueous alumina, obtained by using anionite treatment after one-hour annealing of previous sample at (1) 800°; (2) 800°; (3) 850°; (4) 900°; (5) 950°; (6) 1000 °C.

system dispersion allows single-phase ferrite products to be obtained at much lower temperatures compared to conventional methods of synthesis.

The second way is believed to ensure a higher degree of system organization, since ferrite-forming ions can be occluded by colloidal particles. At the same time, the influence of occluded ions may sometimes result in the formation of larger Fe_2O_3 particles with less reactivity compared with those obtained by the first technique.

Thus, the real structure of products is essentially determined by the choice of cryochemical synthesis technique.

Some features of freeze-drying may be efficiently used in the drying of suspensions, as well as in the so-called 'cryoimpregnation' process.

The freeze-drying of suspensions, which are often formed by the mixing of poorly compatible solutions, prevents the aggregation of separate particles. This makes it possible to form highly dispersed, low aggregated powders, which can be sintered at rather low temperatures yielding small-size crystallites (Shabatin, 1986).

The development of the cryoimpregnation technique was inspired by complications in the preparation of some materials by ordinary cryochemical techniques (freeze-drying, cryoextraction, cryoprecipitation), due to the incompatibility of initial substances in a true solution or incomplete precipitation of components in the last case. To resolve these problems Shabatin *et al.* (1987) proposed to combine different cryochemical techniques (Figure 2.55).

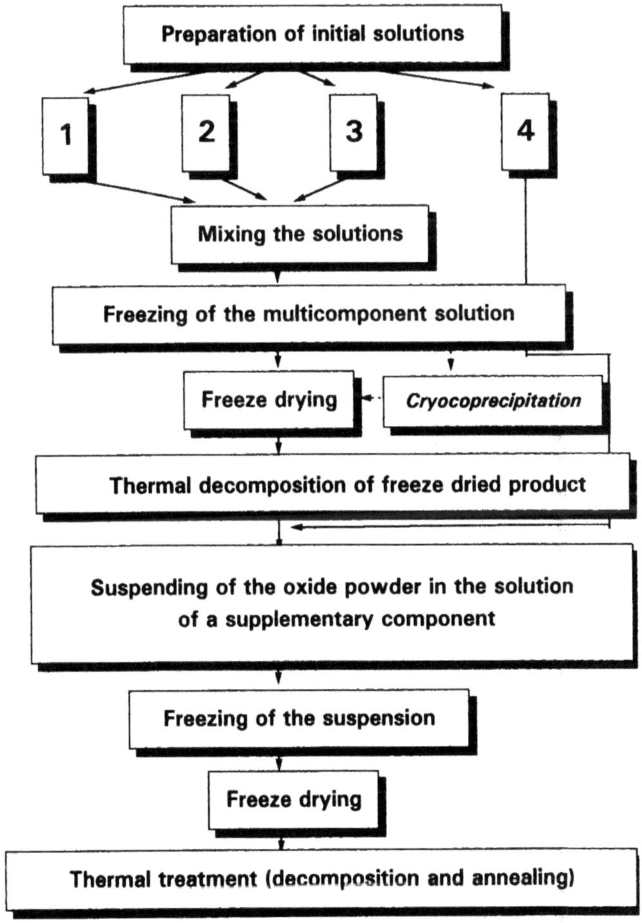

FIGURE 2.55 Complex methods of cryochemical synthesis.

The initial solution, prepared at the first stage, contains all or some of the compatible material-forming components. After freezing, freeze-drying and thermal decomposition this solution yields the product which is further used for the preparation of its suspension in the solution of salts of other components, absent in the initial working solution, which have to be present in the final composition. The obtained mixture is refrozen and freeze-dried; the product is subjected to the following thermal treatment. An undoubted advantage of the proposed scheme is its universality and the possibility of preparing a wide range of pure and doped materials.

A large variety of initial substances which can be used for the synthesis of multicomponent material create additional possibilities of affecting the material structure and its structure-sensitive properties, even in the frames of the same technique. Thus, upon cryochemical synthesis of aluminium–magnesium spinel from oxide powder and sulphate solution, the starting suspension can be prepared in two different ways: by suspending magnesium oxide in the solution of aluminium sulphate or by suspending aluminium oxide in the solution of magnesium sulphate. Studies of the different behaviour of these suspensions during synthesis give reason to favour the second way (Pankov *et al.*, 1982) because of better dispersion of aluminium oxide in magnesium sulphate solution. During ageing such a suspension was transformed into stable gelatinous mass yielding $MgAl_2O_4$ spinel with a more homogeneous ceramic structure. As to the number of possible alternatives, the largest number of them is offered by cryoimpregnation method.

2.2.5 Freeze-drying technique

Freeze-drying process is carried out in installations consisting of three main parts:

1. vacuum chamber where the treated product is placed;
2. condenser intended for removal of vapours from the system;
3. vacuum pumps.

In Figure 2.56 different ways for removal of water vapours from sublimators are represented (Tretyakov *et al.*, 1987). The most common is a technique based on vapour freezing on the condenser. To choose among many commercial sublimators it is necessary to take into account various factors including the nature of the product and its requirements as well as the necessary production scale.

Intensive application of rare earth and transition metal nitrates having rather low eutectic points as well as growing interest to freeze-drying of non-aqueous solutions make more severe demands on the drying equipment than in the food industry, the main consumer of these installations. The first principal demand

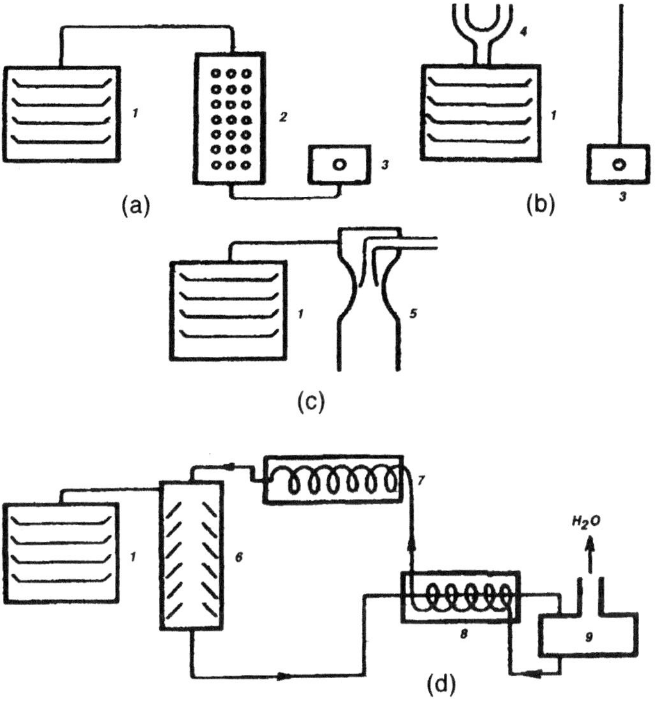

FIGURE 2.56 The main techniques of water vapour removal from vacuum system: (a) adsorption by solid adsorbents; (b) freezing by condenser; (c) direct pumping by flow pumps; (d) absorption by liquid absorbents ((1) drying chamber; (2) adsorbent; (3) vacuum pump; (4) cold trap; (5) vapour ejector; (6) absorber; (7) freezer; (8) heat-exchanger; (9) evaporator).

concerns the geometrical scheme of the equipment. There are two types of freeze-drying installations available on the market: an open type and a chamber type. In the first case the valves with frozen solutions located outside the mainframe of the freeze-dryer are attached by vacuum-flange to the condenser located within the installations. Any kind of heat supply control is often absent in this case.

Installations of the second type are characterized by location of the drying product in trays or vials on the shelves within the vacuum chamber. The shelves are equipped by heat supply control devices for most proposed models. The use of freeze-drying devices of the first group in materials synthesis creates serious complications at least or may be incompatible with the demands of synthesis due to the melting of frozen beads at the early stages of freeze-drying.

Different models of chamber-like freeze-dryers are supplied by controlled heating of shelves only or combined with cooling circuit. A limited number of synthesis tasks like suspension drying, when the formation of eutectics at $T \leq$ 270–265 K is excluded, allows the application of more compact and cheaper installations without cooled shelves. Nevertheless the presence of a cooling circuit is preferable in all cases, taking into account possible changes in the drying object. For the majority of tasks the cooling up to 233 K is enough because further lowering of heat supply leads to a drop in the sublimation rate which is far removed from acceptable values.

Multiple studies showed that a decrease in the condenser temperature lower than 200–210 K is not necessary because a small increase in the freeze-drying rate is accompanied in this case by fast growth of equipment costs. It may be reasonable only for the drying of non-aqueous solutions with low freezing points of solvent. Much more important, a decisive role is often played by the geometry of condenser location within the vacuum chamber. As shown earlier (Figure 2.39) the less distance from the product under drying to the condenser surface, the less p_{H_2O} value near the sublimating layer and hence the higher freeze-drying rate. It means that for the same condenser temperature we can either cut the process duration or improve the product quality by diminishing the amount of liquid phases through arranging the condenser closer to the product. This demand of geometrical closeness is satisfied in the large variety of installations where flat, ribbon-like or, better, surrounding cylinder-shaped condensers are located near drying shelves within the single vacuum chamber.

The power demands of the vacuum system during materials synthesis are rather moderate and adequate for most available models. The range of pressures achieved by conventional fore-pumps (1–10 Pa) approximately correlates with the maximum at the drying rate *vs.* residual pressure curve (Figure 2.32). The achievement of secondary vacuum values is combined with a marked growth in pumping costs and may be reasonable only for the drying of salt compositions with $T_{eut.} \leq 240$ K.

Another important element of the freeze-dryer which is often absent in the open scheme devices is the sensors for product control during the drying process. Along with prevention of the liquid phase formation the main task of these devices is to control freeze-drying completeness. Contact between the underdried product and atmospheric ambient leads to the total degradation of the latter for most of the salt systems used for advanced materials preparation. Methods of drying control – measurement of product temperature, electric conductivity or pressure drop after disconnection of a separately located condenser – are equally valid while the presence of a control device is reasonable for all kinds of synthesis.

One of the primary questions regarding the evaluation of the optimum size of a freeze-drying installation is the total length of the drying cycle which includes:

- preparation of the equipment;
- loading;
- freezing;

- sublimation;
- stoppering;
- unloading;
- defrosting;
- possible sterilization;
- setting up equipment again.

The length of this determines the number of cycles it is possible to achieve per annum. It is desirable (Le Floc, 1986) to reduce this to a minimum by:

- automation of the loading and unloading operation which, on large machines, may need, when done manually, one or two hours per cycle;
- choice of freezing in a separate freezer; in fact, frequently vial filling is a long operation which immobilizes the freezer for a long period;
- envisaging a high performance in the sublimation machine.

Knowing the duration of the cycle and the possibilities offered by the organization within the plant, one can calculate a suitable sublimation cycle and capacity of the machine.

This production can be given by one of a number of machines and lead to a choice which must take into account the following points.

1. The case of several machines of low capacity

Advantages

- easier planning of filling, stock taking and maintenance operations;
- reduction in standstill between each operation;
- less heavy packing materials used all day long;
- staggering of the working day among staff;
- possible halt for maintenance and easy upkeep of machines;
- low rates of use;
- limited loss in the event of major accident (electrical breakdown or accident);
- easier setting up.

Disadvantages

- increase in batches, control and handling;
- dimensions of clean rooms are often larger;
- increase in components (compressors, vacuum pumps etc.) and their upkeep.

2. The case of one high-capacity machine

Advantages

- less onerous product control;
- simplification of operations which take on more significance.

Disadvantages

- less flexibility in planning of manufacture;
- heavy infrastructure of packaging material and personnel as it involves a product which is unstable and needs to be treated quickly but it is not advisable to bring the freeze-drying machine to a halt for too long;
- the implication of large amounts of electrical power and discharge of cooling water.

Any machine efficiency is defined by:

- the sublimation area S, m^2;
- the free space between shelves E, mm;
- and condenser capacity Q, kg.

But equally it should take into account:

- the overall dimensions;
- the weight of components and loading;
- different technical choices which for a large part depend on the price of the equipment (Dalglish, 1990).

A freeze-dryer must meet certain criteria which, although not essential, allow the different machine characteristics to be understood.

Heating power

It is assumed that a standard machine should permit sublimation of 1 kg of water per square metre of surface per hour knowing that heat of sublimation is around 800 kcal/kg of water sublimated. The heating value of kW/hr is equal to 860 kcal/hr. It is estimated that a freeze-dryer be equipped with heating capacity equivalent to 1 kW/m^2 of surface area.

Refrigeration

The machine should be capable of condensing the sublimated vapour with a condenser temperature in the region of −50 °C at the rate of 800 kcal/kg of water and taking into account effective losses, having a theoretical refrigeration capacity of about 1000 kcal/hr at −50 °C/m^2 of surface area.

Fluid circulation

Heating or cooling to the product is brought about by the medium of a special fluid which ought to have the following characteristics:

- low viscosity at low temperatures;
- the highest possible flash point;
- chemical inertness between $-60\,°C$ and $+121\,°C$;
- high thermal conductivity.

Silicon oils possess these qualities but their price is rather high and this is the only factor which limits their use. The oil used has a specific heat of 0.36 kcal/kg per degree. To provide evaporation of 1 kg of water per hour, accepting a temperature variation of $\pm 1.5°$ across the heating plate surface, the necessary output is $Q \times 0.36$ kcal/kg per degree $\times$ 3 degrees $= 800$ kcal/h which gives a total output of $Q = 800$ l/hr per square metre of surface area.

Cycle duration

One can easily estimate the duration of the sublimation cycle if one knows the rate of flow of heat and the power required. It should, however, be noted that for a modern machine with high capacity one should consider in the case of a $10\,m^2$ unit:

- 50 kg of product in vials; or
- 100 kg of product in bulk;
- 1000 to 1500 kg of stainless steel;
- 300 to 500 kg of circulation fluid to refreeze or reheat at each freezing or sublimation stage;
- loaded with 100 kg of product in bulk equipped with 10 kW heating which is used for 80% of the time.

The machines must be equipped with the means for monitoring, recording and alarms.

At the end of secondary drying the refrigerating compressor must be capable of lowering the temperature of the condenser to a suitably low temperature to obtain a satisfactory residual moisture in the product. In short, the freezing requirements can be set for the whole freezing process.

The plates and heat exchangers represent:

- 1000 kg of stainless steel with a specific heat of 0.12 kcal/kg per degree;
- 500 kg of silicone fluid with a specific heat of 0.36 kcal/kg per degree;
- a freezing temperature of $-40\,°C$ and a final temperature of $+40\,°C$.

The amount of energy required will be:

$$Q = 100\,\text{kg} \times 800\,\text{kcal} + 1000\,\text{kg} \times 0.12\,\text{kcal} \times 80^\circ + 500\,\text{kg} \times 0.36\,\text{kcal} \times 80^\circ = 100\,000\,\text{kcal} \quad (2.50)$$

The time required for sublimation will be:

$$T = 100\,000\,\text{kcal}/10\,\text{kW} \times 860\,\text{kcal/kW} \times 0.8 = 15\ \text{hours}$$

to which is added the time for secondary drying according to the amount of residual moisture required in the product.

Efficiency of freeze-drying can be considerably enhanced by a combination of freezing and dehydration processes in a single installation (Brazhnikov, 1983; Kutepov *et al.*, 1989). As it is shown in Figure 2.57 for conventional techniques of freezing using systems with liquid nitrogen the rate of process is not great enough. It may be considerably increased by the use of so-called processes of evaporation vacuum freezing.

The evaporation vacuum technique is characterized by high freezing rates; it is reasonable from the practical, technological point of view, because it makes it possible to obtain products in granulated form and does not require further

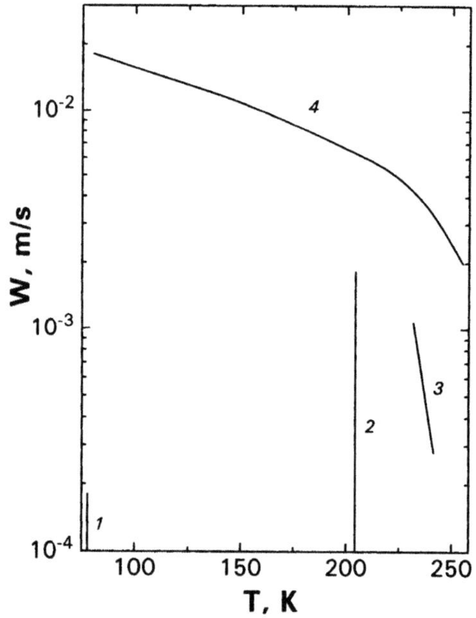

FIGURE 2.57 Process rate for different freezing techniques: (1) liquid nitrogen; (2) hexane; (3) evaporative freezing; (4) freezing on metal surface.

crushing of the product as is the case of products prepared by freezing onto surfaces. The size of most material particles is similar to that obtained by surface freezing but in the last case about 5–10 wt.% of the particles of the outer layer are tenfold greater in size (Karabanov, 1987). Among the disadvantages of the evaporation vacuum freezing technique one can cite the necessity of ensuring that it can work within a narrow range of parameters including pressure in the device, flow rate, viscosity of dispersed liquid during the process of their introduction into a vacuum. The parameters which must be ensured for stable introduction of liquid solutions into a vacuum do not correspond to the optimum parameters for freeze-drying and this complicates the simultaneous realization of the processes of dispersing into a vacuum and of freeze-drying. This also complicates the development of installations working in a continuous regime. On the basis of the discussion above one can conclude that evaporation vacuum freezing and freezing onto cooled surface techniques are effective for large-scale production of disperse powders. An advantage of the evaporation vacuum freezing technique is also the fact that the process of evaporation freezing and freeze-drying can be achieved in a common installation. Therefore in the majority of cases preference can be given to the evaporation vacuum freezing with retention of final product quality (Shatnyi, 1988).

As an example let us consider the evaporation vacuum freezing installation developed by the FMC Corporation (USA) (Figure 2.58).

A liquid precursor is introduced into the granulator. In the flying time the droplets are frozen and fall onto a vibrating platform placed inside the sublimation chamber. The heat supply to moving particles on the platform is produced by the internal electric heating source. The vapour forming upon freeze-drying of the material is absorbed by a special desublimator placed below the platform. Dehydrated particles are directed in the collector and periodically unloaded from there.

Another more common version of the installation for simultaneous freezing and sublimation dehydration was constructed (Ivanov, 1993) on the basis of the sublimator LZ-45.22 (Czechoslovakia) (Figure 2.59).

It includes a vacuum sublimation chamber (3) with internal diameter 900 mm and height 1000 mm, spraying column (2) with internal diameter 600 mm and height 1200 mm, and additional systems such as:

- the system of introducing an initial salt solution and solvent into a vacuum consisting of solution container (8), solvent container (11), two peristaltic pumps (9), (1), connecting tubes and closing valve;
- the system of air pumping consisting of vacuum pump (6), introduction tube and vacuum valve;
- the desublimation system consisting of a desublimator (4), refrigerating machine (7) and connecting tubes;

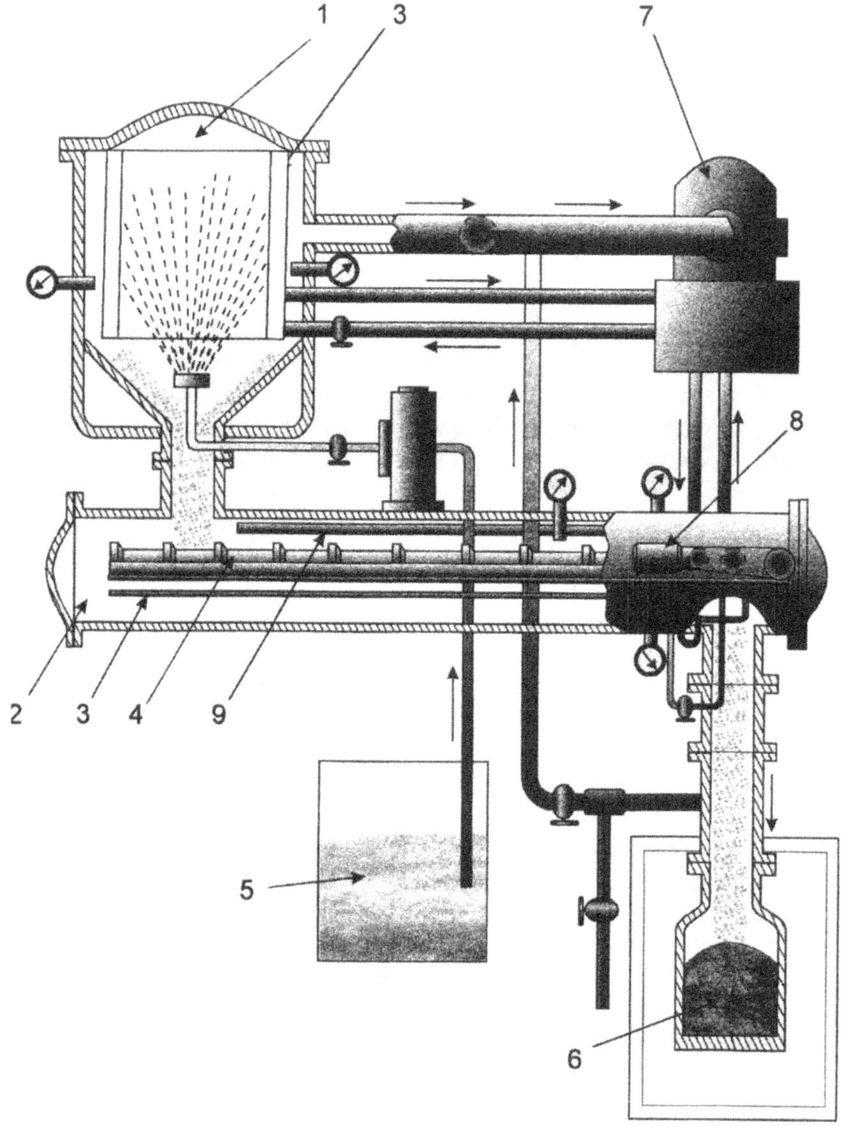

FIGURE 2.58 Apparatus for freeze-drying of continuous moving flow of granules obtained by spraying in vacuum ('FMS Corporation', USA): (1) spraying chamber; (2) sublimation chamber; (3) desublimator; (4) carrier; (5) vessel with liquid to be sprayed; (6) container for dried product; (7) freezer; (8) carrier drive; (9) electrical heater.

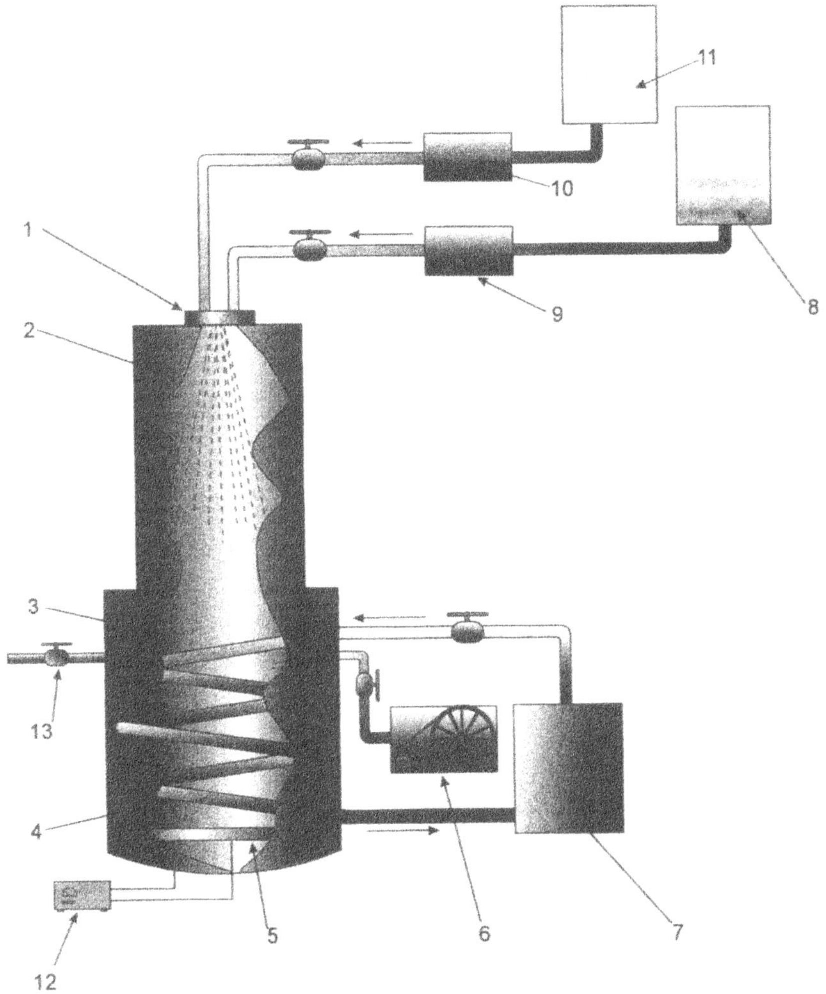

FIGURE 2.59 Experimental installation for combined spraying in vacuum and sublimation: (1) dispergator; (2) dispergation column; (3) vacuum chamber; (4) desublimator; (5) sublimator; (6) vacuum pump; (7) freezer; (8) vessel with salt solution; (9), (10) peristaltic pumps; (11) vessel with solvent; (12) current supply; (13) valve.

- the sublimation system consisting of a sublimator (5) and power source with automation facilities (13);
- the system of air introduction consisting of a tube with a fine calibration orifice (12).

The mode of operation is the following. Firstly, on the control panel the vacuum pump (6), cooling machine (7) and control units should be switched on to ensure the working and monitoring of systems for air pumping and desublimation. After attainment of a steady state corresponding to the pressure no more than 0.05 mm Hg (below the temperature of water triple point) with the temperature of the working surface no more than −40 °C (233 K) the salt solution is injected into a vacuum chamber through the special sprayer (1). Under the force of gravity the droplets of sprayed liquid move towards the sublimator (5). At the same time they are rapidly frozen due to intensive evaporation of the liquid. Frozen droplets fall onto the level surface of the sublimator (5). After spraying a certain amount of solution the closing valve is shut off and dispergation stops. Then, the heat source with automatic units (13) of the sublimation system is switched on and the heating of the sublimator (5) occurs. The heating temperature on the surface of the sublimator is ensured automatically in the range of 60–90 °C (depending on the nature of the treated product) (Hello, 1995). Unloading of the dried product is carried out after switching off the vacuum pump (6) and devacuuming the chamber with air through the fine calibrated orifice (14).

The ultradisperse powder obtained with this equipment is highly hygroscopic and for this reason its unloading and waterproof packing must be effected very quickly. Moreover such powder is kept in hermetic packages with a quantity of calcined silicagel.

2.3 CRYOEXTRACTION PROCESSES

Products of cryocrystallization are most commonly dehydrated by freeze-drying. However, sublimation processes require cumbersome equipment and large energy expenses, and the sublimation time is commonly equal to 10–20 hours. An alternative technique free of the above-mentioned disadvantages is the so-called cryoextraction technique – extraction of ice from cryogranules due to its dissolution in organic solvents at temperatures below the melting temperature of cryogranules. The salt phase of frozen solution should be insoluble in used extractants. The term 'cryoextraction' completely reflects the nature of the physico-chemical processes which only occur in the case when both ice and salt are in a crystal state because extraction is the technique for the separation of solids mixtures based on their treatment by various solvents in which the mixture components dissolve at a different rate. However, cryogranules can sometimes have a more complex composition and include, besides crystal phases, the vitreous ones which exist even upon heating below their melting temperatures in the metastable liquid state. As was mentioned above, this liquid contains practically overall dissolved salt. In the last case the low-temperature process of treatment of the cryogranulate with organic liquids should be considered as salting out and it will be more correct to use the term 'cryosalting out' rather than 'cryoextraction'. Nevertheless, we will only use the term 'cryoextraction' because the most common cryogranulate product contains crystalline and amorphous phases simultaneously if the process is carried out at temperatures when crystallization occurs very slowly (230–240 K) or in the presence of cryoprotectors.

2.3.1 Thermodynamic analysis

Let us consider the process of cryoextraction on a simple example in which cryogranules represent a mechanical mixture of crystals of ice and salt in the form of equilibrium crystallohydrate at a given temperature. Ethanol is used as an extractant. Strictly speaking, such a system is a three-component one, but it is possible to use just the phase diagram 'water–ethanol' (Figure 2.60) if the solubility of the salt in the organic or water–organic liquid phase is negligible. At the initial moment (after introduction of cryogranules in the solvent cooled down to the temperature *T*), the composition of liquid phase corresponds to point *A* in Figure 2.60.

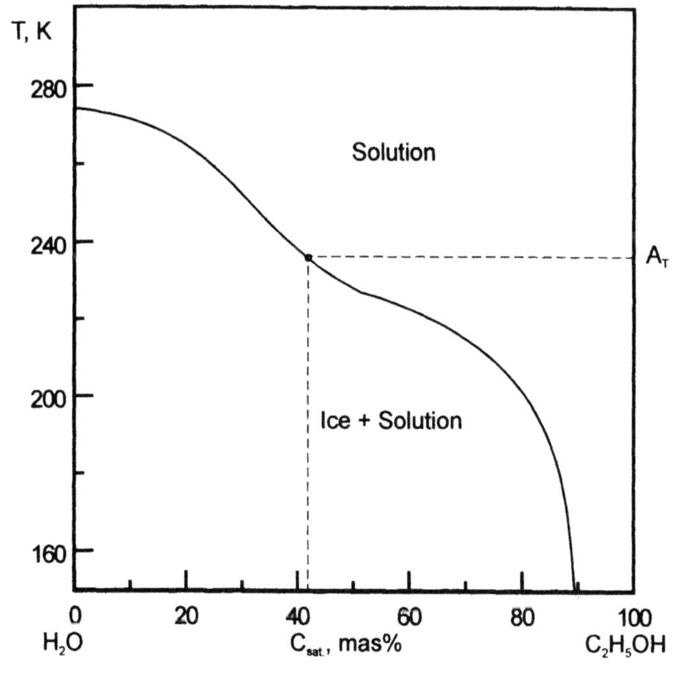

FIGURE 2.60 Phase diagram of the system 'water–ethanol'.

Ice crystals can dissolve in ethanol up to the attainment of equilibrium in the system (point *B*). In this case the composition of the liquid phase corresponds to the solution saturated in respect to ice (C_{sat}). Carrying out the cryoextraction at a higher temperature increases the quantity of dissolved ice, i.e. the extractive capacity of the solvent increases. Therefore, to choose the conditionş for carrying out the process it is necessary to know at least the phase diagram of the system water-extractant.

Requirements to solvents which are potential extractants of ice have been formulated (Mozhaev, 1988). One of these requirements is the high solubility of water in extractants at low temperatures according to the necessity of maximum ice extraction degree from cryogranules. The low temperature of the process only allows extractants with low freezing temperatures to be used. The necessity of fast and easy separation of salt and extractant requires high vapour pressure of extractant (high volatility), but for convenient work with such extractants their boiling temperatures should not be too low. Because the majority of extraction processes take place in a diffusional regime the efficient extractants should have high values of diffusion coefficients of water molecules in the liquid water–organic phase at low temperatures.

It is known that the diffusion rate is highly influenced by the solvent viscosity whose increase leads (according to the Stokes–Einstein equation) to proportional decrease of diffusion coefficient. Hence it is expedient to use the least viscous solvents. It is desirable to take into account such properties of solvents as low toxicity, explosion security, accessibility and low cost as well as other technological parameters.

The potential extractants may be inorganic as well as organic compounds. It is known that substances with high values of dielectric constant (water is among them) are highly soluble in polar solvents. Among such solvents can be alcohols, aldehydes, esters and amines. The presence of polar groups in their structure determines their ability to form hydrogen bonds with water molecules leading to sufficiently high solubility of water in the organic phase even at low temperatures. The requirements of low viscosity and low freezing temperatures allow the first representatives of indicated classes of compounds to be used (Stairs, 1980). Among polar solvents the least viscous is methanol and among low polar solvents the least viscous is acetone. Physical and chemical properties of the most expedient solvents are listed in Table 2.16.

Indicated requirements allow a very restricted number of organic solvents to be considered as suitable for cryoextraction. The list of extractants is even more restricted if we use the most common and technologically expedient organic solvents (Table 2.17).

The absence of literature data on thermodynamic parameters of ice dissolution in organic extractants at low temperatures was the reason to carry out the complete thermodynamic calculation of the process under study.

TABLE 2.16 Melting and boiling temperatures of some organic solvents which dissolve in water at low temperatures.

Solvent	*Formula*	T_m, K	T_b, K
Allyl alcohol	CH_2CHCH_2OH	144	369.9
n-Amylamine	$C_5H_{11}NH_2$	218	377.4
2-Amino-2-methylbutan	$CH_3C(CH_3)(NH_2)CH_2CH_3$	168	–
Acetaldehyde	CH_3CHO	149.5	293.2
Acetone	CH_3COCH_3	178.4	329.2
Acetonitrile	CH_3CN	227.3	354
n-Butylamine	$CH_3(CH_2)_2CH_2NH_2$	222.5	350.8
Dimethylformamide	$(CH_3)_2NCHO$	212	426
Isobutylamine	$(CH_3)_2CHCH_2NH_2$	187.5	341
Isopropanol	$CH_3CH(OH)CH_3$	187.2	355.4
n-Propanol	C_3H_7OH	146	370.2
Propylamine	$C_3H_7NH_2$	190	320.8
Methanol	CH_3OH	175.2	337.5
Methylvinylcarbonyl	$(CH_2CH)(CH_3)CHOH$	193	370
Methylisobutylketone	$CH_3CO(CH_2)_2CH_3$	189.5	391
Methylpropylketone	$CH_3CO(CH_2)_2CH_3$	195.2	374.7
Methylethylketone	$CH_3COC_2H_5$	186.6	352.6
Ethanol	C_2H_5OH	161.0	351.4
Ethylamine	$C_2H_5NH_2$	191.8	289.6
Ethoxyethanol	$C_2H_5OC_2H_4OH$	203	408.6

TABLE 2.17 Physico-chemical properties of some organic solvents.

Solvent	*Freezing temperature,* K	*Dynamic viscosity,* $\times 10^2$, Poise	*Composition of the liquid phase in equilibrium with the ice (wt.% of the solvent) at the temperature* 215 K	243 K	233 K	223 K
Acetone	178.4	0.32	47.0	71.0	87.0	93.0
Methanol	175.2	0.58	25.0	33.0	40.0	46.0
Ethanol	161.0	1.19	53.0	64.8	75.5	85.5
Isopropanol	187.2	2.39	42.0	73.0	77.0	–
Propanol	146.0	2.2	81.0	87.0	91.0	–
Ethoxyethanol	203.0	2.05	39.0	49.0	55.6	61.0

The variation of free energy on dissolving ice in organic extractants ($\Delta G_{sol,T}$) was determined from the relationship:

$$(\Delta G_{sol,T}) = \Delta G_{m,T} + \Delta G_{mix,T} \quad (2.51)$$

where $\Delta G_{m,T}$, $\Delta G_{mix,T}$ are free energy changes of ice melting and mixing of water with organic solvent, respectively, at $T = 273$ K. In the right-hand side of the equation $\Delta G_{m,T} = \Delta H_{m,T} - T\Delta S_{m,T}$, the necessary values of $\Delta H_{m,T}$ and $\Delta S_{m,T}$ were calculated using Kirchhoff's Law. Values of $\Delta H_{m,273}$ and $\Delta S_{m,273}$ were taken as equal to 6.009 kJ/mole and 22.01 J/(K*mole), respectively (Rossini *et al.*, 1950).

To calculate $\Delta G_{mix,T} = \Delta H_{mix,T} - T\Delta S_{mix,T}$ the values of $\Delta H_{mix,T}$ and $\Delta S_{mix,T}$ were found using literature data on mixing enthalpies, mixture heat capacities ($C_{p,x}$) and partial vapour pressures of mixture components (P_1, P_2) at several temperatures near to room temperature (Timmermans, 1960; Belousov and Panov, 1973). To determine $\Delta H_{mix,298}$, P_1, P_2 and $C_{p,x}$, corresponding to various molar ratios of components the linear interpolation procedure was used. The calculation of $\Delta H_{mix,T}$ and $\Delta S_{mix,T}$ was also carried out on the basis of Kirchhoff's Law. Values of $\Delta G_{m,T}$, $\Delta H_{m,T}$ and $\Delta S_{m,T}$ for ice are given in Table 2.18.

As one can see from Table 2.19 the results of calculations of enthalpies of mixing of water with organic compounds correlate satisfactorily with experimental values of ΔH_m obtained using the Calvet-type low-temperature calorimetry (Naumov *et al.*, 1982a, 1982b; Kuleshov *et al.*, 1983, 1986).

2.3.2 Kinetic analysis

A study of kinetic regularities of ice dissolution in different organic extractants was carried out in the temperature range 233–248 K for ice granules with diameter 3.0 ± 0.3 mm. Experimental procedures by Telegina *et al.* (1983) allowed kinetic curves to be obtained, characterizing the dependency of undissolved ice portion (w) on the process time (t), temperature (T), stirring intensity (N) and water content in the liquid water–organic phase (C_{H_2O}) for all studied organic solvents.

TABLE 2.18 Thermodynamic parameters of ice melting at low temperatures.

T, K	ΔG_m, J/mole	ΔH_m, J/mole	ΔS_m, J/K mole
268	108.2	5817.3	21.30
263	212.9	5625.3	20.58
258	313.8	5433.9	19.85
253	411.1	5243.0	19.10
248	504.6	5052.9	18.34
243	594.3	4863.6	17.57
238	608.2	4675.3	16.79
233	762.1	4488.1	15.99
228	839.9	4302.1	15.18
223	913.8	4117.5	14.37

TABLE 2.19 Enthalpy of mixing of water with organic solvents at 248 K.

	$\Delta H_{mix,248}$, J/mole *at molar ratio*		
Components	1:25	1:10	1:5
Water–Methanol	−128.5 ± 4.3	–	–
	−116.0	−250.9	−405.5
Water–Ethanol	−109.5 ± 4.2	–	−316.5 ± 15.9
	−90.5	−180.4	−302.8
Water–Acetone	64.7 ± 10.2	147.1 ± 13.6	60.3 ± 19.2
	65.2	154.2	177.9
Water–DMFA	−190.6 ± 3.5	–	−829.0 ± 15.3
	−177.9	−420.6	−736.5

It was found that the efficiency of extractants evaluated by the time necessary to accomplish the process in identical conditions increases in the following sequence: propanol > iso-propanol > ethoxyethanol > ethanol > methanol. Some typical kinetic curves obtained in a given experimental conditions are represented in Figure 2.61 as examples.

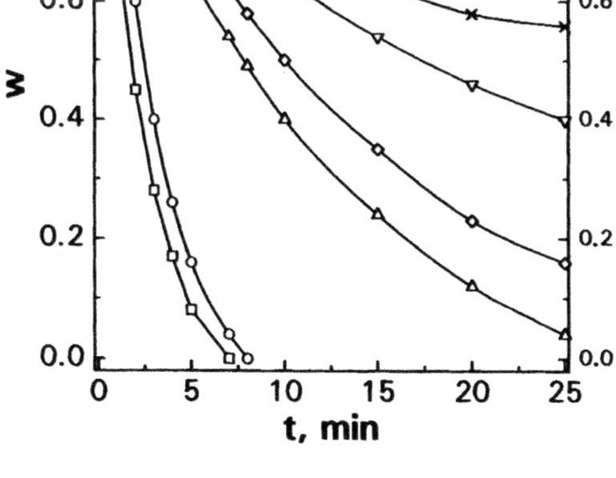

FIGURE 2.61 The amount of undissolved ice (w) *vs.* duration of cryoextraction process for different organic solvents ($T = 248$ K, $N = 73.3$ rad/s). (□) methanol; (○) acetone; (△) ethanol; (◇) ethoxyethanol; (▽) isopropanol; (×) propanol.

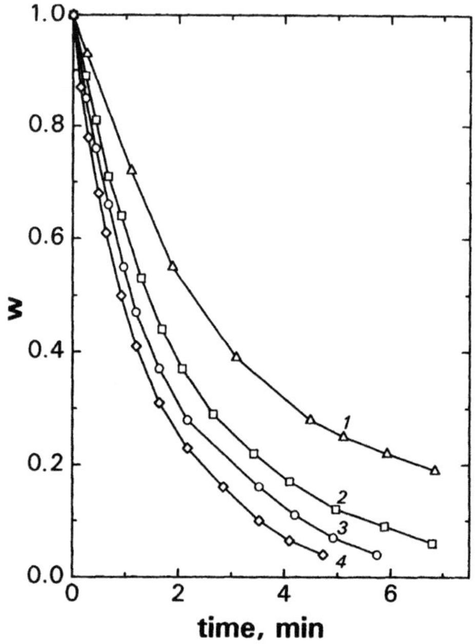

FIGURE 2.62 Kinetic curves for dissolution of ice granules in acetone as a function of mixing intensity at 248 K ((1) 41.9; (2) 73.3; (3) 94.2; (4) 104.7 rad/s).

The limiting step of the process is normally elucidated by studying the dependence of the dissolution rate on the intensity of stirring. Such dependence proves that the process under study takes place according to a diffusional mechanism or combined diffusional–kinetic one (Akselrud and Molchanov, 1977).

It was established that the decrease in stirring intensity in the temperature range 248–233 K causes the decrease of ice dissolution rate in all studied extractants. As an example, in Figure 2.62 the results obtained with acetone as extractant are given.

It has been shown experimentally that other extractants behave in much the same way. This has enabled calculation of the values of mass transfer coefficient β (a constant which specifies the dissolution process) using general kinetic relations (Akselrud and Lysiansky, 1974).

$$dm/dt = \beta S(C_{sat} - C_i) \tag{2.52}$$

where m is the ice mass, t is the process time, S is the phase boundary (interface) which is the surface of dissolving ice particles, C_i, C_{sat} is the water concentration in liquid phase at a given moment and in equilibrium with ice, respectively.

In our calculations the variation of surface of spherical particles in the process of dissolution was approximated using the relation:

$$S = (6m_0/\rho_T d_0)(1 - m/m_0)^{2/3} \tag{2.53}$$

where m_0 is the mass of initial ice sample, m is the mass of substance passed in solution to a given time moment, ρ_T is the ice density, d_0 is the initial diameter of particles. In Table 2.20 the values of β at 248 K ($N = 73.3$ rad/s) for all studied solvents are presented.

Further analysis of dependencies $\beta = f(N)$ showed that upon decreasing the stirring rate the β value tends to zero testifying pure diffusional mechanism of ice dissolution in all extractants.

Therefore, the process of dissolution of ice in studied extractants proceeds by unified mechanism and the limiting stage of this mechanism is the diffusion of water molecules from the surface of dissolving granules into the solution volume. This makes it possible to estimate average diffusion coefficients of water (D_{H_2O}) under different experiment conditions of cryoextraction. The calculation of D_{H_2O} was carried out by equation $D = 0.5\ \beta d_0\ \mathrm{Nu}^{-1}$. The Nu value was estimated on the basis of the empirical relation of Barker and Traiball (Akselrud and Molchanov, 1977):

$$\mathrm{Nu} = 0.052\ \mathrm{Re}^{0.833}\ \mathrm{Pr}^{0.5} \tag{2.54}$$

where Nu, Pr, Re are the hydrodynamic criteria of Nusselt, Prandtl and Reynolds, respectively.

It should be noted that the particle size does not appear in the last equation. It means that in the motion of ice granules in a turbulent regime inertial phenomena do not play a significant role (the so-called automodel regime). In other words, the indicated equation may be used only in cases where the mass transfer coefficient does not depend on particle size. Extra experiments (Table 2.21) showed that variation of average diameter of ice granules from 3.0 to 0.015 cm

TABLE 2.20 Coefficients of mass transfer (β) and diffusion coefficients of water molecules (D_{H_2O}) for different extragents at $T = 248$ K, $N = 73.3$ rad/s.

Extragent	$\beta \times 10^4$, cm/sec	$D_{H_2O} \times 10^6$, cm²/sec
Methanol	6.7 ± 0.5	4.0 ± 1.2
Acetone	8.8 ± 1.5	2.1 ± 0.9
Ethanol	2.0 ± 0.3	0.33 ± 0.11
Ethoxyethanol	1.08 ± 0.18	–
Isopropanol	1.12 ± 0.17	0.26 ± 0.11
Propanol	1.20 ± 0.17	0.21 ± 0.10

TABLE 2.21 Coefficients of mass transfer (β) at the dissolution of ice grains with different initial diameter (d_0) in acetone at 238 K and $N = 73.3$ rad/s.

d_0, mm	3.0	0.024	0.016
$\beta \times 10^4$, cm/s	6.33 ± 0.28	6.20 ± 0.80	5.95 ± 0.87

had no influence on β values within the limits of experimental errors. It will be noted that these experiments were carried out at 238 K because the rate of dissolution of ice granules with small diameter proved to be too high at 248 K to obtain reliable kinetic curves.

Calculated values for water diffusion coefficients in various extractants are given in Table 2.20. Comparing D_{H_2O} values one can see that the increase of water diffusion coefficients occurs in the same sequence is that for extractants efficiency, and this supports the model of the unified diffusional mechanism of ice dissolution.

Let us return to the fact that solid phase dissolution can take place according to a kinetic mechanism or to a diffusional one (Telegina, 1983). In the first case the relationship between the process rate ($\mathrm{d}w/\mathrm{d}t$) and its conditions is commonly expressed as

$$\mathrm{d}w/\mathrm{d}t = -z(w)f(T)v(c) \tag{2.55}$$

where $f(T)$ and $v(c)$ are functions describing the dependence of process rate on the temperature and the solute concentration (usually $f(T) \sim \exp(-E_{act}/RT)$ and $v(c) \sim c^a$, where E_{act} and a are apparent activation energy and reaction order, respectively). Function $z(w)$ is directly connected with the undissolved portion of the component and reflects the surface activity of the solid phase, shape of particles and their size distribution.

For diffusional mechanism of dissolution the similar dependence proved to be more complicated, namely

$$\mathrm{d}w/\mathrm{d}t = -z(w)f(T')v(c)\beta\Delta c \tag{2.56}$$

Here β is the mass transfer coefficient and Δc is the difference of concentrations of one of the components determining diffusion (moving force of the process).

If the process rate is limited by diffusion of water molecules from the surface of granules into a solution volume then $\Delta c = c_{sat} - c_i$, where c_i is the current concentration of water in the solvent, and c_{sat} is the concentration of solution saturated at the given temperature. It was shown (Vigdorchik and Sheinin, 1971) that from relation (2.56) the following integral equation can easily be obtained:

$$(t_1/t_2)_w = (\beta(N_1, T_1)\Delta c_1)/(\beta(N_2, T_2)\Delta c_2) \tag{2.57}$$

where t_1 and t_2 are the times necessary to achieve the same w value in two parallel experiments with different masses. If the limiting process is the diffusion of water molecules from the granules surface into the solution volume then $\Delta c_1 \neq \Delta c_2$ because $c_{sat,T1} \neq c_{sat,T2}$ and the relation (2.57) is non-linear. Results of experimental data treatment using the relation (2.57) are represented in Figure 2.63.

Non-linearity of the curves $t_1 = f(t_2)$ proves that in studied temperature range and stirring conditions the rate of ice granules dissolution is limited by the diffusion of water molecules from the surface of ice granule into solution volume and near the granule surface a water–acetone layer exists with a thickness of δ in which water concentration varies from c_{sat} (at the given temperature) to c_i.

The set of observed experimental data made it possible to calculate the concentration function $v(c)$ describing the process rate relative to water concentration in the water–acetone mixture (Figure 2.64). Knowledge of concentration function allows calculation of the kinetics of the ice dissolution process in hypothetical conditions where the water concentration in the solvent remains constant during the whole process.

An advantage of similar representation of kinetic data consists of the fact that kinetic function is invariant in respect of concentration, temperature and hydrodynamic conditions of the process and it enables simulation of the process

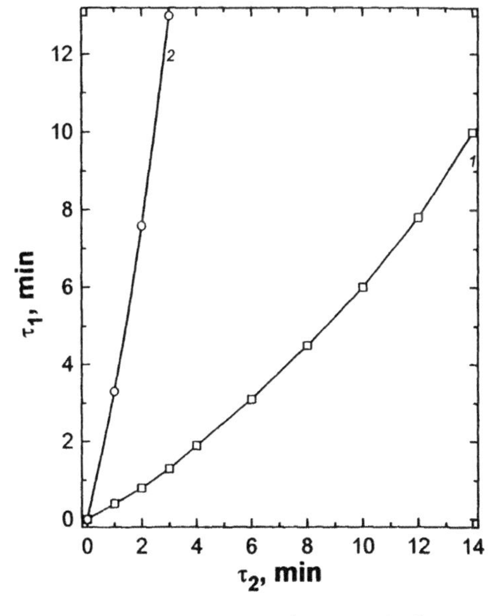

FIGURE 2.63 Correlations between durations of ice granule dissolution in acetone upon similar transformation degree at different temperatures and mixing conditions: (1) τ_1 ($T = 248$ K, $N = 41.9$ rad/s), τ_2 ($T = 233$ K, $N = 73.3$ rad/s); (2) τ_1 ($T = 233$ K, $N = 41.9$ rad/s), τ_2 ($T = 248$ K, $N = 73.3$ rad/s).

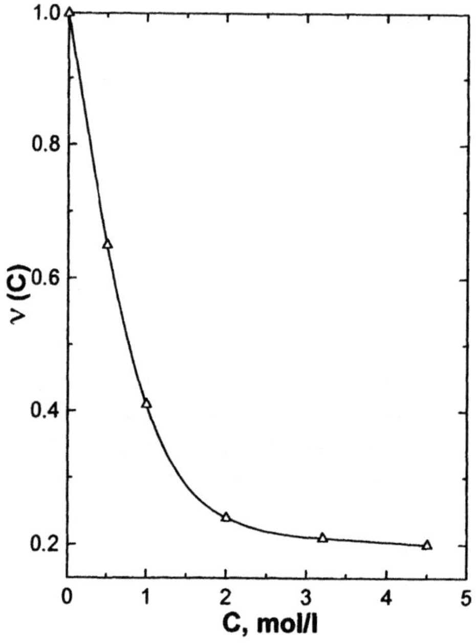

FIGURE 2.64 Concentration function $v(c)$ of ice granule dissolution in water–acetone mixtures.

in various real conditions. To find the kinetic function it is necessary to convert to the normalized time x which is determined by the following relationship:

$$x = \frac{t_i}{t_{tot}} = \frac{\int_0^{t_i} v(c)\Delta c \, dt}{\int_0^{t_{tot}} v(c)\Delta c \, dt} \tag{2.58}$$

The last one represents the ratio of the time necessary for attainment of a given transformation degree $w_i(t_i)$ to the time of complete (or practically complete) transformation (t_{tot}) upon hypothetical (concentrations of reactants are constant during the process) conditions.

In the case under consideration we assumed that complete transformation corresponds to the state with $w_{0.1} = 0.1w_{tot}$, and $w = 1 - 0.9x$. It means that the process of ice dissolution in acetone should have a constant rate if process parameters (T, C_{H_2O}, N) do not change during the experiment.

The diffusional mechanism of ice granules dissolution in organic extractants enable clearing up of the decrease of process rate with the decrease of temperature and stirring intensity or with the increase of water concentration in the extractant and particle size.

It is obvious that the diffusional mechanism of ice dissolution should remain in the cryoextraction conditions. However, due to the presence of the salt framework in cryogranules one can expect the advent of internal diffusion limiting the whole process. In this case, the additional factors affecting the rate of cryoextraction are the concentration of initial solution, the phase composition of cryogranules and the chemical nature of the salts forming a granule framework.

The study of the behaviour of cryogranules with different concentration and chemical nature of salts (Mohr salt, aluminium, lithium, nickel, cobalt, zinc and cadmium sulphates, ammonium trioxalatoferrate ATOF) showed that the increase of salt concentration leads to decrease of ice cryoextraction rate. With the increase of concentration the porous structure of granules becomes more complex (Rudobashta, 1980) which makes it more difficult to transfer water molecules from the reaction zone to the granule surface. At high cooling rates the formation of the salt framework was observed even in diluted (5–9 wt.%) solutions. It means that cryoextraction is limited by internal diffusion impedance in all solutions used in practice (15–30 wt.%). Moreover, with the use of highly concentrated solutions the whole cryoextraction process develops in internal diffusion conditions. Actually, experimental data prove that the rate of ice cryoextraction from frozen droplets of 15% ATOF solution is considerably increased with the increase of stirring intensity. At the same time the rate of ice cryoextraction from frozen solutions with high salt concentration does not depend on the stirring intensity, i.e. the process occurs exclusively in internal diffusional conditions. It is obvious that destruction or absence of salt framework of cryogranules should lead to increase of process rate due to elimination of internal diffusion.

It was established that the relative efficiency of extractants is analogous with the efficiency established for ice extraction, if all conditions of cryoextraction (temperature, quantity of extracted ice, size of cryogranules, hydrodynamical conditions in the reactor, salt concentration in cryogranules) remain unchanged. It can be illustrated by the data obtained for cryogranules of 15% ATOF solutions and represented in Figure 2.65.

For this experimental series we have estimated the contributions of internal and external diffusion processes on the basis of the well-known relation:

$$1/k = 1/\beta + 1/\beta_v \tag{2.59}$$

where k is the cryoextraction rate constant, β_v is the mass transfer coefficient during salt dissolution, β is the mass transfer coefficient during ice dissolution.

It is obvious that this kinetic description of cryoextraction, based on the concept of quasistationarity, is not strict enough because the process really proceeds in non-steady conditions. Nevertheless the similar approach is widely used if the values of coefficient k are averaged. In our case such averaging was realized over the range from 0.8 to 0.2.

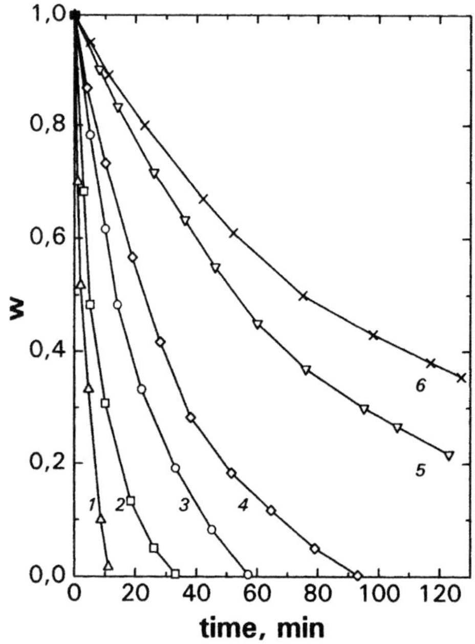

FIGURE 2.65 Kinetic curves for ice cryoextraction from granules of frozen ATOF solutions (15 wt.%) by different extragents: (1) methanol, (2) acetone, (3) ethanol, (4) etoxyethanol, (5) iso-propanol, (6) propanol; $T = 248$ K, $N = 73.3$ rad/s).

Average k values were obtained using the relation $dm/dt = kS(C_{sat} - C_i)$. To evaluate β_v values the β values calculated from kinetics of ice granules dissolution were used. Obtained β_v values enabled the water concentration near the surface of cryogranules to be estimated ($C_s < C_{sat}$ due to water concentration gradient in liquid phase inside salt framework) $dm/dt = \beta_v S(C_{sat} - C_s)$. The results of calculations are listed in Table 2.22, and mechanisms of ice dissolution and cryoextraction are compared in Figure 2.66.

The influence of the chemical nature of salts on the rate of ice cryoextraction has a sufficiently complex character. Experimental relations obtained under conditions where the quantity of ice extracted from cryogranules was the same (concentrations of used solutions were equal to 16.8 wt.%) are represented in Figure 2.67.

Due attention should be given to the fact that the maximum process rate equal to the rate of dissolution for pure ice is observed for cryogranules containing lithium sulphate. It is connected with the fact that lithium sulphate, in contrast to all other used salts, does not form crystal products by cryocrystallization. Hence, in this case there is no salt framework and internal diffusion cannot limit the process, so the rate is maximum under given conditions.

TABLE 2.22 **Values of rate constant and concentrations at the cryoextraction of ice from ATOF cryograins by different organic solvents ($C_{salt} = 16.8$ wt.%, $N = 73.3$ rad/s, $T = 248$ K).**

Extragent	$k \times 10^5$ cm/s	$\beta_v \times 10^5$ cm/sec	C_s, g/cm^3	C_{sat}, g/cm^3
Methanol	24.3 ± 6.7	38.3	0.27 ± 0.05	0.66
Acetone	20.8 ± 3.8	27.2	0.12 ± 0.01	0.35
Ethanol	10.8 ± 2.0	22.8	0.21 ± 0.02	0.36
Ethoxyethanol	4.05 ± 0.09	6.8	0.23 ± 0.02	0.53
Isopropanol	2.95 ± 0.05	4.0	0.11 ± 0.01	0.31
Propanol	5.92 ± 0.05	11.4	0.09 ± 0.02	0.19

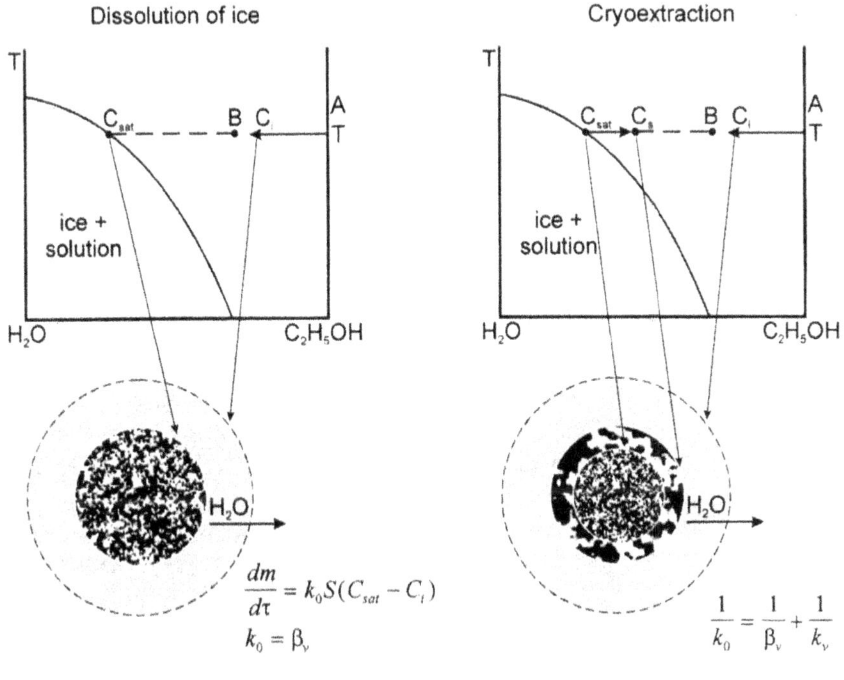

$$\frac{dm}{d\tau} = k_0 S(C_{sat} - C_i)$$
$$k_0 = \beta_v$$
$$\frac{1}{k_0} = \frac{1}{\beta_v} + \frac{1}{k_v}$$

FIGURE 2.66 Schematic diagram of systems behaviour during dissolution of ice granules and cryoextraction of frozen salt solution.

For all other salts, kinetic curves are different. Similar differences seem to be caused by changes in characteristics of microstructure of cryogranules of different salts. To elucidate the influence of microstructure of granules on the ice cryoextraction rate, the last was removed using the freeze-drying technique, and the total volume of pores was determined using the absorption technique with subsequent calculation by the Thomson–Kelvin procedure.

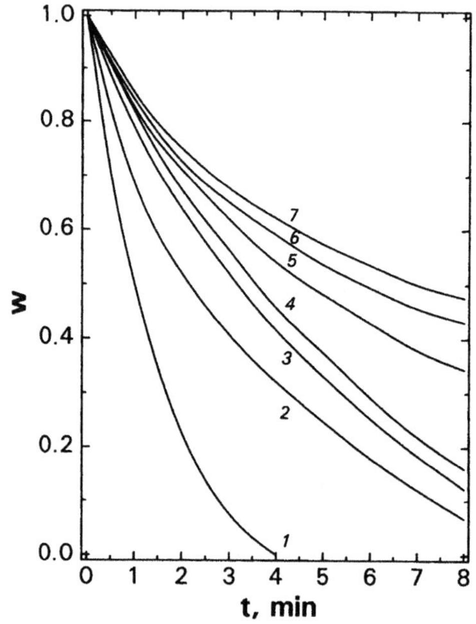

FIGURE 2.67 Kinetic curves for ice cryoextraction by acetone from granules of frozen salt solutions ((1) Li_2SO_4, (2) ATOF, (3) $Al_2(SO_4)_3$, (4) $(NH_4)_2SO_4$, (5) $NiSO_4$, (6) $ZnSO_4$, (7) $CoSO_4$; $c_{salt}^* = 16.8$ wt.%, $T = 248$ K, $N = 73.3$ rad/s, $d_{mixer} = 1.2$ cm).

Results of the analysis of salt granules microstructure are represented in Table 2.23 and in Figure 2.68. As one can see in granules containing $ZnSO_4$, $CoSO_4$ or $NiSO_4$ and having minimum ice cryoextraction rate (Figure 2.67) the volume of pores is several times greater than that for $Al_2(SO_4)_3$, $(NH_4)_2SO_4$ and $(NH_3)_3Fe(C_2O_4)_3$ with higher cryoextraction rate. It seems that the increase of micropores volume reflecting ramification of the salt framework impedes water diffusion in cryogranules leading to a decrease of cryoextraction rate but this correlation is not unambiguous.

In the example of $CdSO_4$ solutions with high salt content (40 wt.%), it was shown that cryoextraction occurs in the region with internal diffusion and the process rate ceases to depend on stirring intensity and thereby the possibility to control the cryoextraction duration is lost. This difficulty in the way of intensification of the process was overcome by conservation of salts in cryogranules as a metastable supercooled solution. This was achieved by the introduction of a cryoprotector – glycerine.

The experimental data obtained on granules of frozen ATOF solutions containing 5 vol.% of glycerine are given in Figure 2.69. Using the low-temperature DTA technique we have established that this quantity of glycerine is sufficient to prevent crystallization of the salt existing in experimental conditions as supersaturated solution even though there is simultaneous presence of ice crystals. In this case the process rate is the same as that for dissolution of pure ice granules.

It should be noted that in reality this process represents cryosalting out – segregation of crystallohydrates by exchange of solvent at low temperatures complicated by simultaneous process of ice dissolution.

Hence, the variations in phase composition of cryogranules due to the introduction of cryoprotectors accelerates the process considerably and allows it

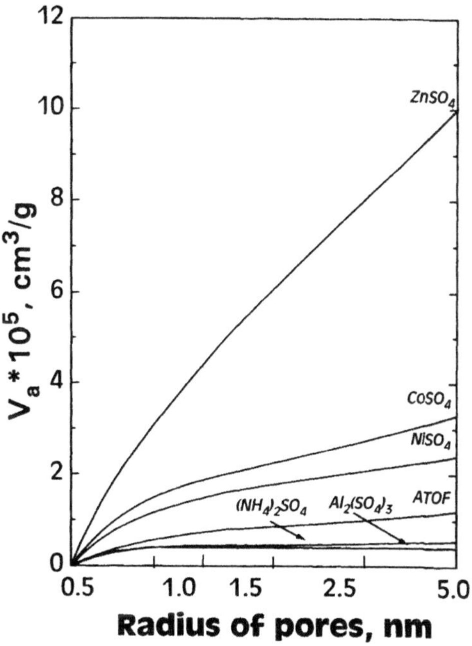

FIGURE 2.68 Correlation between total volume of pores and their radii for granules of different salts obtained by freeze-drying.

TABLE 2.23 Pore size in grains of different salts, prepared by the freeze-drying method (C_{salt} = 16.8 wt.%).

Salt	*Total volume of pores,* cm^3/g × 10^3
$Al_2(SO_4)_3$	0.66
$(NH_4)_2SO_4$	1.35
$(NH_4)_3Fe(C_2O_4)_3$	1.27
$NiSO_4$	1.67
$CoSO_4$	3.33
$ZnSO_4$	11.05

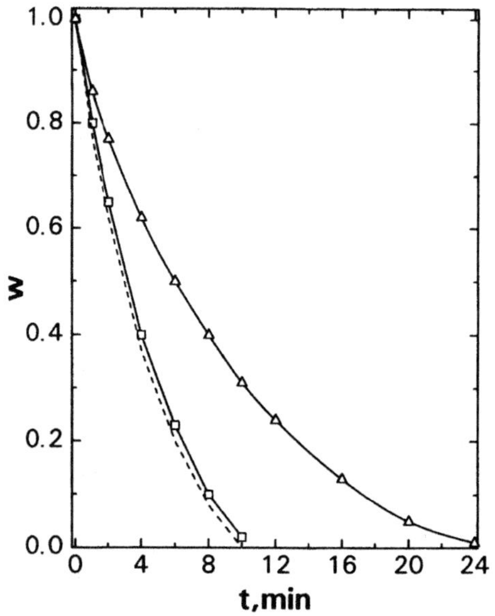

FIGURE 2.69 Kinetic curves of ice cryoextraction from granules of frozen ATOF solutions with different phase composition ($T = 248$ K, $N = 73.3$ rad/s): crystalline ice and crystalline salt (triangles); crystalline ice and metastable salt solution with 5 vol.% of glycerine (squares); dissolution of ice granules (dashed line).

to be carried out it for poorly crystallizing solutions. However, in this case the formation of salt crystallites is determined not by freezing conditions but by parameters of the process of interaction between the granules and organic solvents. This interaction can influence the properties of salt and oxide powders.

2.4 CRYOPRECIPITATION PROCESSES

Different versions of precursors' coprecipitation in the form of insoluble or poorly soluble hydroxides, carbonates, sulphates, oxalates are widely used for preparing multicomponent solid-state materials. In spite of apparent simplicity, coprecipitation proved to be a rather complicated and insufficiently investigated process. Because of this it is difficult to choose the most efficient technique as well as to understand and eliminate the reasons for the unsatisfactory properties of precipitates.

The traditional method of precipitation consists of the addition of a certain volume of reactant solution to a precipitant solution taken in excess as a rule. This simplest route is characterized by continuous change of physico-chemical

TABLE 2.24 Composition of solid phase forming during interaction of $ZnSO_4$ and NaOH solutions.

Molar ratio NaOH/$ZnSO_4$ *in starting mixture*	*Composition of solid phase*
0.40–1.60	$ZnSO_4$–$4Zn(OH)_2$
1.84–2.00	$ZnSO_4$–$4Zn(OH)_2$ + $Zn(OH)_2$
2.02–2.16	$Zn(OH)_2$
2.40–10.00	$mZn(OH)_2$–NaOH (solid solution)

conditions with time leading to the instability of precipitates properties. As an example we can consider the $ZnSO_4$–NaOH–H_2O system represented in Table 2.24. One can see that formation of $Zn(OH)_2$ precipitate is possible only in restricted range of NaOH/$ZnSO_4$ ratios (2.02–2.16). Especially wide variation of precipitates properties occurs upon obtaining multicomponent products because in this case the rate and completeness of individual components' precipitation differ considerably.

The same situation takes place at high supersaturations created by mixing of sufficiently concentrated solutions containing Me^{n+} and $(C_2O_4)^{2-}$ ions in view of the fact that different oxalates are precipitated in different ways. The first portion of sediment is enriched by easily crystallizing components, and the last one is enriched by poorly crystallizing substances. Besides this, the whole crystallization cycle takes several hours. These phenomena are most commonly found in the synthesis of products whose salt precursors substantially differ from their solubility product values. Thus, while obtaining nickel–zinc ferrites by hydroxide precipitation one can expect a great difference in precipitation rates of components because solubility products of $Ni(OH)_2$ and $Zn(OH)_2$ ($2*10^{-15}$ and $3*10^{-15}$, respectively) differ greatly from the solubility product of $Fe(OH)_3$ ($6.3*10^{-35}$). From this point of view the temperature decrease is a favourable factor since it decreases solubility products of all individual components for several orders and hinders their separate precipitation. Besides this, because one of the reagents (namely cryogranules of frozen solution) is solid, there appears to be a possibility of localization of chemical precipitation in microvolumes. Precipitates formed at low temperatures possess high dispersion and chemical homogeneity, which is favourable for the decrease of decomposition temperatures necessary for further interaction between oxide components with the formation of more complicated compounds.

At present the process of cryoprecipitation is usually carried out in the following way. Cryogranules obtained by dispersion and freezing of solutions are first placed in a solution cooled to 230–250 K of precipitant with intensive stirring. After completing the chemical interaction (1–2 hours) the precipitate is separated from the mother liquor by cold filtration, washed to remove precipitant ions (or soluble intermediates) and dried.

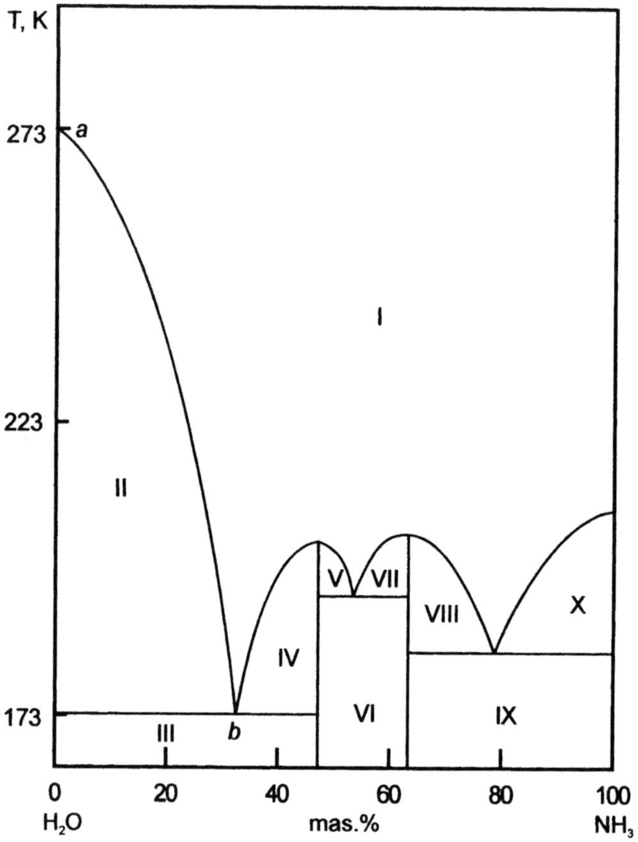

FIGURE 2.70 Phase diagram for the system ammonia–water: (I) solution; (II) solution + ice; (III) ice + NH_3*H_2O; (IV) solution + NH_3*H_2O; (V) solution + NH_3*$0.5H_2O$; (VI) NH_3*H_2O + NH_3*$0.5H_2O$; (VII), (VIII) solution + NH_3*$0.5H_2O$; (IX) NH_3 + NH_3*$0.5H_2O$; (X) solution + NH_3.

At present a lot of data is accumulated confirming the efficiency of the cryoprecipitation technique for preparing piezo- and ferroelectrics, ferrites, granulated oxide powders and many other materials. However, there are no systematic studies of the mechanism of the processes. At first it is connected with the necessity to use diagrams of multicomponent systems which are not yet studied in the low-temperature range. Even in the simplest case when only one kind of cations precipitates, the system should be considered as a three-component one. But in real cases there are at least two cations which correspond to a four-component system.

Aqueous ammonia solutions are the ones most widely used as precipitants. In Figure 2.70 the phase diagram of the system ammonia–water is represented.

The existence of a wide range of liquid solutions at low temperatures and the possibility of obtaining a large number of insoluble hydroxides makes this method promising for the synthesis of many advanced materials. But there are disadvantages as well. Firstly, for the fast and complete precipitation of many hydroxides it is necessary to maintain high pH values which are accessible only by the use of strong alkalis (NaOH, KOH). Secondly, ions of many transition elements (Co(II), Ni(II), Cu(II) and others) form exclusively stable complex ammoniacates readily soluble in water and this leads to losses of one or several cations due to the incomplete precipitation of one or any of them.

For these reasons, in choosing the right kind of precipitant the following criteria should be taken into account:

1. the precipitant should be liquid at low temperatures;
2. all precursors should form insoluble compounds with the given precipitant;
3. any precursors should be very soluble in the same conditions;
4. the initial precipitant should be able to dissolve ice from cryogranules. This process is completely determined by the difference of concentrations in initial and saturated at given temperature solutions (liquidus *ab* in Figure 2.70).

Properties of precipitants efficient for cryoprecipitation process are listed in Table 2.25.

TABLE 2.25 Some properties of aqueous solutions of precipitants.

Precipitant	T_{eut}, K	C_{eut} (wt.%)	C_{sat}* (wt.%) at 253 K
NH_4OH	173	33	14.0
NaOH	244.8	19.0	16.0
KOH	195	34.2	18.0
Na_2CO_3	271	5.7	–
K_2CO_3	236.5	40.8	32.9
HCOOH	224	70.0	39.0
CH_3COOH	246.3	60	49
$H_2C_2O_4$	271.5	–	–

* The concentrations of the solutions in equilibrium with ice at 253 K.

It is obvious that the use of only aqueous solutions considerably restricts the possibilities of the technique. For example, it is impossible to use aqueous solutions of oxalic acid because they crystallize at 271.5 K whereas much lower temperatures (250–230 K) are required. As the process temperature should not exceed the melting temperature of cryogranules the use of non-aqueous solutions of precipitants is the most promising method. In the first place it involves alcohol solutions. The most efficient may be the first representatives of monoatomic alcohols – methanol, ethanol and propanol. Their advantages are obvious:

- many potential precipitants (KOH, $H_2C_2O_4$ and others) are very soluble in these alcohols;
- alcohols have low freezing points and therefore their solutions may exist at sufficiently low temperatures;
- alcohols are also good extractants and able to dissolve considerable amounts of ice with high rates.

Nevertheless non-aqueous solvents have limited application because of a deficiency of data on the low-temperature solubility of precipitants. It does not allow evaluation of the optimum ratio of reagents upon cryoprecipitation.

It is necessary to note that the solubility of many salts in pure water, even for very soluble ones, decreases sharply in water–alcohol mixtures formed on ice extraction (the so-called salting-out effect).

The physico-chemical regularities of cryoprecipitation processes when using alcoholic solutions of precipitants are well illustrated by kinetic investigation (Shabatin *et al.*, 1984) of precipitation as well as cryoprecipitation of cobalt oxalate with water–alcohol solution of oxalic acid (Figure 2.71).

The first stage of interaction between $CoSO_4*6H_2O$ particles and dry ethanol, methanol and water–alcohol mixtures was studied in detail. Dissolution rates of solid particles and concentrations of saturated solutions were measured to elucidate the mechanism of precipitation. Visual observations of the dissolution

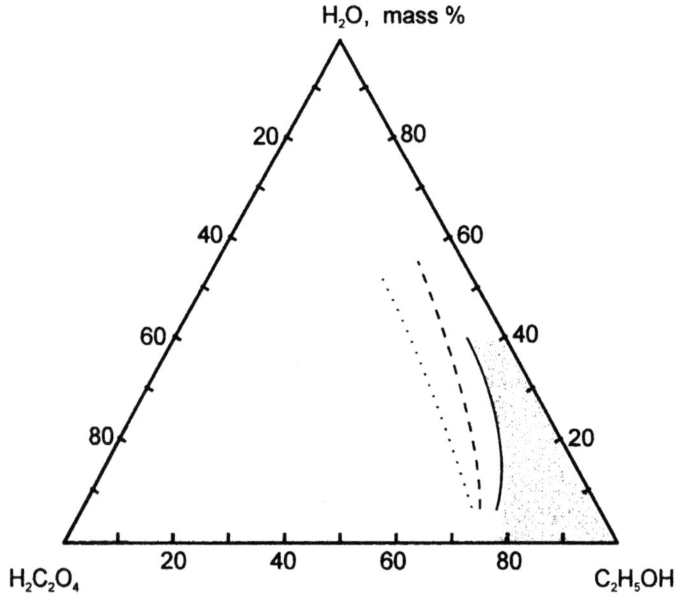

FIGURE 2.71 Solubility limit of $H_2C_2O_4$ in the mixture of water and ethanol at 20 (. . . .), − 15 (- - -) and − 30 °C (—).

of $CoSO_4*6H_2O$ in absolute ethanol showed that, when stirring the mixture, an intensely coloured homogeneous solution is formed, after which a finely dispersed monohydrate of cobalt sulphate precipitates. Metastable solutions proved to contain cobalt sulphate in amounts considerably exceeding the equilibrium ones (36 times).

It should be pointed out that temporal change of electrical resistance of solutions formed upon the interaction of $CoSO_4*6H_2O$ particles with alcohols (Figure 2.72) is also anomalous, but such an anomaly is not observed during the dissolving of anhydrous $CoSO_4$.

In the case under consideration the formation of metastable supersaturated solutions changes the kinetics of precipitation of cobalt oxalate considerably. To take it into account, the solubility diagram for the CoC_2O_4–C_2H_5OH–H_2O system was investigated (Shabatin, 1986). This study enabled the concentration range to be set off for subsequent kinetic analysis of precipitation and cryoprecipitation of cobalt oxalate by a water–alcohol solution of oxalic acid. According to this investigation (Mozhaev, 1988), there is no influence of stirring intensity on the process rate at $T = 293$–263 K for 8.3 to 21.2 mass % water content in water–

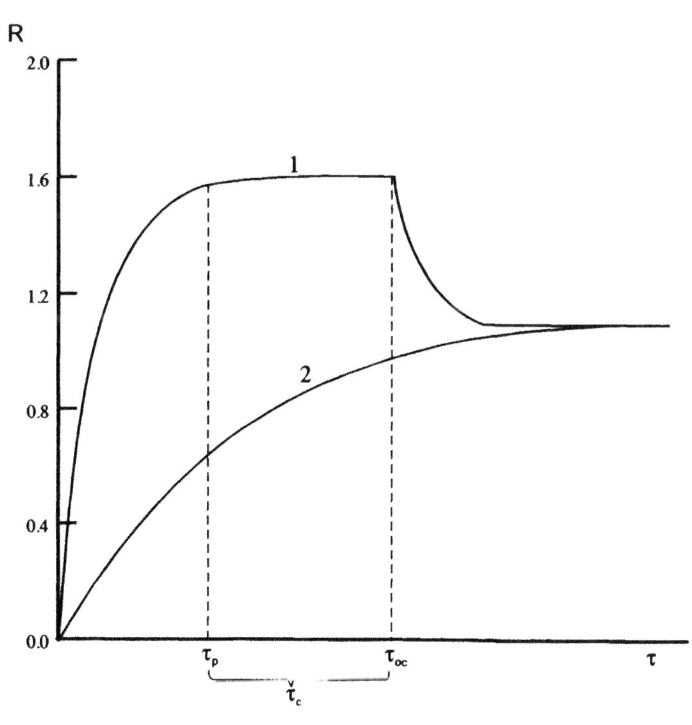

FIGURE 2.72 Changes of electric resistance of solution during interaction with alcohol (1) $CoSO_4*6H_2O$; (2) anhydrous $CoSO_4$.

ethanol solutions. Such behaviour is typical for kinetic regime of precipitation when the limiting stage of the process is the interaction between precipitant and the substance to be precipitated at the interface of liquid and solid phases. It was also shown that the rate of formation of CoC_2O_4 precipitate is decreased with an increase of water content in the precipitant solution. Activation energy of the process and some values of rate constant k_{sr} of the interface reaction between liquid and solid phases calculated using the 'shrinking sphere' equation (Brown *et al.*, 1980):

$$1 - (1 - \alpha)^{1/3} = k_{sr}\tau \tag{2.60}$$

are given in Table 2.26.

Another type of behaviour is observed for 60 wt.% water content in the solvent. At $T = 293$ K the variation of stirring intensity changes the rate of the process considerably whereas at 263 K the rate remains unchanged. These data indicate the mixed diffusional–kinetic regime of precipitation when the contribution of each process changes considerably depending on hydrodynamic conditions and temperature. In this case the interface reaction of liquid and solid phases is the limiting stage of the process only at $T = 263$ K where k_{sr} values can be determined (Table 2.26).

Micromorphological analysis of cobalt oxalate precipitates formed by interaction of $CoSO_4*6H_2O$ crystals with solutions of oxalic acid indicates substantial influence of solvent composition on the size and morphology of precipitated particles. Precipitation by aqueous solutions of oxalic acid resulted in

TABLE 2.26 The values of the kinetic parameters for the reaction of precipitation of cobalt oxalate by water–ethanol solutions of the oxalic acid.

Precipitation C_{H_2O}, wt.%	T, K	N, rad/s	$k_{sr} \times 10^4$, s^{-1}	$E_{(act)}$ kJ/mol
	293	47.1	2.17	
8.3	293	36.7		40 ± 2
	278	47.1	1.21	–
	263	47.1	0.33	
15.3	293	47.1	0.77	–
21.2	293	47.1	0.23	
	293	36.7		–
62.6	263	47.1	0.15	
	263	36.7		

needle-like crystals formation while at low water content (<15 wt.%) irregular polyhedrons are formed.

To understand the reason for this influence we should take into account the great role played by the rate of ice removal from cryogranules during cryoextraction. It can be expected that at fast extraction of the ice the particles of precursor salt remain in contact with the solvent for a rather short time while the water content and salt solubility limit in the last one are quite low. In contrast, low rates of cryoextraction correspond to the mixed diffusion–kinetic regime characterized by continuous contact of the precursor salt with the solvent having much higher values of salt solubility and, therefore, promoted recrystallization processes.

The influence of the nature of organic solvent on the character of precipitation can be illustrated by cryoprecipitation of CoC_2O_4 from water–methanol solutions of oxalic acid having the same water activity values with the above-mentioned water–ethanol solution (14.4 wt.% H_2O in the water–methanol mixture corresponds to 8.3 wt.% of H_2O in the water–ethanol mixture). Formation of metastable supersaturated solutions does not occur in the concentration region under investigation. Kinetic experiments showed the direct correlation of stirring intensity and precipitation rate throughout the whole temperature range. Therefore, apart from water–ethanol solution the precipitation with water–methanol solution of oxalic acid takes place according to the diffusion mechanism. We explain this fact by recalling that the solubility of cobalt sulphate in methanol is higher by a factor of ten than its solubility in ethanol.

Formation of metastable supersaturated solutions is a rather general phenomenon, often occurring during cryoprecipitation. To control the influence of metastability on the precipitation different factors should be taken into account. First of them is a ratio 'precursor salt/precipitant'. Studies of compositional dependence of the metastable solution lifetime for the 'cobalt sulphate–dried ethanol' system (Table 2.27) showed the falling stability of solution for the increasing $CoSO_4{*}6H_2O$ content in the initial mixture.

High content of $CoSO_4{*}6H_2O$ results in crystallization of monohydrate from supersaturated solution prior to complete dissolution of hexahydrate. The quantity of crystallohydrate water in solid phase decreases continuously during dissolution up to the moment when the precipitate falls out from formed metastable solution (Table 2.28).

TABLE 2.27 The variation of the lifetime of cobalt sulphate metastable solution in ethanol versus the composition of the starting mixture.

$C_{CoSO_4{*}6H_2O}$, mol.%	0.175	0.218	0.262	0.306	0.349	0.392	0.436
τ_c, min	119	87	44	36	31	17	15

TABLE 2.28 The content of crystallohydrate water in the solid $CoSO_4$ nH_2O versus the reaction time with ethanol.

τ, min	19	29	39	59
n ($\pm$0.05)	5.08	3.49	2.33	1.45

TABLE 2.29 The influence of the water concentration in the ethanol on the lifetime of cobalt sulphate metastable solution.

C_{H_2O}, mol.%	1.697	2.247	2.916	4.044	4.723	5.324
τ_c, min	370	82	29	10	7.5	2.2

TABLE 2.30 The dependence of the lifetime of cobalt sulphate metastable solution in ethanol on temperature.

T, K	293	303	313
τ_c, min	161	93	58

Another important factor influencing the stability of supersaturated solutions is the water content in the solvent. For mixtures with the same salt/solvent ratio the buildup of water content in alcohol decreases the stability of metastable solutions sharply (Table 2.29). When water content in alcohol exceeded 5.3 mol.% no metastable solution was observed.

The influence of temperature on the stability of metastable cobalt sulphate solution was studied by measurement of the induction period τ_c before the falling of residue from the metastable solution ($\tau_c = 0.133$ mol.%) at a given temperature (Table 2.30). Calculation of the apparent activation energy of the process according to Arrhenius Law gives the value of 38.9 ± 2.9 kJ/mole.

Finally, it is necessary to note that variations of hydrodynamic conditions in the reactor have considerable influence on the value of τ_c. For example, for the mixture with composition 0.31 mol.% at stirring intensities 26.2, 44.2 and 47.1 rad/s the experimental values of τ proved to be equal to 74, 31 and 28 min, respectively. Besides this, crystallohydrate water content in the precursor salt ($CoSO_4$*$6H_2O$ or $CoSO_4$*$3.8H_2O$) also influences the lifetime of metastable solution ($\tau_c = 38$ and 58 min, respectively, for solutions with the same $CoSO_4$ content).

If water concentration in the initial water–ethanol mixture exceeds 5.32 mol.% the formation of the metastable solution is not observed. In this case the interaction of the solid phase with dehydrating liquid causes the dehydration of solid salt without dissolution of solid salt.

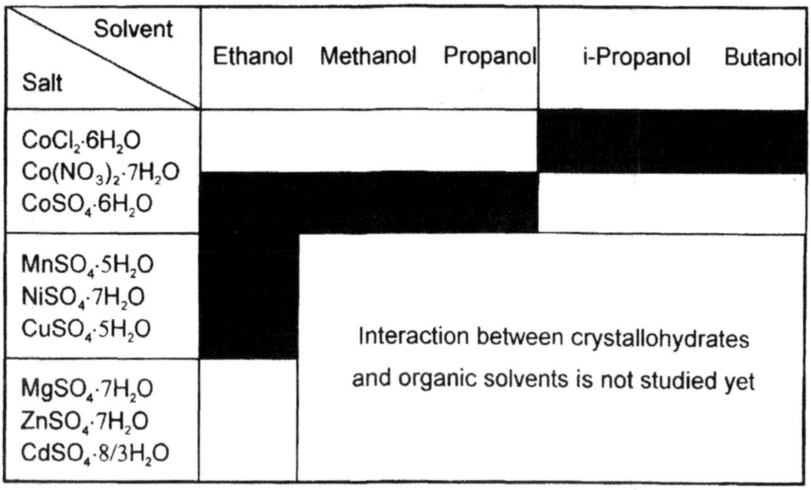

FIGURE 2.73 Interaction of crystallohydrates with organic solvents. Dark area = metastable solution formation.

It was established that the formation of metastable solutions is observed during interaction of $CoSO_4*6H_2O$ with CH_3OH and n-C_3H_7OH but does not occur in iso-C_3H_7OH and n-C_4H_9OH (Figure 2.73).

This difference can be attributed to the falling solubility of cobalt sulphate in alcohols when coverting from methanol to butanol. At the same time the stability of the formed metastable solution decreases sharply from methanol to ethanol and propanol at the same salt concentration. The values of t_c for different salt contents in initial mixtures for methanol and propanol are listed in Table 2.31. The ability of different solvents to form metastable solutions of $CoSO_4$ can also be illustrated by critical values of salt content in the initial mixture enough to form metastable solutions being stable for 30 min. For methanol, ethanol and propanol the corresponding concentrations are equal to 0.992, 0.338 and 0.062 mol.%.

Investigation of the interaction of $Co(NO_3)_2*6H_2O$ and $CoCl_2*6H_2O$ with methanol proved to be impossible because of the high solubility of these salts in

TABLE 2.31 The dependence of the composition of the starting mixture on the lifetime of $CoSO_4*6H_2O$ metastable solutions in alcohol.

Methanol		*Propanol*	
C_{mix}, mol.%	t_c, min	C_{mix}, mol.%	t_c, min
0.763	50	0.057	35
0.914	31	0.085	23
–	–	0.115	10

this solvent. The dissolution of these crystallohydrates in isopropanol and butanol is characterized by a short (5-minute) initial decrease of electric resistance followed by a fast increase of the last one up to the constant value. These results illustrate the intermediate situation when the formation of metastable solutions is complicated by the enhanced solubility of salts which substantially shortened the lifetime of the metastable state.

Studies of interaction of absolute ethanol with transition metal sulphates showed a remarkable difference in their ability to form metastable solutions. The formation of metastable solutions is found to occur for Mn^{2+}, Ni^{2+} and Cu^{2+} salts while for Zn^{2+}, Cd^{2+} and Mg^{2+} sulphates it is not observed.

The outlined data may be clarified by the model developed by Shabatin (1986) and Malykh (1985) for the interaction of $CoSO_4*6H_2O$ with methanol. Dissolution of $CoSO_4*6H_2O$ particles in CH_3OH causes formation of $(SO_4)^{2-}$ ions and hydrated $[Co(H_2O)_6]^{2+}$ ions. Then interaction of these ions with alcohol leads to a stepwise exchange of five water molecules in aqua-complex for alcohol molecules so that Co^{2+} ion holds its octahedral coordination in solution up to the moment of precipitation. The sixth water molecule possessing the highest energy of $H_2O–Co^{2+}$ bond is not involved in this exchange. Lower solubility of newly formed alkohydrate in the alcohol results in destabilization of metastable solution and precipitation of cobalt sulphate monohydrate.

According to these considerations the possibility of formation and stability of metastable solution are determined by the number of water molecules that are able to be exchanged for alcohol molecules in aqua-complexes. This conclusion correlates quite well with the fact that, among crystallohydrates under investigation, the greatest ability to form metastable solutions is observed for compounds of 3d transition elements (Co^{2+}, Mn^{2+}, Ni^{2+}, Cu^{2+}) having a remarkable tendency to form different complex compounds.

2.5 PROCESSING OF CRYOPRODUCTS

2.5.1 Structure and ageing of cryoproducts

The salt or hydroxide products of freeze-drying, cryoextraction or cryoprecipitation processing are considerably far from the equilibrium state, mainly because of their very high degree of dispersion. Although the conditions of all processes mentioned above can vary substantially, the comparison of product microstructures suggests that dispersion increases in the raw: cryoprecipitation–cryoextraction–freeze-drying. Thus, the very high dispersion of freeze-drying products formed during cryocrystallization can be ascribed to their inherent characteristic.

As mentioned above (section 2.2.2), for the most part, the products of freeze-drying can be obtained in one of the following forms: equilibrium lower hydrates;

'vacancy structures'; amorphous hydrates. Let us consider the structural peculiarities of such salt products (Shlyakhtin, 1985).

Lower hydrates are quite stable dehydration products of higher hydrates. For example, the dehydration of heptahydrates $MeSO_4*7H_2O$ (Me = Mg, Mn, Fe, Co, Ni, Zn) results in stable monohydrates $MeSO_4*H_2O$. These hydrates can be produced under harsh conditions of freeze-drying – at intensive heat supply and considerable hampering of vapour diffusion out of the cryogranule. This situation resembles the conditions of equilibrium monohydrate synthesis. To give an example, the temperature of sublimator heat plates no less than 300 K and the salt layer no less than 10 mm thick are quite necessary to prepare the freeze-dried iron sulphate monohydrate.

At the same time, there are some specific features of 'cryochemical' monohydrates making them dissimilar to 'equilibrium' (well-crystallized) ones:

- higher dispersity (dispersion degree may account for differences in thermoanalytical data for 'cryochemical' and equilibrium monohydrates (Oleinikov *et al.*, 1985b); for example, dehydration of 'cryochemical' monohydrates can proceed in a wider temperature range and generally at 15–20 K lower than dehydration of well-crystallized samples); and
- different proportion of diffraction maxima compared to test samples. (Mössbauer spectra of 'cryochemical' and well-crystallized $FeSO_4*H_2O$ are identical within experimental errors and this suggests that the nearest surrounding of Fe(II) cation is much the same).

The so-called 'vacancy structures' are much more complex at this point (Mutin *et al.*, 1979). According to chemical analysis they contain 1.6–1.8 moles of H_2O/mole of salt. Nevertheless, their X-ray patterns are substantially different from those of equilibrium crystallohydrates with low water content and are much more similar to patterns of higher crystallohydrates $MeSO_4*4H_2O$ and $MeSO_4*6H_2O$ (Figure 2.74).

Mössbauer spectra indicate that the nearest surrounding of an iron atom in a vacancy structure of $FeSO_4*xH_2O$ is quite similar to that in monohydrate. Therefore, vacancy structures can be regarded as having intermediate properties – the properties of higher tetra- or hexahydrates at macrolevel (XRD) and of monohydrates at microlevel (Mössbauer spectroscopy).

Thermal analysis curves for vacancy structures are similar to those for 'cryochemical' monohydrates. This fact implies that the rearrangement of unstable vacancy structures during dehydration may occur and more stable equilibrium monohydrates would form. Sometimes such rearrangement can proceed in an avalanche-like fashion. This may account for a distinct exothermic effect on DTA curves at the beginning of dehydration. In this case, the DTA curves would have two minima: at low (120–150 °C) temperature, elimination of adsorbed water, and at high (300 °C) temperature, elimination of water from

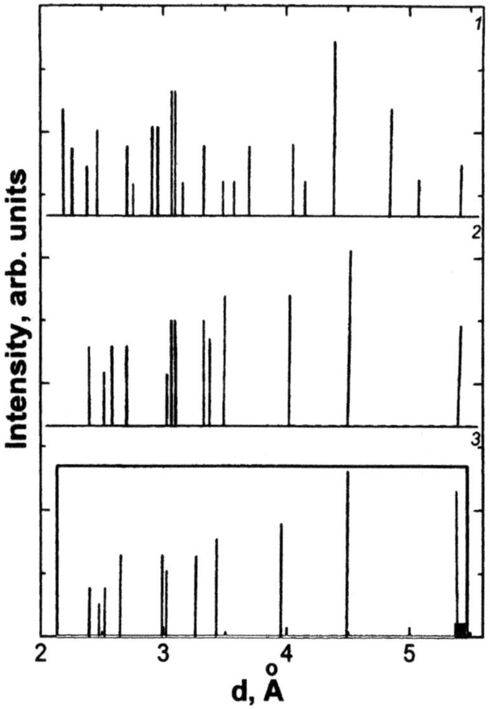

FIGURE 2.74 XRD patterns of: $FeSO_4{*}6H_2O$; $FeSO_4{*}XH_2O$ with vacancy structure; $FeSO_4{*}4H_2O$ (top to bottom).

monohydrate $MeSO_4{*}H_2O$. However, usually this is not the case, so dehydration and structural rearrangement should take place simultaneously. Thus, 'vacancy' structure $MeSO_4{*}xH_2O$ seems to be thermally unstable.

The structural peculiarities of 'vacancy' compounds may point to the conditions under which they are prepared. Thus, dehydration resulting in vacancy monohydrate would occur only if the crystal lattice elements have low mobility. Besides the factors connected with the salt nature, the mobility may be influenced by external conditions. It was shown that at vapour pressure (P_{H_2O}) up to 10^2 Pa some vacancy structures remain stable at room temperature (Mutin and Watelle, 1979). To preserve significantly more stressed vacancy structures formed during freeze-drying, much lower values of P_{H_2O} are essential. So, there is no wonder that vacancy monohydrates are usually formed at low heat input when diffusion of water vapours out of the cryograin does not limit the process.

In the absence of destabilizing factors the vacancy structures can be kept for a long time. The estimation of their stability in vacuum ($P \leq 10\,Pa, T = 260\,K$) has shown (Shlyakhtin, 1985) that the dried products do not change their structure for two weeks at least. At higher vapour pressure the vacancy monohydrate would decompose as illustrated by the following experiment. The freeze-drying of frozen solution of zinc sulphate was carried out in two identical reactors. The conditions differed by the presence of disperse ice layer in one of the reactors; the sample to be sublimed was placed upon the layer. The X-ray pattern of the latter product has indicated that the mixture of tetrahydrate and monohydrate in equal quantities was formed (Figure 2.75).

The structure of the other corresponded to hexahydrate '$ZnSO_4*6H_2O$' with some admixture of monohydrate. These results once more call attention to the primary role of water vapour in the process of freeze-drying product formation.

In conclusion, it is worth noting that vacuum dehydration of recrystallized $ZnSO_4*7H_2O$ has resulted in a product, whose composition was $ZnSO_4*xH_2O$ ($x = 1.4$–1.6), but whose structure corresponded to crystallohydrate $ZnSO_4*6H_2O$. Therefore, the formation of vacancy structures during vacuum dehydration seems to be quite common.

Now, let us consider in more detail the structure and properties of X-ray amorphous products of freeze-drying. X-ray amorphous hydrates have been known for half a century; however, until now all the data concerned only their identification.

Shlyakhtin (1985) has shown that proper synthesis conditions make it possible to prepare amorphous salt products from sulphates of nickel, magnesium, cobalt

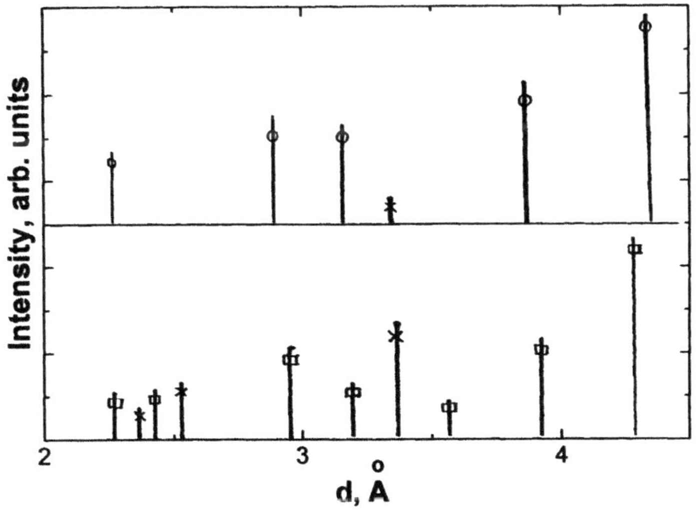

FIGURE 2.75 XRD patterns of: frozen $ZnSO_4$ solutions after freeze-drying at different P_{H_2O} ($P_{II} > P_I$). Squares – vacancy structure similar to $ZnSO_4*4H_2O$; circles – vacancy structure of $ZnSO_4*6H_2O$ type; crosses – $ZnSO_4*H_2O$.

and iron, but there appear to be difficulties if zinc sulphate is used. It should be noted that the properties of synthesized amorphous hydrates, for the most part, are similar. According to data of chemical analysis the water content is approximately equal for all these compounds and lies in the range of 2.2 to 2.5 moles H_2O per mole of salt. The parameters of thermal dehydration process for these compounds are also the same – the elimination of water occurs continuously in a temperature range of 300–600 K. According to the data of differential thermal gravimetry (Figure 2.76) the dehydration of these compounds is not accompanied by the crystallization of intermediate products.

This fact is implied by the absence of sharp peaks at $T = 470–620$ K which would appear in the case of crystalline monohydrate formation. At the end of dehydration of amorphous $CoSO_4{*}xH_2O$ the distinct exothermic effect (Figure 2.76A) was detected by DTA sensors even if the plate-like crucible was used. In this temperature range such an effect can only suggest spontaneous crystallization, and its position at the end of the thermoanalytic curve (Figure 2.76B) points to the crystallization of the anhydrous product. If dehydration proceeds with a diffusion hampering, DTA curves are similar to those described earlier for crystalline vacancy structures (Figure 2.76C). Significant exothermic effects were not observed, suggesting simultaneous water elimination and crystallization. Therefore, thermal stability of X-ray amorphous hydrates is relatively

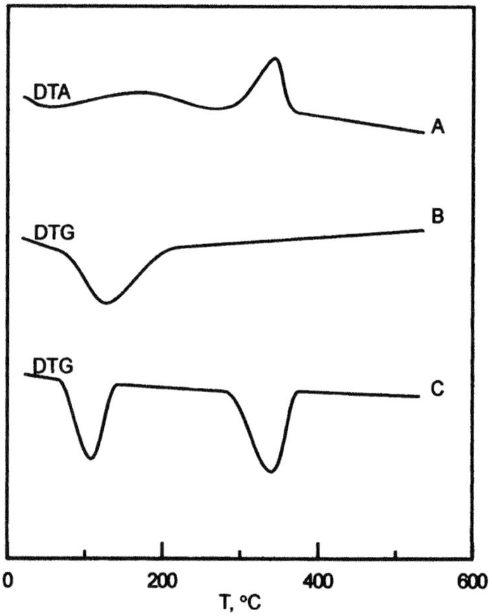

FIGURE 2.76 Thermoanalytical curves for thermal dehydration of amorphous $CoSO_4{*}XH_2O$. (A) DTA, flat crucible ($H < 1$ mm); (B) DTG, flat crucible; (C) DTG, cylindrical crucible ($H = 10$ mm).

low and decreases in the presence of water vapour. Most amorphous hydrates are quite stable at room temperature, especially in the absence of moisture.

It seems that the reasons for such stability could be realized from consideration of structural peculiarities of amorphous hydrates. Mössbauer spectroscopy data, in this case, enable the situation to be understood more clearly, because of its applicability both to crystalline and amorphous salts. The spectra analysis of amorphous $FeSO_4{*}xH_2O$ (Figure 2.77) has led to the following results (Shlyakhtin *et al.*, 1988):

1. The spectrum corresponded neither to vacancy structure $FeSO_4{*}xH_2O$ nor to the structure of known hydrates of iron sulphate and was characterized by the specific parameters: $\Delta = 2.80$ (77 K); 2.50 (290 K) and $\delta = 1.48$ (77 K); 1.45 (290 K) mm/s.

2. The spectrum lines are considerably broadened, complicating the analysis; however, one of two main models of amorphous phases seems to be more preferable (Mott, 1987; Gubanov, 1965). The first model – a statistical one – assumes random distribution of atoms in space, i.e. the absence not only of a long-range order, but of a short-range order as well. The chemical bonds still ensure the necessary distances between atoms, but their space orientation is quite random. The second model, called the microcrystalline one, assumes the

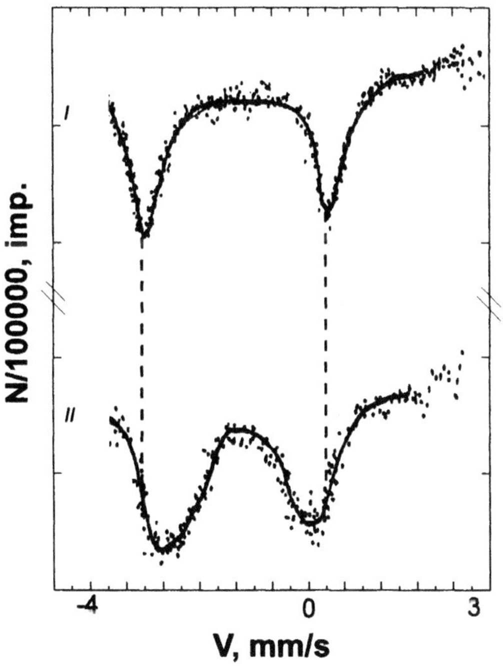

FIGURE 2.77 Mössbauer spectra of amorphous hydrate $FeSO_4{*}XH_2O$ at (I) 77 K; (II) 290 K.

existence of a great number of areas with near-range order randomly distributed in the sample and separated by disordered areas. The spectrum with distinct absorption maxima proves certain preferential surrounding of iron atoms, i.e. brings evidence for the second model. This model also agrees well with the data of thermal analysis (Figure 2.76).

Freeze-drying and thermal decomposition are usually carried out in different devices. Most of the modern industrial sublimators are installations with a discrete working cycle and a cycle time of 20–30 hours. Thus, more or less long-time conservation of freeze-dried salt products after their removal from sublimators is quite necessary. The short-time interaction of dehydrated products with water vapour is inevitable anyway, because of the current design of sublimators where after opening the chamber is filled with atmospheric air. So, it would be reasonable to divide the processes involved into two types:

(a) the processes of short-time interaction between the product and atmosphere moisture at the moment of sublimator opening;
(b) the processes of long-time interaction between salt samples and water vapours during conservation.

In any case, the product of freeze-drying undergoes considerable changes during conservation. For example, rapid crystallization of an initially amorphous product prepared by the freeze-drying of a frozen solution of yttrium, barium and copper nitrates (1:2:3) (Figure 2.78).

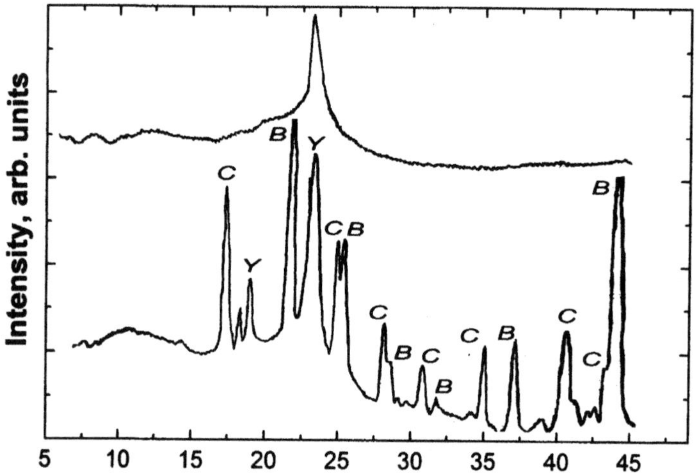

FIGURE 2.78 XRD patterns of freeze-dried solution of Y, Ba and Cu nitrates (Y/Ba/Cu=1/2/3) (a) immediately after completion of freeze-drying (top curve); (b) after storage for 30 min at $P_{H_2O} <$ 300 Pa. Y, B, C – Y, Ba and Cu nitrates, respectively.

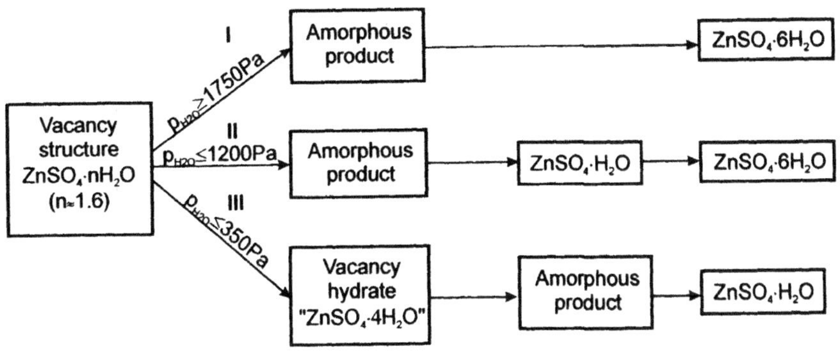

FIGURE 2.79 Transformations of lacunary $ZnSO_4{}^*XH_2O$ (vacancy structure) during storage at different P_{H_2O} values.

Figure 2.79 provides a generalized scheme for the experimental results on hydration of the freeze-dried crystalline product with the vacancy structure $ZnSO_4{}^*nH_2O$ at various values of P_{H_2O}.

It is striking that the hydration process is accompanied by decay of the initial structure, even though the arrangement of the salt fragments corresponds to the structure of the hexahydrate. Thus, the restoration of water to the vacant positions in the structure is hindered, although the high dispersity of the product provides significant amounts of water to be adsorbed and chemisorbed. Indeed, for $P_{H_2O} < 350$ Pa, the total water content in the samples actually does not change and the observed phase transitions may be ascribed only to an increase in the amount of adsorbed water.

The growth of P_{H_2O} leads to a considerable increase in the water content in the samples (up to $n = 6$). However, the phase transition results firstly in the formation of the monohydrate and not until then, the hexahydrate, which is stable at high P_{H_2O}, would form. It has been shown that at $P_{H_2O} < 1200$ Pa, the processes involving the relaxation of vacancy structures initially predominate and, after that, the hydration of the product actually arises. Our results are in good accordance with the proposal of a low rate for the conversion of water into the crystal hydrate form.

The effect of adsorbed water on the phase transitions in the system may be explained by the loss of strength upon adsorption (Rebinder effect) (Shchukin *et al.*, 1982). The dehydrated product with the structure of the vacancy crystallohydrate can undergo significant mechanical strain. The loss of stability upon adsorption leads to the decomposition of surface layers. Then the vacancy structures transform into the less-strained defective tetrahydrate or even into the monohydrate which is stable at the initial water content. Thus, the possibility of the monohydrate formation increases in regard to the growing amount of adsorbed water in the sample. The amount of water, in its turn, increases with the growth of P_{H_2O}. Apparently, this could account for the difference in hydration mechanisms at the initial stages of the process at P_{H_2O} greater than 1750 Pa and less than 350 Pa.

The decay of large crystallites of the dehydrated crystallohydrate into many fine crystallites during hydration, which was observed by SEM, brings evidence for such mechanism. The amorphous phase apparently forms in the same manner; the amount of this phase increases considerably during the phase transitions.

Neither monohydrate nor tetrahydrate phases form upon hydration of the product at $P_{H_2O} > 1750$ Pa. This failure may be related both to the greater rate for this process than in the previous case and the greater amount of capillary-bound water which leads directly to recrystallization effects.

It has also been noted that similar effects are observed in coarsely crystalline zinc sulphate after dehydration in vacuum at low temperatures. This was illustrated by the SEM data obtained for single crystals of dehydrated $ZnSO_4$*$7H_2O$ before and after interaction with water vapour. Thus, the dispersity of the cryocrystallization product is not the major factor determining the mechanism of the dehydration and following hydration of freeze-drying products. The course of these processes is determined by the simultaneous influence of vacuum and low temperatures on the crystallohydrate and, thus, by the formation of vacancy structures not only for zinc sulphate, but also for a number of other compounds differing greatly in their chemical nature (Oswald *et al.*, 1975; Mutin *et al.*, 1979; Chupakhin, 1983). On the other hand, the decomposition of vacancy structures during interaction with water is hardly related to the chemical reactions of the system and seems to be the result of the significant stresses in the vacancy structure. This indicates that the decomposition of vacancy structures during hydration is rather common and should occur upon the interaction of water vapour and the vacancy crystal hydrates of any chemical nature.

The universality of the scheme given in Figure 2.80 can also be proved by the results obtained while investigating the ageing of the vacancy product (see p. 100) prepared by freeze-drying, which contained nickel, zinc and iron sulphates together and corresponded to one of the commercial ferrites.

The initial precursor was a solid solution of (Ni,Zn,Fe)SO_4*$4H_2O$ containing 1.7–1.8 moles H_2O per mole of salt. Its ageing was carried out during a week at 298 K and at different values of P_{H_2O} (50–2400 Pa spanning an interval from 2 to 90% of absolute humidity).

X-ray analysis of products kept at $P_{H_2O} < 1.5*10^3$ Pa indicated that they were completely amorphous. The conservation of samples at $P_{H_2O} = 1.5*10^3$ Pa (the most common value of atmospheric humidity) and at room temperature led to transformation of products into crystalline monohydrates with relatively low content of amorphous phase. According to data of thermal analysis, this transformation was not accompanied by phase degradation and the monohydrate was still the solid solution of three sulphates (Ni,Zn,Fe)SO_4*H_2O.

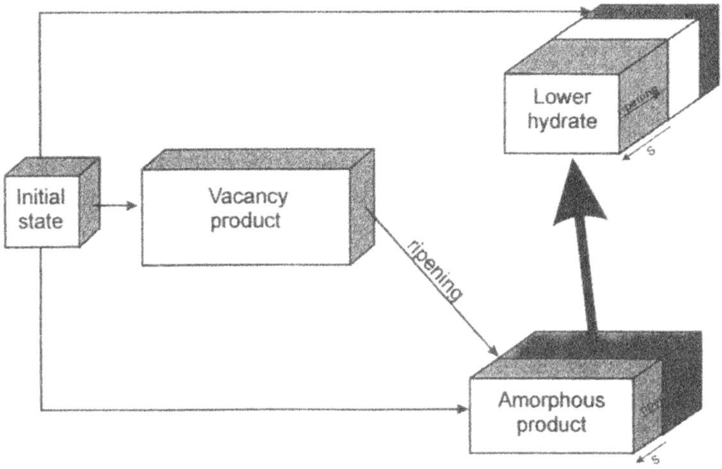

FIGURE 2.80 Interrelations between different types of freeze-drying products.

Turning back to the problem of ageing mechanism for the freeze-drying products, it is worth noting its practical consequences. The most common freeze-drying products of frozen sulphate solutions produced in commercial devices are crystalline vacancy hydrates (see p. 153). Their extreme instability to atmosphere moisture makes them inappropriate for storing. Therefore, it is necessary to stabilize one of the possible products of their ageing. Formation of higher hydrates leads, as was mentioned above, to phase degradation of multicomponent system and for this reason is undesirable. In the case of synthesis of nickel–zinc ferrites, monohydrates appear among ageing products usually when the P_{H_2O} value is sufficient for deeper hydration; the attempts to stabilize this kind of product were not successful. Now, it seems that the amorphous hydrates are the most promising for long conservation. Along with high resistance against the outer influence, these products also have a high chemical homogeneity.

The scheme in Figure 2.80 suggests that keeping the vacancy products at sufficiently low P_{H_2O} is the best method for obtaining amorphous hydrates. To avoid the catalytic effect of water vapour leading to undesirable shrinking of salt powders, the pressure should be minimized. However, it was shown that, in this case, the formation of amorphous phases proceeds quite slowly with simultaneous monohydrate formation. The reason for the observed phenomenon seems to be a considerably lower rate of amorphous phase formation at lower P_{H_2O}, compared with the rate of crystalline monohydrate formation. Obviously, the formation of heterogeneous product is also unacceptable in this case.

One of the possible ways can be a short-time (1–2 hours) exposure of the initial vacancy product to atmospheric moisture, ensuring the necessary water content in the system (2.3–2.5 moles H_2O per mole of salt). After that, the obtained product should be hermetically packed. The amorphous hydrates obtained in this way are appropriate for long-time storing in dry boxes.

To sum up all the information above, we can conclude that:

1. the pressure of water vapour determines not only the composition of the final ageing product, but also the mechanism of the rearrangement taking place;
2. at any P_{H_2O}, the ageing of vacancy hydrates is accompanied by their degradation;
3. the amorphous hydrates are relatively stable against ageing.

In addition, it should be pointed out that long-time contact of freeze-dried salt products with water vapour leads to considerable shrinking (collapse).

2.5.2 Dehydration and thermal decomposition of cryoproducts

The main feature of the sublimation-drying product is its very porous salt framework. The highly developed internal surface between the solid and gaseous phases is usually a pattern of porous structures in topochemical processes of degradation and thermal decomposition, and the nucleation of the new phase can arise at any point of this interface.

According to the classic description of topochemical processes, their initial stage extends only to a small depth, corresponding to conversion degree from 3 to 5%. It involves mainly the elimination of gaseous products from the most energetically stressed regions, where the bond energy is lower than in the volume. Gases eliminated across the crystallite surface soon get removed from reaction zone, and the next stage, which is called the **induction period**, comes about. At this stage the conversion degree does not change remarkably, but ‘maturation’ and emergence of the interface take place. The induction period itself represents a very complicated process consisting of several elementary stages. The main ones of these seem to be the drawing near of the defect regions. This process leads to their cooperation and results in formation of a defect region larger than of critical size, enabling the nucleation of a new phase and formation of a new interface.

Quite easy formation of defect regions at any point of the porous product obtained by freeze-drying leads to the saturation of salt framework internal space with gaseous products even at low ($\alpha < 0.01$) decomposition degree. As a result, the processes of dehydration and thermal decomposition can be considerably hindered. The process of gaseous product transfer from the separate crystallite surfaces to the geometrical boundary of the salt granule is called intradiffusional transfer. The term was chosen for the following reasons:

1. there is true external diffusion determined by transfer of gaseous products from the geometrical boundary of porous salt granule into the inert carrier-gas flow;

2. there is negligible internal diffusion in the salt crystallites assembled in the salt framework, because of their small size.

It should be noted that changes in hydrodynamic parameters of the reactor, affecting the conditions of external diffusion, do not promote the gas transfer inside the porous products obtained by the sublimation-drying technique.

Hence, the topochemical reactions involving cryochemical salt products are characterized by the perceptible contribution of internal diffusion. This contribution considerably increases with advancing the reaction zone inside the solid. On the other hand, the internal diffusion in the sublimation-drying products is either constant (if the products formed by dehydration and low-temperature decomposition do not crystallize) or considerably decreases due to less molar volume of the oxide product compared to that of initial salt (if the products formed in topochemical reactions do crystallize).

These features of sublimation-drying products clearly manifest themselves in dehydration and thermal decomposition processes at low temperatures of heating. While dehydration of such products, a little dependence on heating rate is usually observed. As shown in Figure 2.81, the dehydration curves of $Fe(OH)_3$ obtained by the cryochemical technique are almost coinciding for different heating rates

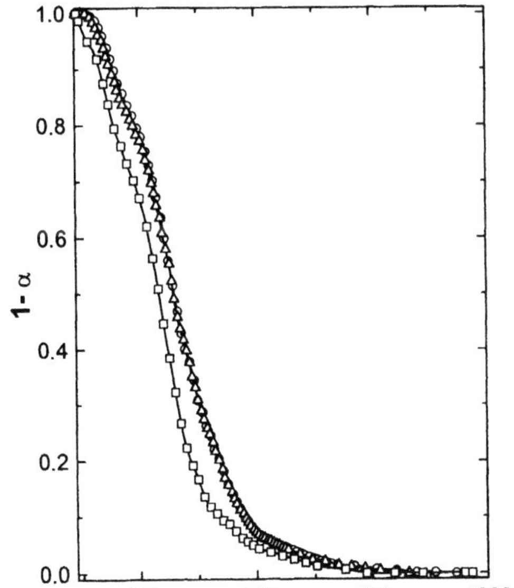

FIGURE 2.81 Thermal dehydration of freeze-dried $Fe(OH)_3$ at different heating rates. Squares – 2.5 K/min; triangles and circles – 5 and 10 K/min respectively.

from 5 K/min to 10 K/min. Similar behaviour can be observed for the salt products of freeze-drying.

To elucidate the correlations between cryoproduct structure and the process of its thermal decomposition, the behaviour of the model substance – ammonium tri-oxalatoferrate $(NH_4)_3[Fe(C_2O_4)_3]*3H_2O$ (ATOF) – was studied (Sokolov, 1978; Mozhaev, 1988). It was found that kinetic curves for each sample involved had the induction period, which value τ was determined graphically (Figure 2.82). The further analysis of dependencies $\ln \tau = f(1/T)$ and evaluation of apparent activation energy values for α-Fe_2O_3 nucleation showed that the latter depends on the salt prehistory.

For an ATOF sample obtained by crystallization from the solution saturated at 333 K (sample I) E_{act} was equal to 117.5 ± 10 kJ/mole, and for the sample obtained by freeze-drying of cryograins with diameter 3 mm (sample II) E_{act} was equal to 152.5 ± 10 kJ/mole. Using the grains with smaller size for freeze-drying (grains with diameter 130 μm, sample III) to lessen the internal diffusion resulted in a decrease of apparent activation energy up to $E_{act} = (123.7 \pm 9)$ kJ/mole.

It is worth noting that for the ATOF sample obtained by slow isothermal evaporation of solvent from the solution (sample IV), the value of E_{act} does not differ from sample I within the range of evaluation. At the same time, the induction period for sample IV is considerably shorter than for sample I. It was shown that the duration of the induction period is related to defects emerging:

1. as a result of decreasing the separate crystallite size, for example, by mechanical grinding; or

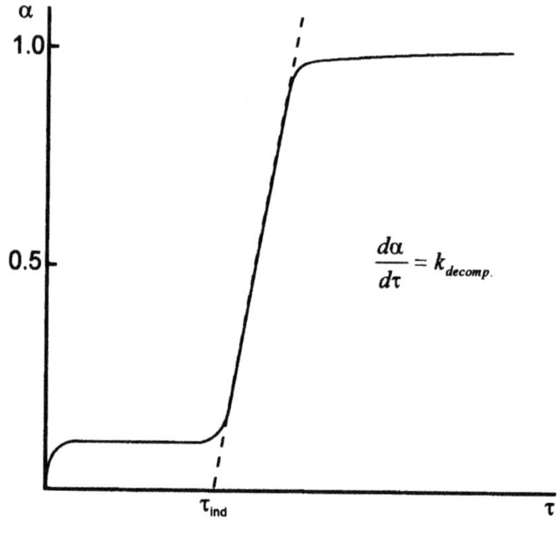

FIGURE 2.82 Typical TGA curve of dehydrated ATOF decomposition at isothermal conditions.

2. while formation of the salts with particular microstructure where each crystallite actually is an ensemble of smaller particles (for example, formed during recrystallization under the condition that the rate of crystal growth due to ionic mass transfer is of the same range as the rate of crystal growth due to coalescence of smaller particles).

In crystals with high concentration of growth defects, the latter can partially anneal or be annihilated while the temperature increases; and this results in a decrease of E_{act}. The higher initial concentration of defects, the lower the activation energy value. Thus, in the row of similar extractants 'methanol–ethanol–propanol', the first extractant was found to have the best capability for salt product recrystallization (Mozhaev, 1988). For this reason, the ATOF sample, obtained by extraction of ice from cryograins with methanol (under high supersaturation), was characterized by the highest defect content and, as a result, by minimal τ_i values and $E_{act} = (64.5 \pm 5)$ kJ/mole.

For samples VI and VIII obtained by extraction with propanol and ethanol the apparent activation energies are equal to (123.5 ± 10.0) kJ/mole and (94.0 ± 7.0) kJ/mole, respectively.

Therefore, microstructure of salt products obtained by cryoextraction technique differs considerably from that of products obtained by freeze-drying. Porous salt framework formed during cryocrystallization undergoes substantial changes during cryoextraction. Finally, it represents an ensemble of fine salt crystallites, where the concentration of defects and microstructure are mainly determined by the properties of the extractant–water system and by the duration of the process. Varying the conditions of powder preparation leads to differences in kinetic properties of substances under treatment. For example, the dehydration and thermal decomposition of cryoextraction products begin at the temperatures 50–100 K lower than for well-crystallized products. It is a direct result of more rapid formation of interface in salt systems with defects. In contrast, in the products of freeze-drying the formation of interface is complicated, because of the low defect content and internal diffusion hindrance. However, the highly developed surface and rapid increase of diffusion rate with temperature growth result in the fact that the dehydration and thermal decomposition occur in the same temperature range as for products of cryoextraction.

However, the mechanism of dehydration and thermal decomposition for freeze-drying and cryoextraction products can differ considerably from the mechanism of such processes for products obtained by equilibrium crystallization. This difference is especially noticeable if dehydration and thermal decomposition are carried out step by step.

As an example, let us consider the thermal decomposition process for lead nitrate $Pb(NO_3)_2$ obtained by equilibrium crystallization:

$$Pb(NO_3)_2 \rightarrow xPbO{*}Pb(NO_3)_2 \rightarrow 2PbO{*}Pb(NO_3)_2 \rightarrow 3PbO{*}Pb(NO_3)_2 \rightarrow 4PbO{*}Pb(NO_3)_2 \rightarrow PbO$$

The main reason of multistage decomposition of $Pb(NO_3)_2$ is diffusion hindrance; the accumulation of gaseous products would impede the process on one or another stage. Multistage mechanism is less characteristic while decomposition of sublimation-drying products

$$Pb(NO_3)_2 \rightarrow PbO{*}2Pb(NO_3)_2 \rightarrow 2PbO{*}Pb(NO_3)_2 \rightarrow PbO$$

and cryoextraction products

$$Pb(NO_3)_2 \rightarrow 2PbO{*}Pb(NO_3)_2 \rightarrow 4PbO{*}Pb(NO_3)_2 \rightarrow PbO$$

Such differences in mechanism were observed in the case of other salts forming several intermediates during dehydration and thermal decomposition.

For quantitative analysis of such topochemical reactions as dehydration and thermal decomposition of cryochemical products, the formal kinetic relations are usually applied (Young, 1966; Barret, 1973). They are classified into three main groups and their arrangement is determined by the limiting stage of the process.

Belonging to the first group are nucleation models which can be represented by the Avrami–Kolmogorov equation (Brown *et al.*, 1980):

$$\alpha = 1 - \exp(-k\tau^n) \tag{2.61}$$

where α is a conversion degree in the reactions of dehydration or thermal decomposition; k is the constant of the process; τ is the duration of the process; n is the parameter depending on the reaction mechanism, nucleation rate and geometry of nuclei.

The last equation is often represented in logarithmic form

$$\ln\frac{1}{1-\alpha} = k\tau^n \tag{2.62}$$

or

$$\ln\ln\frac{1}{1-\alpha} = \ln k = n\ln\tau \tag{2.63}$$

Equation 2.63 is usually applied for handling the experimental data in coordinates $\mathrm{lb}(1/1-\alpha) - \ln\tau$ (lb indicates double logarithm).

TABLE 2.32 Value of *n* in kinetic equation versus the mechanism of reaction and growth conditions of nuclei.

Mechanism of reaction, growth conditions and geometry of nuclei	*n*
I. Polymorphous transformations, eutectoid reactions, nuclei growth, reaction controlled on the phase separation	
Conditions	
1. Increasing velocity of nuclei formation	> 4.0
2. Constant velocity of nuclei formation	4.0
3. Decreased velocity of nuclei formation	3–4
4. Nuclei formation takes place on the face planes of the crystals after saturation	1.0
II. Diffusion growth of nuclei	
Conditions	
1. Nuclei of different forms grow from limited initial volume and form with the increasing rate	> 2.5
2. Nuclei of different forms grow from limited initial volume and form with the constant rate	2.5
3. Nuclei of different forms grow from limited initial volume and form with the decreasing rate	1.5–2.5
4. Nuclei of different forms grow from limited initial volume and form with the zero rate	1.5
5. Nuclei of different forms grow from considerable initial volume	< 1.5
6. Nuclei have a needle shape or plates with limited length, where size is small compared with the dividing space	1.0
7. Broadening of long cylinders (needles) at the constant length	1.0
8. Broadening of very large flat plates at the constant square	0.5
9. Precipitations on the dislocations	< 0.5

The *n* values, as a function of reaction mechanism, growth conditions and nuclei geometry, are listed in Table 2.32. Comparison of presented values with *n* values calculated according to Equation 2.63 makes it possible to draw a conclusion about the mechanism of investigated process.

Another approach to determination of process mechanism is developing the nomograms with axes 'transformation degree α – reduced time $\tau/\tau_{0.9}$' for various *n* values (Figure 2.83).

Reduced time can be estimated as the ratio of the process duration to the time necessary to reach the conversion degree $\alpha = 0.9$. Nomograms similar to those given in Figure 2.83 allow the comparison of processes in different time intervals, estimation of *n* value, and, therefore, determination of the limiting stage and mechanism of the process.

The second group of kinetic equations comprises the relations drawn from the law of mass action. The general cxpression is

$$-\frac{d(1-\alpha)}{d\tau} = \frac{d\alpha}{d\tau} = k(1-\alpha)^n \qquad (2.64)$$

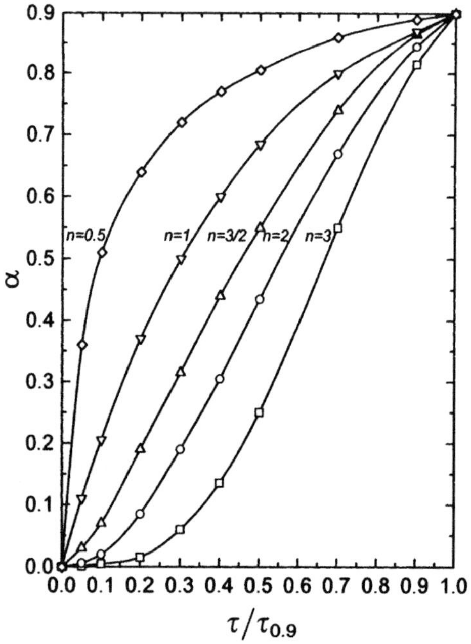

FIGURE 2.83 Simulation curves of solid-state reactions limited by nucleation process and described by equation $\ln(1/1-\alpha)=k\tau^n$ for $n=0.5$; 3/2; 2 and 3.

It is supposed that Equation 2.64 describes topochemical processes, in which the limiting stage is the reaction at the interface, i.e. true chemical interaction.

By separation of variables in Equation 2.64 and subsequent integration in limits from 0 to τ, the following expression can be obtained:

$$\frac{1}{1-n}[1-(1-\alpha)^{1-n}]=k\tau \qquad (2.65)$$

where k is a constant and n is the apparent order of solid-state reaction.

In this case, it is also possible to compose a nomogram for different models of growth including shrinking sphere ($n=\frac{2}{3}$) and reaction at plane surface ($n=0$).

Finally, the third group of kinetic equations comprises the so-called diffusion equations. We shall consider in detail the derivation of only one of them – the Jander equation.

It was found experimentally that

$$\frac{dx}{d\tau}=\frac{k_s}{x} \qquad (2.66)$$

where x is the thickness of planar layer of the product, τ is time and k_s is a constant depending on the properties of reactants and process conditions.

Later, assuming local thermodynamic equilibrium at any point of reaction zone, Equation 2.66 was obtained theoretically. It was shown that $k_s = kD$, where D is the diffusion coefficient of ions limiting the process. By integration of Equation 2.66 with boundary conditions ($x = 0$ if $\tau = 0$) we obtain

$$x^2 = 2kD\tau = k_p\tau \tag{2.67}$$

where k_p is the rate constant for parabolic-law growth of the product.

To apply Equation 2.67 for the description of process kinetics in powder systems, Jander considered the model based on the following premises (Figure 2.84).

1. component A consists of uniform spherical powder particles with initial radius R_0;
2. due to high surface diffusion, component B readily forms a continuous layer of reaction product on the surface of particles A (this assumption is true if the ratio $R(A)/R(B)$ is sufficiently great and component B is in excess compared to component A);
3. solid-phase interaction is limited by volume diffusion of component B through the layer of product to component A (unidirectional diffusion);
4. the reaction product does not form a solid solution with reactants;
5. the ratio of the product volume to the volume of reactant consumed during reaction (component A) is close to 1;

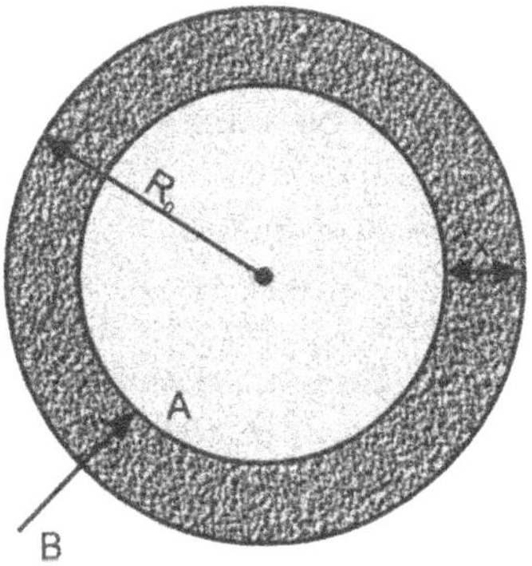

FIGURE 2.84 Scheme of solid-state reaction according to Jander's model: x – thickness of product layer; R_0 – initial radius of reagent particle A.

6. the diffusion coefficient of transferred ions does not change with time and activity of reactants on the surface of the reacting layer is constant;
7. the thickness of product layer changes with time according to parabolic law.

The volume of each A particle remaining at the τ moment is expressed by equation

$$V = \frac{4\pi}{8}(R_0 - x)^3 \tag{2.68}$$

On the other hand, if at the τ moment the conversion degree is equal to α, then

$$V = \frac{4\pi}{3}R_0{}^3(1 - \alpha) \tag{2.69}$$

Combination of Equations 2.68 and 2.69 leads to the relation

$$x = R_0[1 - \sqrt[3]{1 - \alpha}] \tag{2.70}$$

which with Equation 2.67 gives the relation

$$(1 - \sqrt[3]{1 - \alpha})^2 = \frac{2kD\tau}{R_0{}^2} = k_{\mathrm{Ja}}\tau \tag{2.71}$$

called the Jander equation.

Applicability of the last equation can be verified by graphic linearization of experimental data in coordinates

$$(1 - \sqrt[3]{1 - \alpha})^2 = f(\tau)$$

Experimentally, it has been shown that for the most part of solid-phase reactions the Jander equation works well only at low conversion degree $\alpha < 0.2$–0.4.

Changing the premises used in derivation of the Jander equation enables additional kinetic equations to be obtained for different diffusion mechanisms of solid-state interaction (Table 2.33) assuming local thermodynamic equilibria at any point of reaction zone. This implies that kinetic equations in Table 2.33, at close inspection, can be used to describe the reactions in mixtures of monocrystalline powders or in polycrystalline samples, in which concentration of non-equilibrium defects determined by chemical, thermal or mechanical prehistory is negligible. However, the active-state powder reactants with high concentration of non-equilibrium defects are commonly used. This increases diffusion mobility of lattice components considerably and, therefore, speeds up

TABLE 2.33 The basic kinetic equations for diffusion models of solid-state interaction, assuming local thermodynamic equilibria at each point of the reaction zone.

The name of the equation accepted in publications	*Time dependence of the thickness of the product layer or of the volume of unreacted substance*	*Changed or introduced assumptions*	*Kinetic equation*
Jander equation	$\frac{dx}{d} = \bar{k}$	–	$(1 - \sqrt[3]{1-\alpha})^2 = k_J \ln \tau$
anti-Jander equation	–	Single direction diffusion of the reagent A through the product layer to reagent B	$(\sqrt[3]{1+\alpha} - 1)^2 = k_{AJ} \ln \tau$
Ginstling–Brounstain equation	$\frac{dx}{d\tau} = \frac{kR_0}{x(R_0 - x)\tau}$		$-\frac{2}{3}\alpha - (1-\alpha)^{2/3} = k_{GB} \ln \tau$
anti-Ginstling–Brounstain equation	–	Single direction diffusion of the reagent A through the product layer to reagent B	$+\frac{2}{3}\alpha - (1+\alpha)^{2/3} = k_{AGB} \ln \tau$
Zhuravlev–Lesokhin–Timpelman equation	–	Variation of the concentration of the diffusion reagent during reaction	$\left[\left(\frac{1}{1-\alpha}\right)^{1/3} - 1\right]^2 = k_{ZLT}\,\tau$
Carter–Valency equation	$\frac{dV_\tau}{d\tau} = -\frac{4\pi kDR_\tau}{R_2 - R_\tau}$	Variation of the product volume V_1 in comparison with the equal volume of covered reagent V_2	$\frac{z}{z-1} - (1-\alpha)^{2/3} - \frac{[1+\alpha(z-1)]^{2/3}}{z-1} = k_{CV}\tau (z = V_1/V_2)$
anti-Carter–Valency equation	–	Single direction diffusion of the reagent A through the product layer to reagent B	$\frac{1}{z} + \frac{2}{3}\alpha - \frac{(1+\alpha z)^{2/3}}{z} = k_{AKV}\tau$
Dunvald–Vagner equation	Fick's Law	Diffusion of the non-steady-state flow of the reagent from the stationary source into sphere radius R_0	$\ln \frac{6}{\pi^2(1-\alpha)} = k_{DV}\tau$

the interaction. At the same time, heat treatment (heating or isothermal exposition) facilitates annihilation of non-equilibrium defects whose concentration decreases with time. Therefore, during reaction of active reactants the local equilibrium does not actually take place both in product volume and on the interface between phases. The equations derived with respect to this fact are listed in Table 2.34.

The complex behaviour of solid-state reactions in real systems often makes it impossible to apply the kinetic equations based on model concepts. In this case, the experimental data processing is facilitated if the **generalized kinetic model (GKM)** using a set of rate constants k_i depending on conversion degree is applied. In fact, for the homophase reaction taking place in the gas or liquid, the relation (2.64) based on the mass action law is true. At the same time, for the most part of solid-state reactions, the relation (2.64) in logarithmic form is non-linear (Figure 2.85).

The reasons for this seem to be the multistage character of the process, close values of rate for different stages, non-uniform distribution of reaction zone in material volume, polydispersity of reaction mixture etc. In general, Equation 2.64 can be represented as follows:

$$\frac{d\alpha_i}{d\tau} = k_i{*}(1 - \alpha_i)^n \tag{2.72}$$

where k_i is a hypothetical constant of homophase reaction at the provided conversion degree, i. Experimental dependence $\alpha = f(\tau)$ can be described by the power function

$$\alpha = a\tau^b + c \tag{2.73}$$

TABLE 2.34 The basic kinetic equations for diffusion models of solid-state interaction between active (non-equilibrium) phases.

The name of equation accepted in the literature	*Time dependence of the thickness of product layer or of the volume of unreacted substance*	*Kinetic equation*
Tamman equation	$dx/d\tau = k/\tau$	$1 - \sqrt[3]{1-\alpha} = k_T \ln \tau$
Kroeger–Ziegler equation	$dx/d\tau = k/x\tau$	$(1 - \sqrt[3]{1-\alpha})^2 = k_{KZ} \ln \tau$
Ginstling–Brounstain Halbert equation	$\frac{dx}{d\tau} = \frac{kR_0}{x(R_0 - x)\tau}$	$1 - \frac{2}{3}\alpha - (1-\alpha)^{2/3} = k_{GBH} \ln \tau$

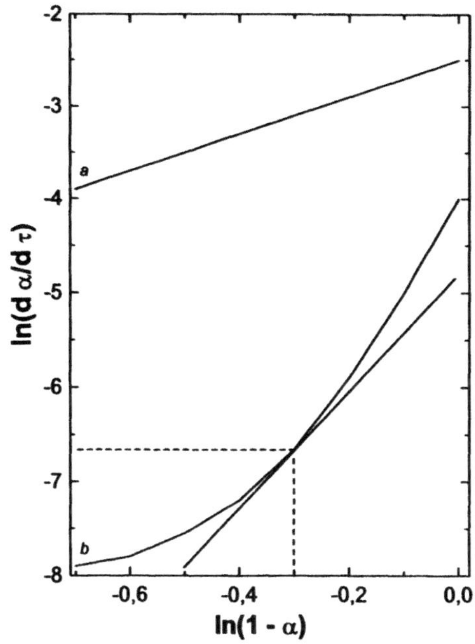

FIGURE 2.85 ln(dα/dτ) *vs.* ln(1 − α) dependencies for homogeneous (a) and heterogeneous (b) reactions.

where a, b and c are constants. By differentiation of Equation 2.73 we can find the process rate

$$\frac{d\alpha}{d\tau} = ab\left(\frac{\alpha - c}{a}\right)^{(b-1)/b} \tag{2.74}$$

If we take a logarithm and differentiate the latter equation by $\ln(1 - \alpha)$ we obtain

$$n = \frac{b-1}{b}\,\frac{1-\alpha}{c-\alpha} \tag{2.75}$$

$$\ln k^* = \ln b + \frac{1}{b}\ln a + \frac{b-1}{b}\ln(c - a) - \frac{b-1}{b}\,\frac{1-\alpha}{c-\alpha}\ln(1-\alpha) \tag{2.76}$$

By the treatment of numerous experimental data with Equations 2.73–2.76, it has been shown that the distribution of hypothetical homogeneous reaction

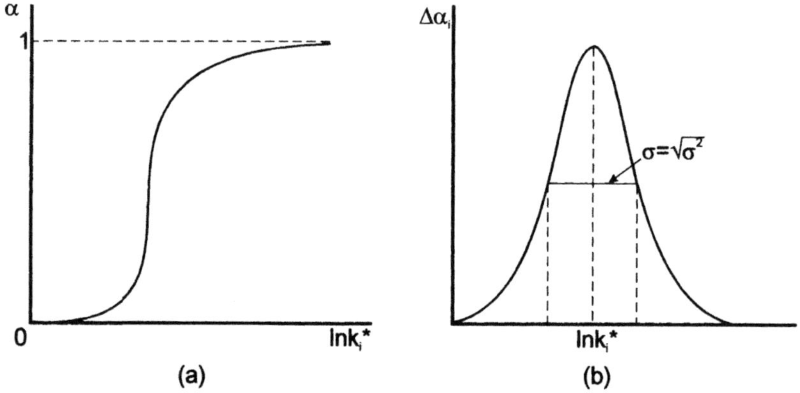

FIGURE 2.86 Distribution of hypothetical homogeneous reaction constants $\ln k^*$ *vs.* conversion degree α applied for description of real heterogeneous process. (a) integral form; (b) differential form.

constants $\ln k_i$ according to conversion degrees obeys normal Gauss distribution law (Figure 2.86):

$$\alpha_i = \frac{1}{\sqrt{2\pi}} \int_{(\ln k^*)_{\alpha=0}}^{(\ln k^*)_{\alpha_i}} e^{-\frac{(\overline{\ln k^*} - \ln k^*)^2}{2\sigma^2}} d \ln k^* \tag{2.77}$$

The most important quantitative parameters of this distribution are average value $\ln \bar{k}^*$ and dispersion of average value σ^2 (or average square deviation $\sigma = \sqrt{\sigma^2}$) (module σ, because σ can have negative and positive values), which can be calculated in the following way. Let us represent Equation 2.77 as

$$\alpha_i = \frac{1}{2\pi} \int_0^{U_i} e^{-(U_i^2)/2} \, dU \tag{2.78}$$

Values of integral function in the right part of Equation 2.78 are tabulated. Therefore, for each value α_i it is possible to determine the value U_i satisfying the relation (2.78). On the other hand, $U_i = (\ln \bar{k}^* - \ln k_i^*)/\sigma$ (compare Equations 2.77 and 2.78). It is obvious that dependence between U_i and $\ln k_i^*$ calculated from Equation 2.76 is linear and the tangent of slope is equal to $1/\sigma$; $\ln \bar{k}^*$ can be calculated by taking into account the fact that $U_i = 0$ ($\alpha_i = 0.5$).

The assumptions of the generalized kinetic model suggest that at $\sigma = 0$ the solid-state process proceeds in kinetic regime (i.e. can be described by the single rate constant) and $\sigma > 0$ proves that the process is limited by diffusion.

This fact enables the use of the value of σ as a relevant parameter for optimizing the dehydration and thermal decomposition processes accompanying the processing of homogeneous salt solution up to the oxide phase. In general, these processes should be carried out in kinetic regime, because if they proceed in the diffusion regime the concentration gradient in solid phase appears not only for the diffusing component (for example, the gradient of water vapour pressure upon dehydration), but also for all components in the system (because of the confined thermodynamical description of its behaviour assuming the local equilibria). As a result of the latter, the decay of obtained solid solutions can occur.

Unfortunately, the low sensitivity of formal kinetic models due to their deterministic character enables the change in mechanism to be detected only after irreversible changes in solid solutions.

Alternatively, GKM is a more sensible tool allowing the detection of the very beginning of mechanism change in solid-state processes and, therefore, optimization of their conditions.

As an example of GKM advantage, we would like to provide the experimental results on dehydration of Mohr salt $FeSO_4*(NH_4)_2SO_4*6H_2O$ in a fluidized bed.

Experimental study of dehydration process and following mathematical treatment of data has led to the following values of optimization criteria σ:

t, °C	50	58	62	70
σ	0	0.12	0.17	0.32

These data point to the change in character of Mohr salt dehydration at a temperature of about 50 °C. At $t < 50$ °C the limiting stage of the process is the chemical interaction ($\sigma = 0$). The deviation of the process from purely kinetic regime ($\sigma \neq 0$) at $t > 50$ °C can at once be detected by GKM. With temperature increasing, the changes in mechanism grow more perceptible and, finally, this leads to degradation of the monophase system.

As shown on Figure 2.87, the parameters of Mössbauer spectrum of Mohr salt subjected to dehydration at the temperature 50 °C are almost constant (chemical shift $\delta = 1.59 \pm 0.5$ mm/s, quadruple splitting $\Delta = 1.71 \pm 0.05$ mm/s).

Spectrum line broadenings can be connected with a defective structure emerging during partial dehydration (these broadenings disappear if spectra are registered at $T = 77$ K). At a higher dehydration temperature (for example, up to $t = 62$ °C), the superposition of two spectra can be observed. One of them is similar to the spectrum of starting Mohr salt and the second one has parameters $\delta = 1.59 \pm 0.5$ mm/s, $\Delta = 1.71 \pm 0.05$ mm/s), which is characteristic for $FeSO_4*nH_2O$ ($n < 7$). Commonly used procedures of formal kinetics allow change in the mechanism to be detected only at $t > 70$ °C.

Thus, the analysis of typical topochemical process – dehydration of Mohr salt – with GKM (a more sensitive model than those used in formal kinetics) enables prediction of the change of reaction mechanism and, thus, the change of structure-sensitive properties of materials. The results obtained can be used to optimize the basic topochemical processes (freeze-drying, dehydration and

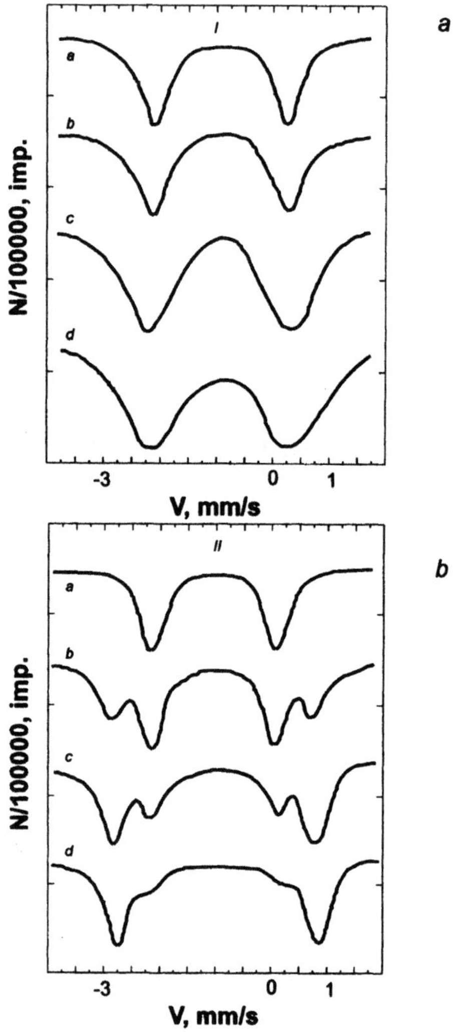

FIGURE 2.87 Mössbauer spectra of Mohr salt at different stages of thermal dehydration in fluidized layer.
(a) $t = 50\,^{\circ}C$, conversion degree $\alpha =$ 0; 0.18; 0.34 and 0.47 (curves a–d, respectively);
(b) $t = 62\,^{\circ}C$, $\alpha = 0$ to 0.5 (top to bottom).

thermal decomposition) accompanying the processing of cryochemical salt products.

The analysis of freeze-drying kinetics with GKM (Table 2.35) leads to the following conclusions:

TABLE 2.35 The results of an analysis of the sublimation process of frozen solution of Mohr salt based on a universal kinetic model.

(a) Influence of concentration. Weight of the dried product = 74 g. Frozen temperature = −95 °C

C, wt.%	9.0	15.3	29
σ	0.48	0.42	0.56
$\sigma/\ln K^*$	0.101	0.095	0.124

(b) Influence of drying regime, $C = 15.3$ wt.%, $T_{\text{froz.}} = -95$ °C, $m = 20$ g

	σ	$\sigma/\ln K^*$
Static	0.76	0.153
Vibrolayer	0	0

1. the influence of initial solution concentration (9 wt.% $< c <$ 29 wt.%) on the homogeneity of sublimation dehydration process is quite low (within the error limits 10–15%). This is proved by close values of average square deviation σ and of its ratio to measured k value $\sigma/\ln \bar{k}^*$ (Table 2.35);
2. permanent agitation not only accelerates the homogeneous process of freeze-drying, but also enables it to proceed without perceptible diffusion hindrance.

Analysis of kinetic results on dehydration (Table 2.36a, b) and thermal decomposition (Table 2.36c) processes also enables discovery of the best method for the transformation of the cryochemical salt product into the final oxide and, meanwhile, to maintain a high chemical homogeneity of products.

Let us consider in detail the problem of high homogeneity preservation during transformation of a frozen multicomponent system into a final oxide product. It was shown that high chemical homogeneity can degenerate when the spontaneously increased diffusion of components becomes more probable. This can occur, first of all, due to phase transitions in the systems under treatment. Similar critical situations have been discussed earlier; they were connected with the appearance of a liquid phase during freeze-drying (see section 2.2.2) and with structure reorganization via an amorphous phase formation during product ageing (see section 2.5.1).

Through the practical experience, it was found that changes occurring on final treatment of the salt product are especially dangerous for chemical homogeneity.

When the salt is heated at a low rate and reaches a certain temperature, the spontaneously increased diffusion of components, usually leading to phase degradation of salt system, can still occur before thermal decomposition. When the salt is heated at a high rate, the thermal decomposition begins before the relaxation processes take place, and high chemical homogeneity of a multicomponent system, thus, can be saved.

TABLE 2.36 Analysis of fluidized layer dehydration and thermal decomposition of Mohr salt based on the universal kinetic model.

(a) Influence of the vibration parameters (weight $m = 100$ g, dehydration temperature $t = 60\,°C$)

Frequency	25			40			55		
Amplitude	2.0	2.6	3.2	1.4	2.0	2.6	0.8	1.4	2.0
$-\ln K^*$	3.80	4.09	4.26	4.25	4.24	4.22	5.01	5.16	5.24
σ	1.09	1.08	1.10	0.72	0.73	0.72	0.45	0.46	0.44

(b) Influence of dehydration temperature ($A = 2$ mm, $f = 55$ Hz)

T, °C	60	70	80
$-\ln K^*$	5.27	5.16	5.02
σ	0.48	0.88	1.35

(c) Influence of the weight of Mohr salt and decomposition temperature ($A = 2$ mm, $f = 55$ Hz)

Weight of the sample, g	*Temperature*, °C	$-\ln K^*$	σ
100	700	3.16	0.84
	725	2.92	0.64
	750	2.74	0.14
	775	2.55	0.01
200	750	2.73	0.40
	800	2.60	0.28
	850	2.46	0.11
	900	2.31	0.02

Let us consider some real examples, taking the system of heptahydrates $MeSO_4*7H_2O$ as a model. These compounds fall into two structural series: rhombohedral (Me = Mg, Ni, Zn) and monoclinic (Me = Fe, Co) (Aslanian and Balarev, 1977). Salts of the same series can produce a continuous series of solid solutions, while interaction of salts of different series is limited. We should take into account the fact that any of the binary sulphate systems $(MeMe')SO_4*7H_2O$ can be obtained in the form of three main products: amorphous, crystallinez vacancy and monohydrate.

Results of thermoanalytical investigation of freeze-drying products in $MeSO_4$–$Me'SO_4$ systems, presented in Figure 2.88 and in Tables 2.37–2.38, suggest that the dehydrated products may be solid solutions as well as mixtures of monohydrates of the components involved.

It should be noted that the dominant factors facilitating preservation of chemical homogeneity of salt solid solutions are the low tendencies of components to dissolve in water and to crystallize from water solutions. Thus, for example, a solid solution $(Ni,Zn)SO_4*7H_2O$ can be obtained by equilibrium co-crystallization, but it does not survive under severe vacuum dehydration

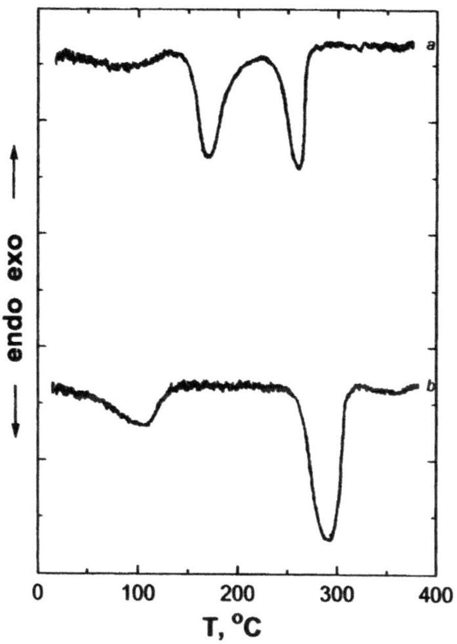

FIGURE 2.88 DTG curves for thermal dehydration of freeze-drying products: $ZnSO_4 + FeSO_4$ (a); $CoSO_4 + FeSO_4$ (b).

simulating the freeze-drying conditions. As a result of vacuum treatment, due to high solubility in water, the solid solution readily decomposes. On the other hand, in the system $MgSO_4$–$FeSO_4$ a metastable solid solution with vacancy structure (see p. 153) forms during thermal treatment even if the components are mixed

TABLE 2.37 **Thermoanalytical estimation of the behaviour of binary freeze-dried salt products during its thermal decomposition (systems forming the continuous sequence of solid solutions).**

Type of crystal lattice of solid solution	*System*	*Type of hydrate*	*Behaviour during dehydration*
rhombohedral	$NiSO_4$–$MgSO_4$	Eq.	+
		Vac.	+
		Amorph.	+
rhombohedral	$NiSO_4$–$ZnSO_4$	Eq.	–
		Vac.	+
		Amorph.	+
monoclinic	$FeSO_4$–$CoSO_4$	Eq.	+
		Vac.	+
		Amorph.	+

TABLE 2.38 Thermoanalytical estimate of the behaviour of binary freeze-dried salt products during its thermal decomposition (systems forming the limited sequence of solid solutions).

System	*Region of separation*	*Type of hydrate*	*Behaviour in the region*	
			of homogeneity	*of separation*
$MgSO_4$–	10–50%	Eq.	+	–
–$FeSO_4$	$FeSO_4$	Vac.	+	+
	–	Amorph.	+	+
$ZnSO_4$–	50–65%	Eq.	–	–
–$FeSO_4$	$FeSO_4$	Vac.	+	–
		Amorph	+	+

Note: +: single stage dehydration; – :dehydration in several stages. Eq.: equilibrium lower hydrate, Vac.: vacancy structure, Amorph. : hydrate of amorphous type.

mechanically. From this fact, the second important conclusion can be drawn: the probability of homogeneity preservation increases at a higher deviation of product from the equilibrium state. From Tables 2.37 and 2.38 it can be seen that in all studied systems the formation of a two-component product in the form of amorphous hydrate facilitates homogeneity preservation; for more equilibrium (than amorphous phase) vacancy hydrates, this is sometimes not the case (for example, for the system of iron sulphate–zinc sulphate). Thus, to preserve high chemical homogeneity in the systems with a tendency to spontaneous segregation, it is necessary to carry out freeze-drying in such a way that the amorphous hydrate would form.

Studies of the product (i.e. of solid solution) homogeneity drew some interesting conclusions. For example, in the 'magnesium sulphate–iron sulphate' system the range of homogeneity corresponds to vacancy solid solutions of two kinds: on the basis of $MgSO_4{*}xH_2O$ and $FeSO_4{*}yH_2O$. Variations of solid solutions composition lead to regular shift of dehydration temperatures. Experimentally observed correlations between composition and temperature is also valuable with regard to metastable solid solutions, corresponding to segregation areas in an equilibrium phase diagram (Figure 2.89).

In this case relatively small changes in the freeze-drying regime can lead to the products with identical cationic composition, but different structures. Certainly, when the 'bifurcations' occur, it is impossible to predict which of two vacancy structures will be the base for salt framework of the obtained product.

Such unusual behaviour of systems can be observed only if a cryochemical technique providing an initially high degree of components mixing is used.

Regarding the dehydration processes in the systems formed from heptahydrates with identical crystallochemical structure and with high resistance to water action, the changes in their cation composition usually lead to continuous displacement of DTA peaks on the temperature scale (Figure 2.90).

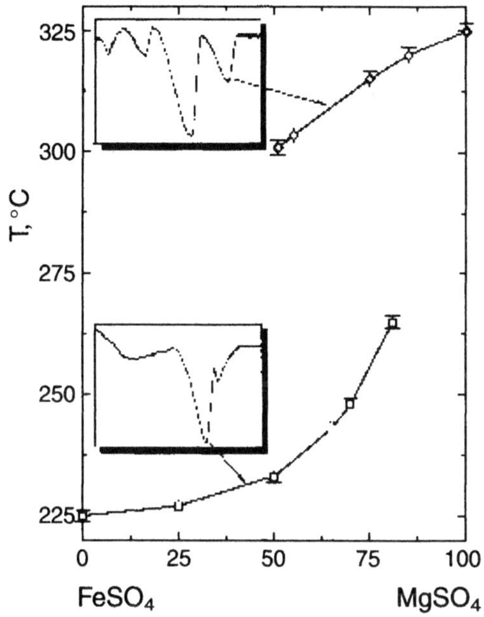

FIGURE 2.89 Compositional dependence of characteristic peaks of dehydration for solid solutions based on $FeSO_4{}^*xH_2O$ (bottom curve) and on $MgSO_4{}^*yH_2O$ (top curve).

The characteristic for the $FeSO_4$–$CoSO_4$ system is the fact that the DTG peak for solid solution containing about 40 mol.% $CoSO_4{}^*xH_2O$ is a little displaced compared with that corresponding to dehydration of pure $CoSO_4{}^*xH_2O$.

Investigations of three-component systems have shown that their properties (as for two-component ones) are additive, in regard to the properties of salts in their composition.

The predicted triple equilibrium diagram $NiSO_4{}^*7H_2O$–$FeSO_4{}^*7H_2O$–$ZnSO_4{}^*7H_2O$ (Figure 2.91) indicates that the area, appropriate for ferrite syntheses (solid line), is close to the area corresponding to decomposition of solid solutions. In this case, the most reliable way to save high chemical homogeneity is freeze-drying, leading to a product in the form of amorphous hydrate.

The synthesis of required compositions in an amorphous state is rather difficult because of the high content of zinc sulphate, and may be carried out only if drops of small size are frozen and low-temperature regimes of freeze-drying are used (Figure 2.92). The application of such regimes considerably increases (2–3 times) the process duration and is inappropriate from a technological point of view. Accordingly, the demands for the new methods to affect the phase composition of freeze-drying products to obtain amorphous hydrates are quite obvious.

The most efficient among these methods seems to be the further lowering of pressure in the sublimation chamber. Thus, the decrease of pressure down to

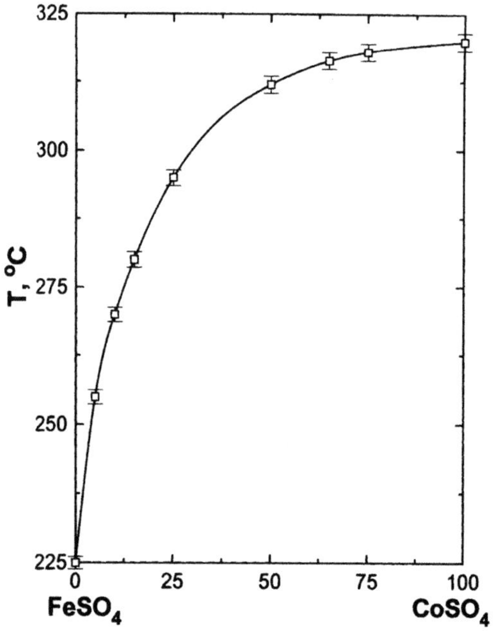

FIGURE 2.90 Characteristic temperature of thermal dehydration for various $Fe_xCo_{1-x}SO_4{*}xH_2O$ solid solutions.

0.05 Pa makes it possible to prepare the amorphous zinc sulphate hydrate $ZnSO_4{*}xH_2O$, which is rather difficult to obtain using other techniques. Such an approach seems to be the most universal for amorphous phase preparation. However, its technological application is considerably complicated, because the majority of commercial sublimators are designed to work at $P > 1$ Pa.

Another important factor affecting the phase composition of freeze-drying products is the pH of the initial solution. In particular, it was shown that at a fixed sublimation regime (including pressure in chamber 3–4 Pa) regular pH decrease of nickel, zinc and iron sulphate solution from 2.2 to 0.9 enables the whole row of the product's states to be obtained: amorphous, crystalline vacancy hydrate and monohydrate. The mechanism of pH influence on the product composition is related, first of all, to the increase of mass exchange between salt crystals and microquantities of liquid phase, which usually appear during freeze-drying, especially if acid solutions are used. More intensive mass exchange leads to the shrinkage of salt framework and enhances the role of diffusion during the following dehydration. Thus, the probability of preparing the crystalline product increases.

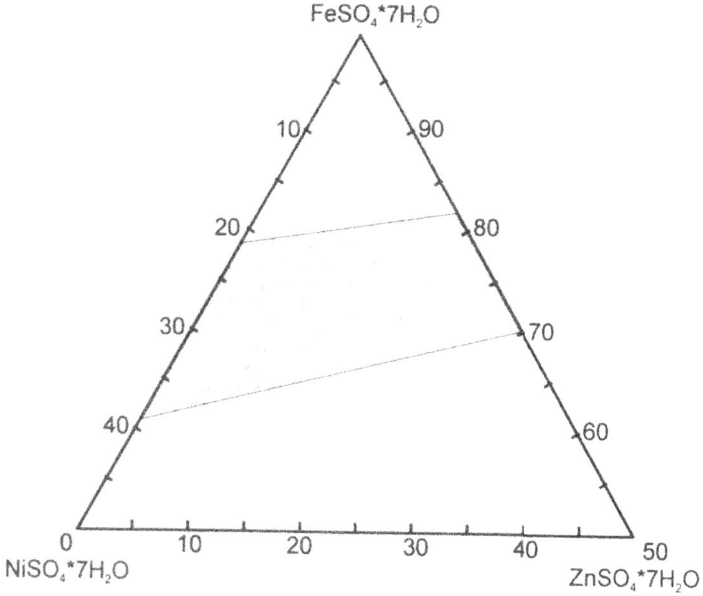

FIGURE 2.91 Presumable ternary equilibrium diagram for the system $NiSO_4{*}7H_2O$–$FeSO_4{*}7H_2O$–$ZnSO_4{*}7H_2O$.

At the same time, the possibility of preparing the amorphous hydrates by the pH increasing at relatively high (1–10 Pa) pressure, usual for commercial production units, is of great interest. However, to make use of it, some information about the behaviour of salt solutions near the limits of their stability (pH > 2) is still needed. Obviously, the problem requires further investigation.

An efficient method of chemical homogeneity evaluation is thermoelectromotive force (thermo-EMF) technique. Its application is demonstrated by the example of hematite doped with lithium oxide. In Figure 2.93 a part of the phase diagram for the Fe–O system is presented in coordinates $\lg P_{O_2} - f(1/T)$.

Line 1 corresponds to the existence of hematite and magnetite; line 2 to transition from intrinsic p-type conductivity to electron conductivity (n–p transition) in hematite. Doping of hematite with lithium oxide increases the concentration of electrons and shifts n–p transition in the region of higher oxygen pressure. The reason for this seems to be that the lithium ions possess donor properties. At the same time, electrical properties of doped hematite depend on uniformity of additive distribution. So, if the impurity concentration is low and intrinsic lithium ions do not interact with one another, the doping process can be described by the quasichemical equation (scheme):

$$m\mathrm{Li_2O}_{\rightarrow \mathrm{Fe_2O_3}} \rightarrow 2m\mathrm{Li}_i + 2me' \qquad (2.79)$$

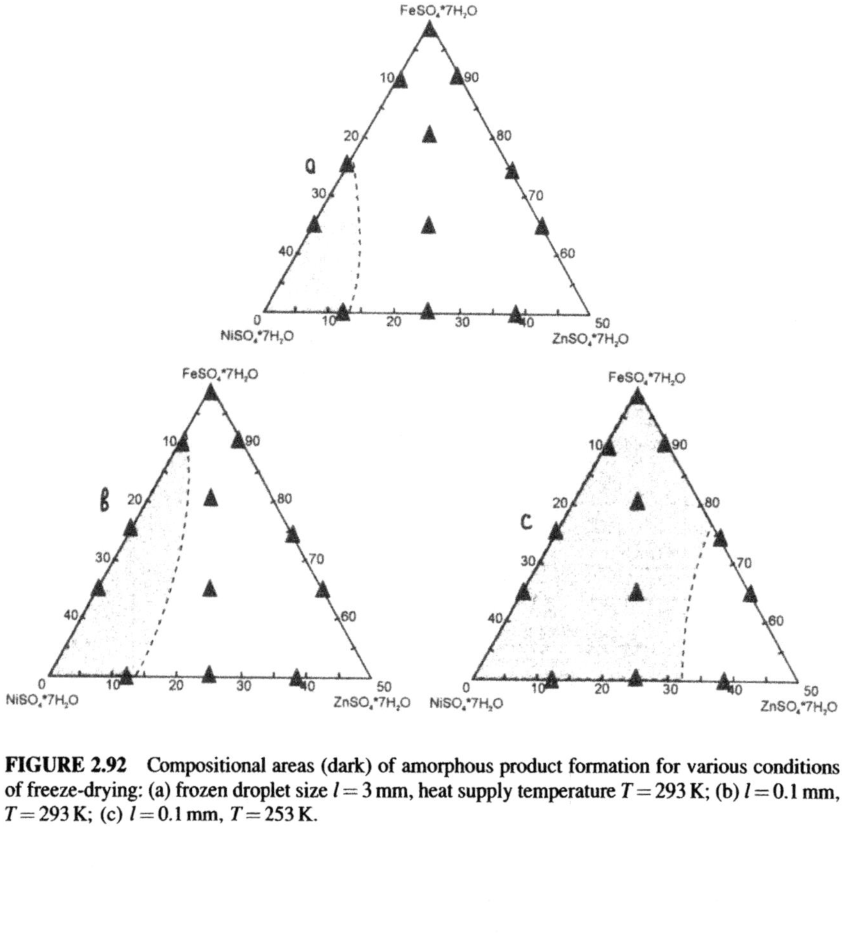

FIGURE 2.92 Compositional areas (dark) of amorphous product formation for various conditions of freeze-drying: (a) frozen droplet size $l = 3$ mm, heat supply temperature $T = 293$ K; (b) $l = 0.1$ mm, $T = 293$ K; (c) $l = 0.1$ mm, $T = 253$ K.

In this case, it is reasonable to expect the maximum influence of additive on the electrical properties of hematite (line 4, Figure 2.93).

Quasichemical equation

$$m\text{Li}_2\text{O}_{\rightarrow \text{Fe}_2\text{O}_3} \rightarrow (\text{Li})_{2m} + e' \tag{2.80}$$

describes another hypothetical state corresponding to the complete association of introduced lithium ions and to minimum variation of electron concentration in hematite. For the sample in this state, the shifts of n–p transition, such as for pure hematite, cannot be observed (line 2, Figure 2.93).

In practice, depending on the heat treatment regime of cryochemical synthesis, the intermediate states can be realized (point a on line 3, Figure 2.93).

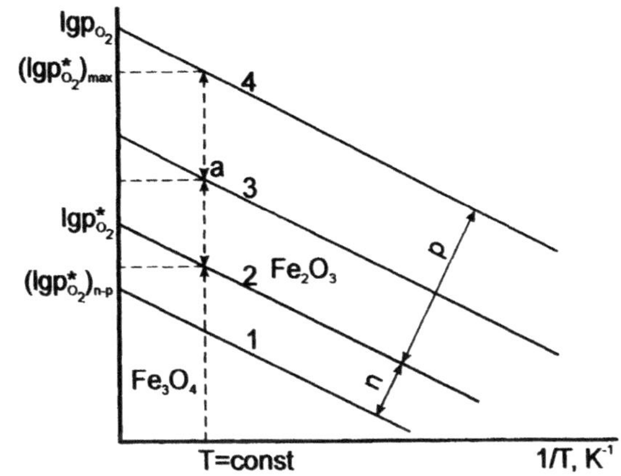

FIGURE 2.93 Displacement of p–n transition area for hematite doped by Li_2O as a function of dopant distribution homogeneity.

At a given doping level, the shift of point *a* on the phase diagram is a good quantitative criterion of chemical homogeneity preservation during the treatment of cryochemical salt products.

2.5.3 Powder processing

Powders are rarely the objective of cryochemical syntheses. More often they are precursors for compact materials obtained by forming and sintering of powders. In respect to this, the main attention should be given to the structure of powders as well as of polycrystalline products of their sintering. It has been noted that structure can be characterized at three hierarchic levels: micro-, meso- and macrolevel (Legasov *et al.*, 1986).

Microlevel corresponds to the crystal structure of the substance which is used to create the material. For crystal structure characterization, the X-ray diffraction (space group and lattice parameters), electron diffraction (high structure organization with formation of Magneli phases (see Collongues, 1971; Magneli, 1953), electrophysical and thermodynamic techniques (type and degree of disordering with formation of point defects and their associates) can be used. Sometimes, the crystal structure of substance can be affected by the process parameters during the synthesis (variation of temperature, total pressure, partial pressure of volatile component etc.) and by additional thermal treatment, or by doping.

Mesolevel – fine crystal structure (FCS) – can be characterized by coherent domain size (CDS), mean square strains and by concentration of extended defects (Tretyakov *et al.*, 1973b). These parameters of FCS commonly are estimated from X-ray data.

Finally, the macrolevel corresponds to rheologic structure (in the case of powders) or to ceramic structure (in the case of sintered ceramics) (Saltykov, 1970). The ceramic structure is often referred to as the microstructure. Microstructure can also be considerably modified by various technological procedures including variation of sintering temperature, gaseous phase composition and doping.

The rheologic and ceramic structures can be characterized by optical and electron microscopy techniques (size, shape and distribution of crystallites, individual powder particles and their aggregates; size, shape and distribution of pores), sieve and sedimentation analyses (distribution of particles by size), absorption techniques (specific surface values), hydrostatic weighting (density and porosity of compact) etc. (Oleinikov *et al.*, 1985a).

The rheologic structure of ceramic powders is usually required to be highly consolidated and provide the maximum green density. To solve this problem various methods can be used; first of all, surface-active substances can be applied to modify the surface properties of powders. In some cases, the powder structure may be considerably changed by mechanical treatment.

It should be emphasized that two structure levels – fine crystal structure and rheologic structure – are often combined and named as the real structure.

During the development and preparation of new materials it is expedient to influence all structure levels. Taking into account that each structure level is characterized by various interrelated parameters, it would be relevant to integrate the parameters into compact groups (Tesker, 1986). It was assumed that correlation of measured parameters is an indirect manifestation of several hypothetical factors. Each factor can be actually observed as a group of measured parameters (compact group), using techniques described in literature (Braverman and Muchnik, 1983). Parameter aggregating was used according to the given quantitative criteria. As such a criterion, the function

$$F = \sum_{k=1}^{L} \frac{1}{m_k(m_k - 1)} \sum_{i,j \in G_k;\, i \neq j}^{M} \alpha_{i,j} \tag{2.81}$$

was chosen, where L and M are the number of factors and parameters, respectively, m_k is the number of parameters in a compact group and α_{ij} is the coefficient of correlation between ith and jth parameters.

The fundamental is the property of function F, that at $m_k = 1$

$$\frac{1}{m_k - 1} \sum_{i,j \in G_k;\, i \neq j} \alpha_{i,j} = 0 \tag{2.82}$$

The suggested approach can be reliably used for the systems with a high degree of internal organization – fine crystal structure, rheologic structure and ceramic structure. Regarding the crystal structure, it has a limited appplication. For example, the concentrations of different impurities usually determined by the precursor preparation technique correspond to the random stochastic values and do not correlate one to another.

It should be noted that the number of direct parameters, i.e. directly characterizing some structure elements on different hierarchic levels, is considerably lower than the number of indirect parameters. Both direct and indirect parameters used to characterize fine crystal, rheologic and ceramic structures are listed in Table 2.39. The choice, certainly, is rather arbitrary. However, it should be pointed out that the number of direct parameters, providing the actual information about some hierarchic level, is restricted, whereas the number of indirect parameters is unlimited. Sometimes, the number of indirect parameters characterizing rheologic structure exceeds 50.

Analysis of direct parameters of fine crystal structure of ferrite powders (Tesker, 1986) has shown that actually there are two factors generating two compact groups of measured parameters. The first one, which includes the CDS and the probability of crystal twin formation, characterizes the mosaic structure of materials; the second, which includes the value of mean square strains and the probability of deformation defect appearance, is connected with deformation distortions. Therefore, when the representative parameters, directly characterizing the FCS, are to be selected, only two parameters can be used – for example, the CDA size (D_{CDS}) and the size of microdistortions.

The situation is quite different when analysis of experimental data on rheologic structure of oxide powders is carried out. In this case, both direct and indirect parameters of rheologic structure are combined in a complicated way. At the same time, the indirect parameters prevail in the experimental data, because they can be more readily obtained and, thus, are more available.

The parameter combination was applied to analyse the experimental rheological parameters of Ni-Zn- and Mn–Zn-ferrite powders obtained by both the standard ceramic and cryochemical methods. To characterize different kinds of powders and ceramics, a set of rheological parameters was used: poured powder density (ρ_{gr}); shaken powder density (ρ_{sh}); fluidity (θ); specific surface according to BET technique (S_{sp}) or measured from velocity of gas flow through pressed pellet (S_{gpm}); mean aggregate size ($\bar{D}_{ag}$) and its mean square deviation (σ_{ag}) according to sieve analysis data; mean aggregate size ($\bar{d}_{ag}$) and its mean square deviation ($\sigma_{\bar{d}_{ag}}$) according to sedimentation analysis data; humidity (ω).

Analysis of the data presented in Table 2.40 leads to the conclusion that parameters characterizing rheological structure of ferrite powders regardless of materials prehistory can be divided into three quite steady compact groups: volume–mass, surface and geometric. So, another important conclusion can be

TABLE 2.39 Direct and indirect diagnostic parameters of different structural levels.

	Real structure	*Rheological structure*	*Ceramic structure*
Direct parameters	1. Coherent domain size (CDS) D 2. Value of microstrains $\sqrt{\xi^2}$ 3. Probability of twins formation β 4. Probability of stacking faults formation γ	1. Average size of particles and their standard deviation (electron microscopy data) d_r, σ_{d_r} 2. Average size of aggregates d_{ag} and their standard deviation $\sigma_{d_{ag}}$ determined by sedimentation analysis data 3. Specific surface S_{sp} (BET method) 4. Powder density ρ_{pik}	1. Average size of crystallites D_{cr} and their standard deviation $\sigma_{D_{cr}}$ 2. Average size of pores d_p and their standard deviation σ_{d_p} 3. Bulk density ρ_i
Indirect parameters	1. Value of excess Gibbs energy, enthalpy, entropy and heat capacity of active phases. 2. Value of excess partial Gibbs energy of oxygen $\Delta G_{O_2}{}^{ex}$, for active phases 3. Value of thermoelectric power (EMF) 4. Parameters of solid-state interaction in model reaction (isopolythermal methods) 5. Parameters of catalytic reactions 6. Shrinkage of compacted samples after sintering 7. Parameters obtained using contact volt–ampere method	1. Green density ρ_{gr} 2. Density after shaking ρ_{sh} 3. Coefficient of fluidity θ 4. Humidity ω 5. Specific surface (gas-permeability method) S_{gpm} 6. Angle of natural slope α (dynamic) 7. Coefficient of internal friction μ_{in} 8. Coefficient of external friction μ_{ex} 9. Average size of aggregates D_{ag} and their standard deviation σ_{ag} (sieve analysis data)	All static and dynamic structural-sensitive parameters, including: coercive force H_c, saturation induction B_m, residual induction B_r etc.

TABLE 2.40 Aggregated rheological parameters of the ferrite powders.

Object	*The result of the parameter aggregation*	*Correlation*
ceramic technology	I ρ_{gr}; ρ_{sh}; $\bar{d}_{ag}$; $\sigma_{\bar{d}ag}$	0.926
+	II S_{sp}; θ ; ω	
mechanical treatment	III $\bar{D}_{ag}$; σ_{ag}	
ceramic technology	I ρ_{gr}; ρ_{sh}; S_{gpm}; $\bar{d}_{ag}$	0.913
+	II S_{sp}; θ; ω	
mechanical treatment	III D_{ag}; σ_{ag}	
cryochemical technology	I ρ_{gr}; ρ_{sh}	0.982
+	II S_{sp}; θ; ω	
mechanical treatment	III D_{ag}; σ_{ag}	

made that at least three parameters from each group are necessary to characterize rheologic structure.

Eventually, the analysis of the most complete set of ceramic structure parameters (mean crystallite size $\bar{D}_{cr}$ and its mean square deviation $\sigma_{\bar{D}_{cr}}$, as well as mean pore size $\bar{d}_p$ and its mean square deviation σ_{d_p}) allows division of them into two compact groups – the first includes mean crystallite (D_{cr}) and pore (d_p) size, and the second their mean square deviations ($\sigma_{\bar{D}cr}$ and σ_{dp}). It is quite obvious that for reasonable description of ceramic structure it is expedient to use the mean size of crystallites and its mean square deviation (Kolotyrkina, 1983; Tesker, 1986).

Accordingly, the parameter combination enables selection of the minimum set sufficient for powder system description from numerous parameters. But after this procedure the system is still rather complex, especially if chosen parameters are independent. Fortunately, there are regularities in solids which enable the situation to be simplified.

One of these regularities is the ability of solid reactants obtained from different initial substances to manifest a kind of 'memory': being synthesized from the same precursors they inherit an amazing ability to be especially active only in certain heterogeneous processes (Tretyakov, 1978). Thus, iron(III) oxides obtained by thermal decomposition of different salt precursors – for example, oxalate, nitrate and sulphate – are identified as the same phase α-Fe_2O_3. Meanwhile, the samples of Fe_2O_3 synthesized from different precursors manifest unequal reactivity. For example, iron oxide obtained by decomposition of oxalate in oxygen has enhanced (compared to other samples) catalytic activity, and the sample synthesized from iron sulphate manifests high ability for sintering. The effect is quite reproducible and stable for a time.

The example above illustrates the ability of chemical systems to keep and transfer the information. It is from this fact that the analogy with memory arises.

The influence of the initial phase composition on the properties of solid products formed during topochemical transformations of the following type:

$$A(s) \rightarrow B(s) + C(g) \tag{2.83}$$

was discovered by H.F. Huettig in the 1930s, when he studied the solid phase reactions, and was called 'matter memory'. However, the term was not widely accepted, and the results of multiple observations on the influence of precursor nature on the properties of solid products were interpreted as the effect of chemical and thermal prehistory of the materials. Taking into account that the properties of solids can also be influenced by mechanical treatment, it was usual to refer to the effect of chemical, thermal and mechanical prehistory (Legasov *et al.*, 1986).

Now, the effects similar to the prehistory influence are named with the more general term: 'topochemical memory'.

Topochemical memory (TCM) is a result of the correlations between the conditions of the topochemical process, leading to real structure formation, and the reactivity of solids, which is determined by the ability of solids to 'remember' their origin and to display this in subsequent topochemical processes. The main property of TCM is an ability to inherit the precursor features up to the final product, in spite of various types of external treatment.

Topochemical memory is very important for practical application; it may result both in positive and negative consequences, and can sometimes be used or should be suppressed during processing of advanced materials (Oleinikov, 1995).

Now let us consider how the topochemical memory arises. It is evident that during a heterogeneous reaction, the real structure of initial reactants changes into the real structure of products. On the basis of general kinetic rules, it is possible to predict the features of the latter structure, paying attention to the conditions of its formation. To simplify the consideration, let us choose two of the most easily controlled parameters: the temperature of topochemical reaction T and the rate of gaseous product elimination W (for the process of the type (1)) (Delmon and Boldyrev, 1979). It should be emphasized that the first parameter considerably determines the degree of thermodynamic deviation from the equilibrium during product formation, while variation of the second affects the dynamic deviation from the equilibrium in the open system.

However, to make the simplest model for the structure formation, these two parameters T and W would be insufficient.

To analyse the temperature influence on crystal lattice formation, we shall consider two temperature ranges, assuming that at low temperatures T_l ($<600\,°C$) the dominant is the process of new phase nucleation, and at high temperatures T_h ($\gg 600\,°C$) the process of nuclei growth. In other words, at low

TABLE 2.41 The relationship between kinetic parameters of topochemical reactions: $A_s \rightarrow B_s + C_g$ and peculiarities of solid-state reagent B_s reactivity.

Temperature	*Velocity of gas evolving as gas product*	*Physico-chemical character of solid-state product*	*Reactivity of solid-state product*
Low	High (W_h)	Individual fine particles, microstrains	High absorption and catalytical properties
(T_l)	Low (W_l)	Ensemble of fine crystallites, separated by dislocations	High activity in solid-state reactions
High	Low (W_l)	Individual large crystallites	Low reactivity
(T_h)	High (W_h)	Ensemble of large crystallites, separated by dislocations	High ability to sintering

temperatures a great number of small particles would be expected to form, while at high temperatures the formation of large particles is much more common.

Regarding the influence of gaseous product elimination rate, it is worth noting that low rate of gaseous product elimination from reaction zone W_l facilitates the consecutive growth of new phase nuclei, while the high rate facilitates their parallel growth. On consecutive growth, every new nucleus, hypothetically, would form after complete formation of the precedent; eventually, the product would consist of separate particles. In contrast, on parallel growth, the nuclei form simultaneously and intensively aggregate, resulting in a giant cluster eroded with dislocation defects.

Combining the limit values of temperatures (T_l and T_h) with limit values of elimination rates (W_l and W_h), as an ultimate case, the products with four different RCS and, therefore, with various structure-sensible properties could be obtained (Table 2.41).

Let us, at first, consider the formation of materials at low temperature T_l and relatively low rate of gaseous product elimination W_l. These kinetic conditions enable the oxide powders to be obtained, consisting of well-faced small particles with internal microstresses and deformations. Such powders usually have enhanced adsorption and catalytic activity. An example of them is oxide powders obtained by slow decomposition of individual oxalates, nitrates and carbonates as well as of their mixtures.

If decomposition proceeds at a high temperature T_h and a high rate of gaseous product elimination W_h, oxide powders consisting of small crystallites with high concentration of dislocation defects, leading to microstresses and deformations in solid, can be obtained. Such features provide high activity of the material in heterogeneous reactions. Actually, to intensify the solid phase reaction, it is necessary to increase the number of contacts between initial oxide particles (high dispersion degree) and facilitate the diffusion through crystal lattice of reactants (the high concentration of dislocations). Such behaviour is characteristic for oxide powders obtained by fast heating of oxalates, nitrates or carbonates to relatively low temperatures.

From these considerations, an important conclusion can be made: the variation of dynamic deviation from equilibrium, i.e. of W parameter, can be used to control the reactivity of oxide phase. Thus, varying the rate of gaseous product elimination during decomposition of the same salt, one can obtain either a satisfactory oxide catalyst or oxide with high reactivity.

Other combinations of reaction conditions, especially high-temperature T_h and low rate of gaseous product elimination W_l, result in the formation of oxide powders consisting of relatively large crystallites (at high temperature the nucleus growth prevails upon the process of nucleation), with an almost complete absence of dislocation defects and microstresses in their structure. Therefore, the material synthesized at such conditions would not manifest enhanced reactivity towards catalytic and absorption processes or solid-state reactions.

At the same time, high-temperature T_h and high rate of gaseous product elimination W_l enable preparation of the oxide powders which consist of relatively large particles with the only defect type, which is the stacking faults – twins. (It should be noted that oxide powders formed in the short time during the fast decomposition of salts hold internal microstresses and deformations, emerged because of the incomplete redox processes, i.e. oxygen exchange.) These powders have the maximum reactivity during sintering. As an example, the oxide powders obtained by decomposition of sulphates and schoenite-like salts can be mentioned. The specific behaviour of such materials is usually accounted for by relatively low density of dislocation defects, which are the main vacancy sources and sinks, providing material stabilization up to relatively high temperatures when the sintering occurs (Geguzin, 1984).

The model above, concerning real structure formation and based on general kinetic rules, requires some general comments.

1. In the model, only four ultimate cases were considered, which corresponded to the formation of four types of real crystal structure. Each of the products can be characterized by a set of dominant defects. However, in any real situation, the RCS of product only resembles the structures discussed to a certain degree. In other words, an infinitely great number of intermediate states can be realized between any two of those situations (for example, between T_l and T_h at a fixed parameter W, or between W_l and W_h at fixed temperature T), i.e. a great number of materials with specific RCS can be synthesized.
2. Under particular conditions of synthesis, the real crystal structure forms in solids which have their own set of spatial defects and structure-sensible

properties. So, in every specific process, precise correction of synthesis conditions are quite necessary to produce the phases with the required type of activity.

3. The use of reactants with different crystal structures as precursors creates doubts about the correctness of 'isotropic' concepts in the model discussed. In other words, according to Curie's principle (Sirotin, 1982), to produce a stricter model of polycrystalline solid formation, not only 'external' synthesis conditions should be taken into account, but also the symmetry of the fields of external and internal actions. The gradient of temperature or of chemical potential of gaseous products along the sample may serve as fields of external action in this case; and the type of initial reactant crystal lattice, as fields of internal action.

Results obtained from model considerations of real structure formation can be represented by a graphical scheme, where the crystallites and the character of their interaction depending on conditions of the topochemical process are shown (Figure 2.94).

It is worth noting that the real structure formed at T_l and W_h resembles dissipative structures well known in non-equilibrium thermodynamics.

Evidently, the topochemical memory would not occur if the system were in a thermodynamically equilibrium state, because such systems are 'dead', i.e. there is no driving force, and this results in strongly determined behaviour, i.e. the

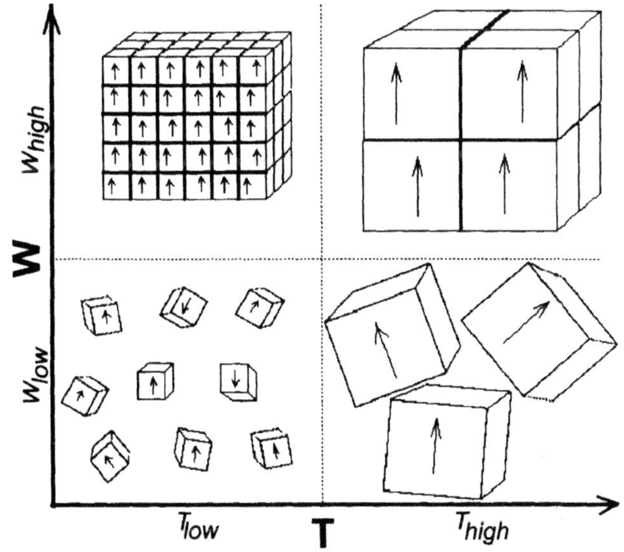

FIGURE 2.94 Hypothetical structure of powdered solid product of topochemical reaction like $A_{(s)} = B_{(s)} + C_{(g)}$ as a function of process kinetics.

result of the process is not influenced by kinetic factors, but is determined only by thermodynamics. It is known that for the systems in a thermodynamically equilibrium state the final state of the system does not depend on the method by which it was reached.

The analysis of correlations between generalized flow I (mass, heat etc.) and driving force of the process $\Delta\xi$ (Figure 2.95) has shown that in any open system the self-organization can occur (Nicolis and Prigogine, 1989). Thus, in the chemical system, the excessive Gibbs energy ΔG can act as a driving force; and if $\Delta\xi = 0$, the system is in equilibrium state. When deviations from equilibrium are small, the dependence of the flow on the driving force is linear. So, the behaviour of the system is strictly determined and can be described by Onsager relations (Nicolis and Prigogine, 1989). When undamped fluctuations play a remarkable role, the relation $I = f(\Delta\xi)$ becomes non-linear, and reaching a certain threshold $\Delta\xi > \Delta\xi_{cr}$, a bifurcation occurs, i.e. instead of one stable path of the process development, the system evolution can follow several stable directions.

Taking into account significant deviations of cryochemical systems from equilibrium, it is reasonable to expect that in the freeze-drying products similar 'bifurcations' can occur rather often.

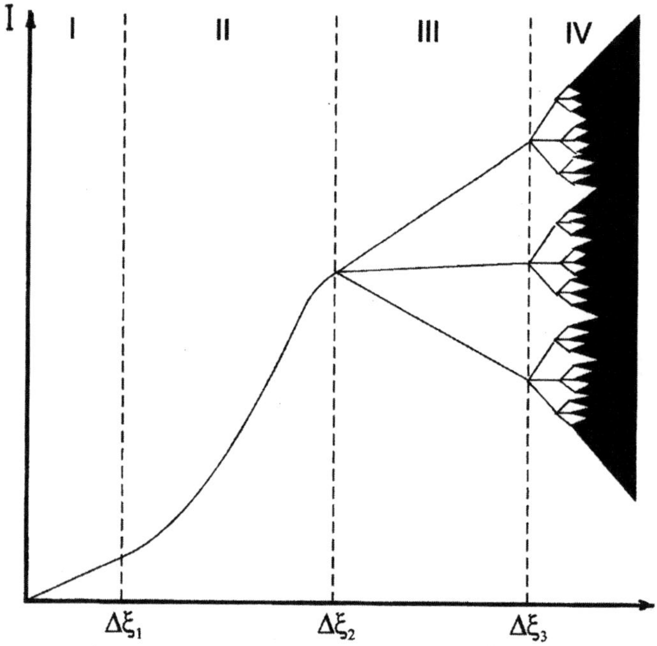

FIGURE 2.95 Dependence of generalized flow *vs.* driving force of the process: appearance of self-organization phenomenon.

Concerning the products of topochemical reactions, the most probable state of the system in low-temperature range is an ensemble of separate elementary volumes (for a non-uniformly heated liquid this state corresponds to the chaotic Brown motion of molecules), aggregated quite weakly. In this case, the nucleation rate is considerably higher than the rate of mass transfer, and the collective behaviour of individual elementary volumes (as elementary volumes, the new oxide phase nuclei would be considered) can be observed only in confined spatial sites (the so-called 'spatial correlation distance' l_k is rather short). Ultimately, at low temperature the l_k value corresponds to the average crystallite size of the salt used as a precursor in topochemical reaction

$$A(s) \rightarrow B(s) + C(g) \qquad (2.84)$$

At high temperatures the product of the topochemical reaction is formed under conditions, whereupon the rate of mass transfer considerably exceeds the rate of nucleation. As a result, a considerable increase of spatial correlation distance l_k (compared to l_k at T_1), sometimes exceeding the average crystallite size of initial reactant A, can be observed (in this case, the aggregation of separate crystallites is possible).

Hence, in real powder systems, instead of two hypothetical states predicted on the basis of kinetic models of real structure formation at a given temperature ($T = \text{const}$) an indefinitely great number of states with different degrees of elementary volume aggregation can occur. This multitude of states occurring usually due to variations in topochemical reaction conditions, provides the ability to accumulate and conserve 'information' in the form of a specific real structure.

The area of partial aggregation (middle part) as well as two hypothetical areas corresponding to the complete aggregation and disaggregation of elementary volumes in the system is presented in Figure 2.96.

In this area, the higher the aggregation degree of elementary volumes (in Figure 2.96 the increase of aggregation degree is indicated by arrows), the higher is the deviation of the system from equilibrium at the moment of new phase formation, i.e. the greater is the rate of gaseous product elimination W.

In this respect, it should be emphasized that with higher deviation of the system from equilibrium at the moment of new phase formation, the 'information' kept by the system will be more stable, i.e. the 'quality of topochemical memory' will be better. Ultimately, the best 'topochemical memory' should exist in the systems with a tendency to self-consistent behaviour appearing in all elementary volumes and facilitating collective transition in the totally new state.

To study the influence of cryochemical processing on the structure of oxide products, the fine crystal and rheologic structures of iron(III) oxide powders prepared by the thermal decomposition of Mohr salt were compared (Ischenko *et al.*, 1997). The Mohr salt was obtained by recrystallization (series I) and by cryocrystallization followed by sublimation drying (initial salt concentrations were 0.03 mole/l and 0.9 mole/l, for series II and III, respectively).

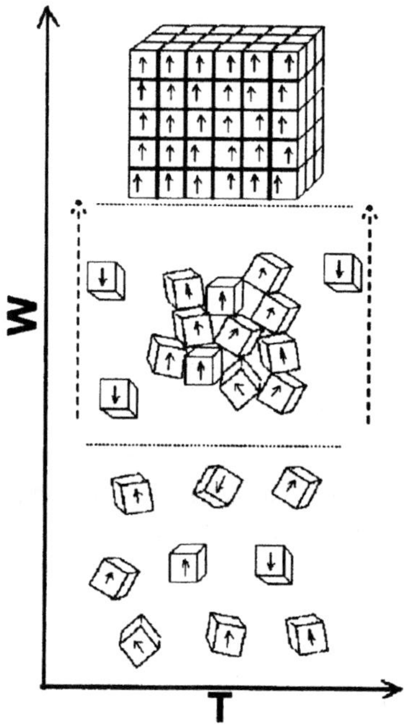

FIGURE 2.96 Schematic of crystallites behaviour at different temperatures and different deviations of the system from the equilibrium.

X-ray analysis data show (Figure 2.97) that all products of thermal decomposition at 700 °C are characterized by the same (within error limits) coherent domain sizes (CDS), which are 20–30 μm. Such a similarity seems to be determined by the critical size of the α-Fe_2O_3 nucleus which would form at this temperature.

At the same time, during the following thermal treatment the CDS values for different iron oxide samples vary in a different way (Figure 2.97). Thus, the samples of series II keep the minimum CDS up to $t = 1000$ °C, while the behaviour of other 'cryochemical' samples (series III) are almost the same as the behaviour of series I, regardless of heating conditions. However, close values of CDS do not mean that parameters of rheologic structure are the same. Actually, the analysis of dependences of mean crystallite size (evaluated by statistical treatment of microphotographs) on the treatment temperature for α-Fe_2O_3 samples (Figure 2.98) shows that the increase of particle size is the same for cryochemical samples, regardless of initial salt concentrations and thermal treatment regimes (slow (10°/min) heating or bringing in heated furnace), as well as for the sample obtained by decomposition of crystalline salt at intensive

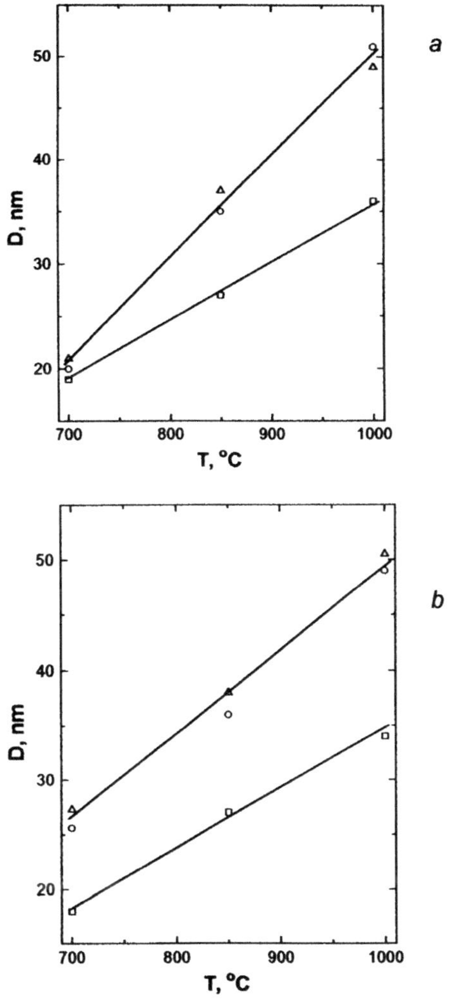

FIGURE 2.97 Mean size of subcrystallites *vs*. heat treatment temperature for crystalline salt (triangles) and cryochemical product obtained from solutions with different concentration (circles = 0.9 M; squares = 0.03 M). (a) low heating rate; (b) high heating rate.

heating. Meanwhile, for the sample obtained by decomposition of crystalline salt at slow heating, a more intense increase of particle size is observed.

These results agree well with the aggregate structure of iron(III) oxide powder formed from individual particles (Figure 2.99). It can be seen that for samples with a similar character of particles growth, the similar aggregate structures with isolated fragments are observed (Figure 2.99a, b, d). In contrast, for the sample showing more intense crystallite growth (Figure 2.98c) the three-dimensional framework of microparticles is characteristic.

Quantitative data on aggregate structure of Fe_2O_3 powders – histograms 'aggregate size distribution' – were obtained by dynamic light-scattering technique. These data show (Figures 2.98 and 2.100) that for samples with slow particle growth aggregates about 1 μm in size are predominant, while for samples with more intense particle growth a large part of the aggregates are about

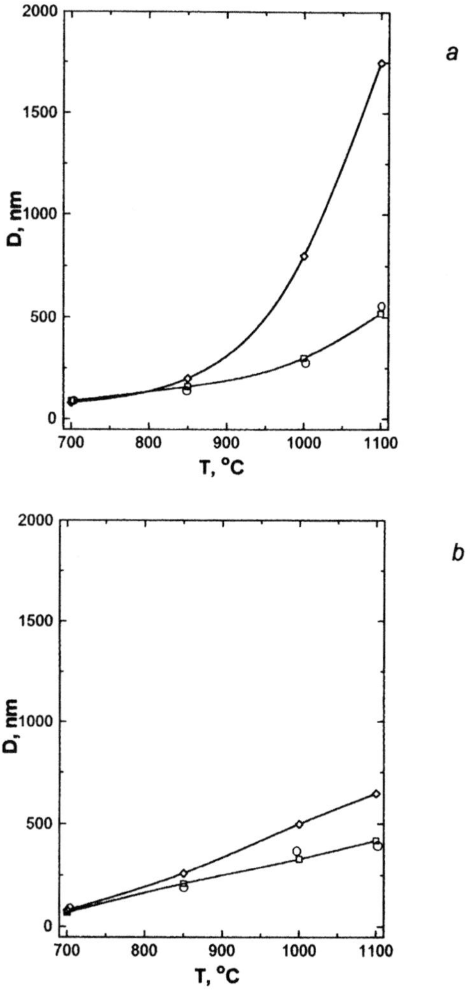

FIGURE 2.98 Mean size of crystallites *vs.* heat treatment temperature for crystalline salt and cryochemical products (see Figure 2.97).

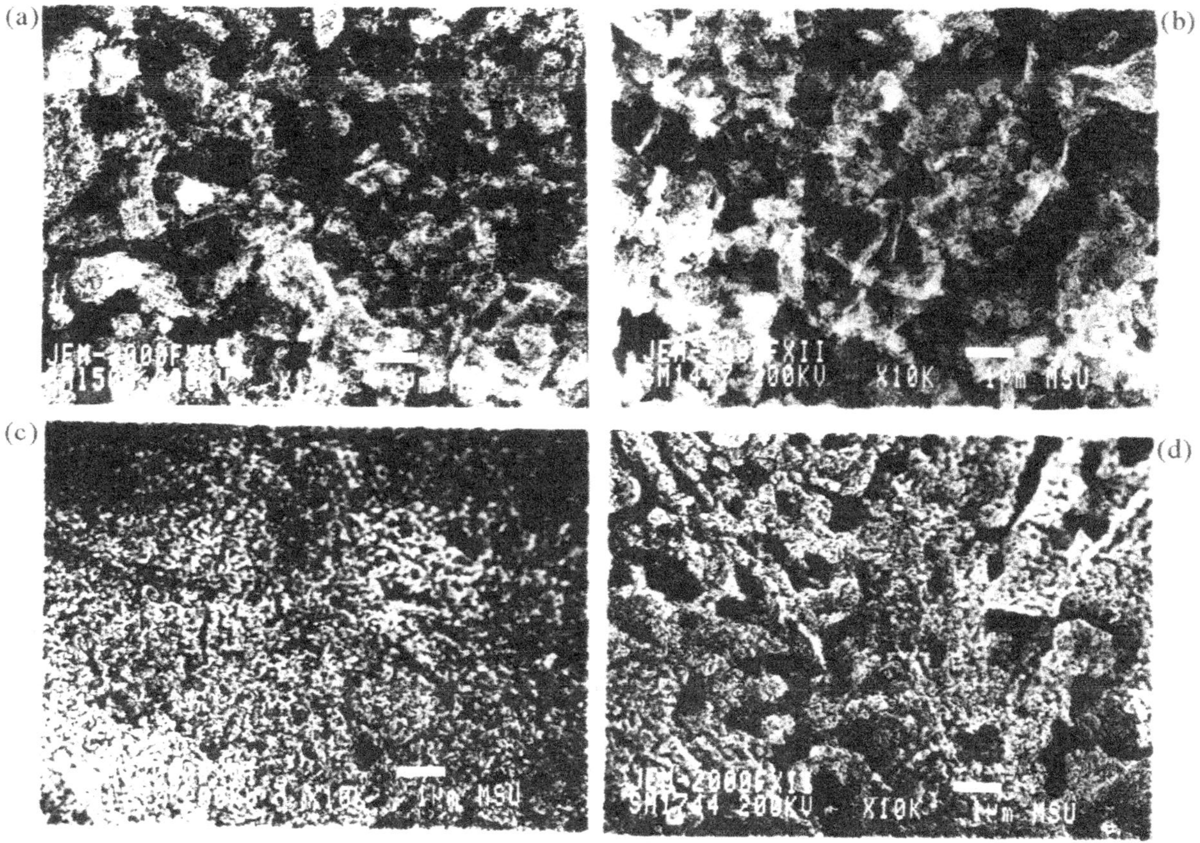

FIGURE 2.99 SEM images of Fe_2O_3 powders obtained by thermal decomposition of large crystals of Mohr salt (c, d) or fine freeze-dried salt powder (a, b). b, d – fast heating to decomposition temperature; a, c – slow heating.

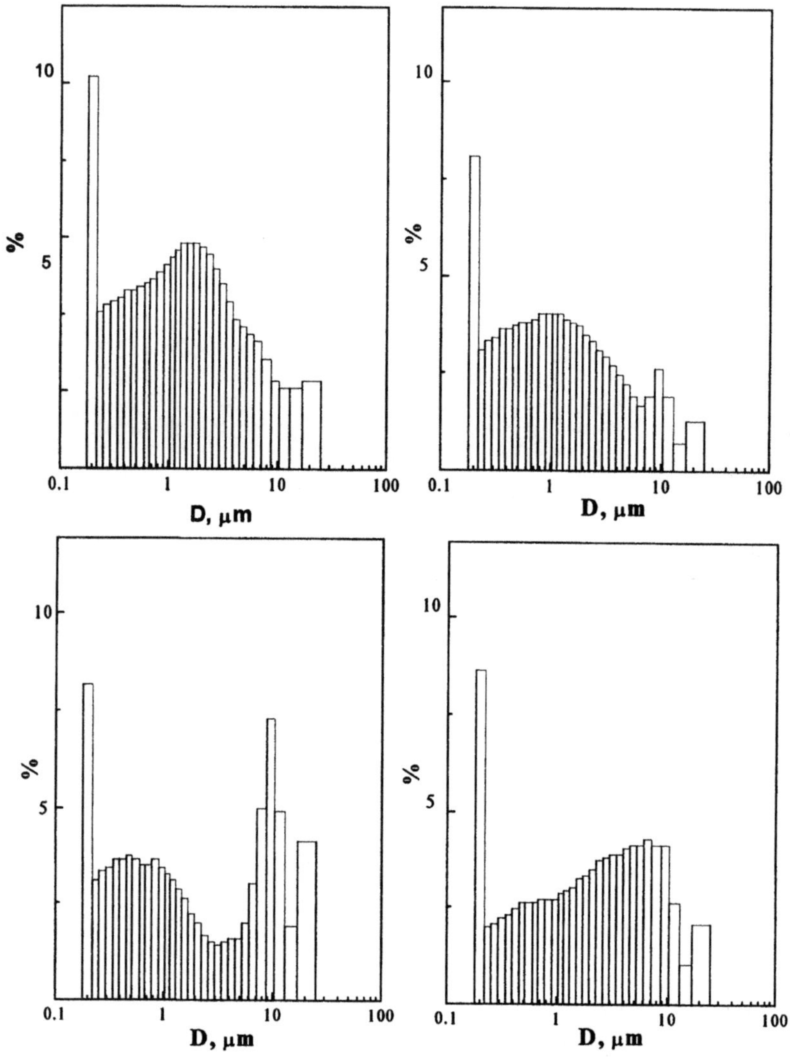

FIGURE 2.100 Size distribution of particle aggregates for iron(III) oxide powders with different thermal and chemical prehistory (see Figure 2.99).

10 μm in size. Thus, it is quite reasonable to assume that one of the main factors determining the particle growth during thermal treatment of oxide powder is the character of particle aggregation (the latter may include aggregate size, density of particle packing within aggregates, as well as the density of larger aggregate formations of interconnected aggregates).

On the whole, reported experimental data make it possible to conclude that at low heating rates the oxide powder microstructure reflects the features of the salt precursor microstructure. At fast heating rate (up to several hundred degrees per minute) the influence of prehistory is considerably diminished. The main difference in microstructure of Fe_2O_3 powders appears in their aggregate structure; the following thermal treatment brings about the differences at a new hierarchic level – the level of crystallites, which are just determined by the difference in initial aggregate structures.

Such correlations between the particle growth and aggregate structure enable particle growth to be controlled through a deliberate change of powder microstructure at the aggregate level. One of the simplest methods to create such changes seems to be its mechanical destruction.

To compare the efficiency of various dispersing techniques (grinding in a ball mill of planetary type and ultrasonic treatment in liquid) for Fe_2O_3 microstructure transformation, the changes in size distribution of particle aggregates with the course of action time have been studied. The results obtained (Figure 2.101) show that ultrasonic treatment is more efficient than grinding, because in a relatively short time the smaller and more uniform aggregates are obtained, while after mechanical treatment a considerable part of the greater aggregates are still observed. For this reason, the ultrasonic treatment was used as the dispersing tool in further experiments.

Taking into account the fact that the aggregate structure of oxide powder is a final stage of microstructure evolution during thermal decomposition and that there are other stages also determining the evolution course, the ultrasonic treatment was used not only towards the oxide product ($t = 700\,^{\circ}C$), but also towards the initial cryochemical salt and semi-product obtained at 500 °C. After the ultrasonic treatment all the samples were heated at the rate of 10 °C/min and annealed at 1100 °C. The sizes of produced crystallites were compared to those of control sample crystallites; the latter was prepared by the same thermal processing but without ultrasonic treatment.

The results obtained showed (Figure 2.102) that ultrasonic treatment is more efficient when the oxide product of decomposition is treated (for this sample the sharp increase of crystallites size is observed).

On the basis of these observations, we can assume that the reason for more intense crystallite growth would be the increase of powder density. The ultrasonic treatment of oxide powder leads to destruction of loose framework formed during thermal decomposition of cryochemical precursors and composed of more dense aggregates. This leads to more dense packing of smaller aggregates, followed by formation of new bonds between them. As a result, a more dense framework forms, providing more intense growth of crystallites during subsequent annealing. The ultrasonic treatment of initial salt precursors or intermediate products turns out to be inefficient, due to the destruction of forming bonds between salt particles during thermal decomposition.

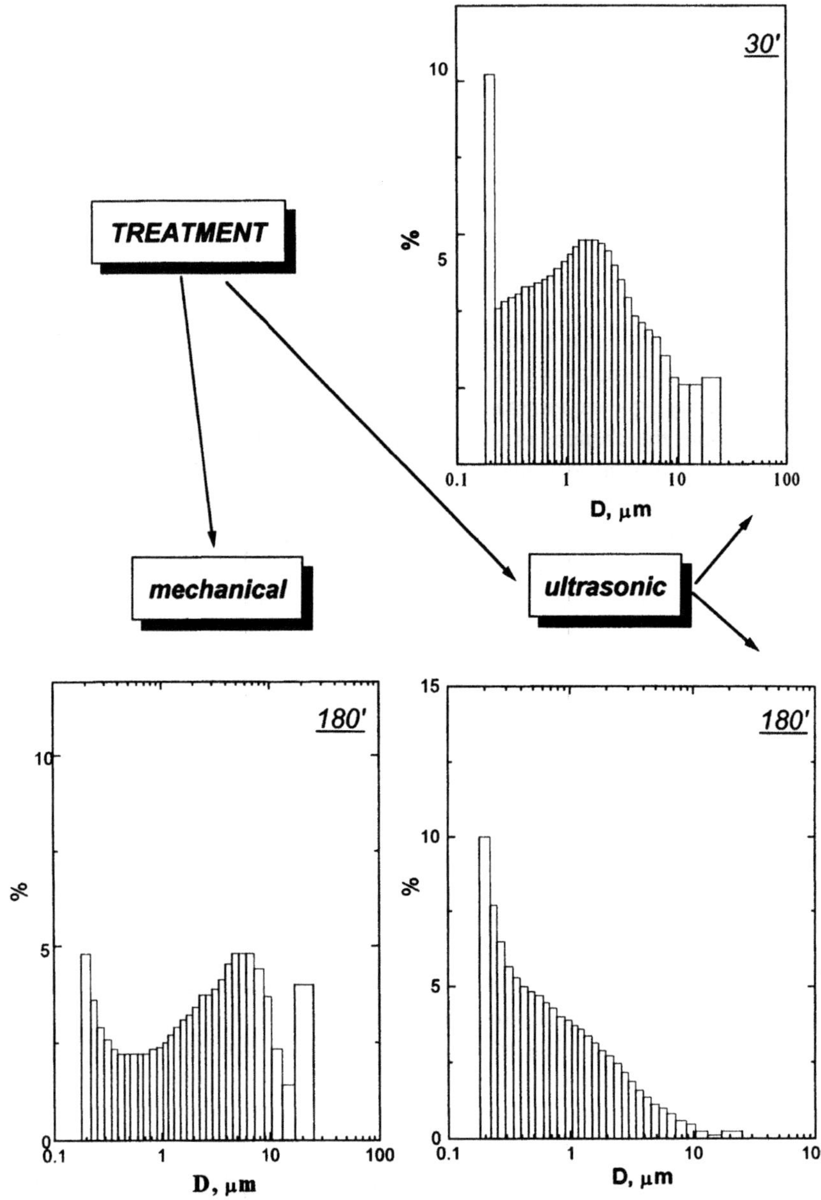

FIGURE 2.101 Comparison of effectiveness of planetary milling and ultrasonic treatment in powder disaggregation.

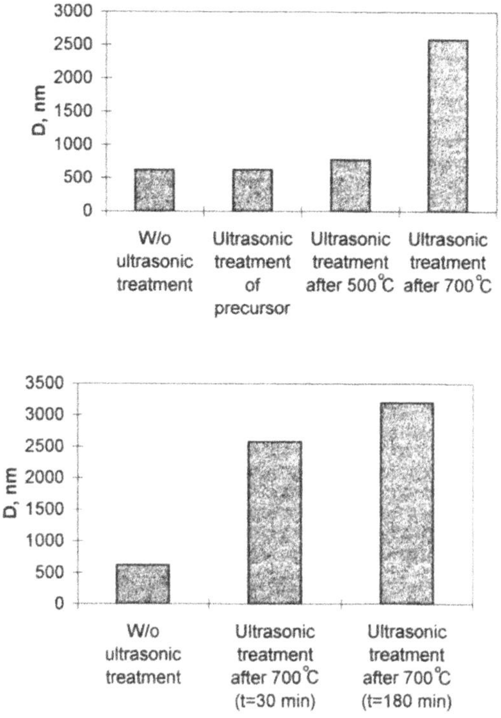

FIGURE 2.102 Effectiveness of ultrasonic treatment application at different stages of powder processing.

Therefore, affecting the aggregate structure of oxide products obtained by thermal decomposition, it is possible to transform it and to change, thus, the character of particle growth during the following thermal treatment.

Another effective technique for the destruction of aggregates formed during thermal decomposition of cryochemical salt products is cryodispersion, which will be discussed in section 2.6.

2.5.4 Sintering processes

The mechanically firm materials are usually obtained from powders by forming (pressing) and sintering. Just the first attempts have shown (Landsberg and Campbell, 1965; Kim and Monforte, 1971) that powders with cryochemical prehistory can be successfully used for the preparation of ceramics with high density ($\rho_{rel} > 0.97$, where ρ_{rel} is the density of ceramic relative to its X-ray density). The main feature of such ceramics is their specific microstructure, which is sufficiently homogeneous and consists of relatively small crystallites.

The reasons for microstructure homogeneity for ceramics with cryochemical prehistory can be obtained from the following concepts of sintering (Levin *et al.*, 1978).

Pressed powder represents a non-equilibrium system, wherein: (1) there are pores, i.e. open internal surfaces; (2) there are interfaces between crystallites: (3) the lattices of some crystallites are deformed during pressing or powder preparation.

According to these three sources of free energy in the system, in the course of sintering three processes can take place: the sintering itself, recrystallization and relaxation. The sintering itself refers to consolidation and strengthening of particle aggregates and elimination of pores from bulk material. Recrystallization involves the growth of new crystallites, accompanied by the lessening of others. Relaxation refers to elimination of crystallographic deformations in the lattice. The final microstructure of ceramics, certainly, will depend on the relative intensities of sintering, recrystallization and relaxation at different stages of the process. The dominance of one or another will be determined by the chemical composition of the system, by the degree and character of its deviation from equilibrium.

The process of sintering may be divided into three to six stages (Figure 2.103).

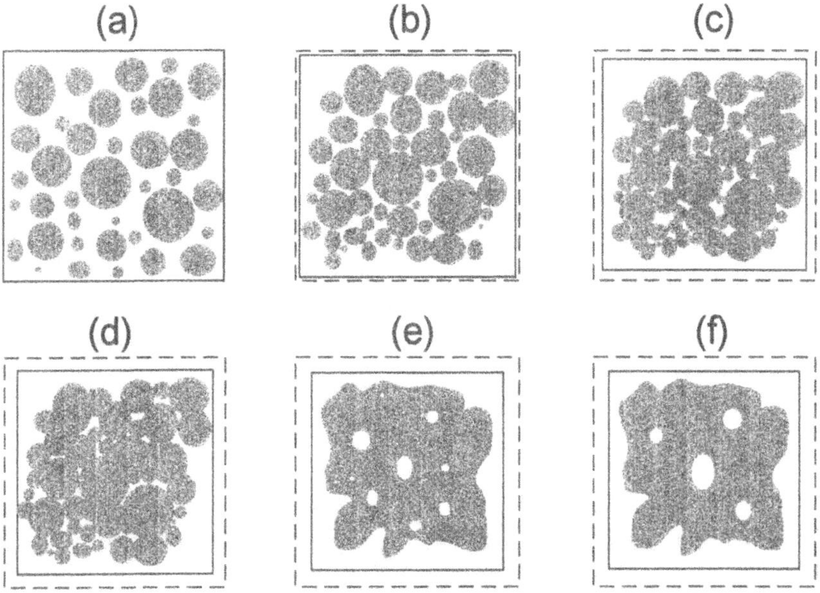

FIGURE 2.103 Stages of sintering the compact of spherical particles: (a) initial state; (b) mechanical contact between particles; (c) formation of necks; (d) formation of closed pores; (e) spheroidization of pores; (f) elimination of pores.

1. Initial stage

At this stage the porosity of products is so high that the idea of 'pore' has no practical sense. The peculiarity of this stage is considerable shrinkage of pressed material. At the same time, the sharp increase of contact surface between particles (the developing necks) takes place. It should be noted that only at some sites are contacts between compressed particles quite perfect (i.e. realized at atomic level) and continuous. For the most part, the contact surface includes various imperfections typical for solids: micropores, cracks and voids. So, Van der Waals's forces are of great importance for such contacts. At initial stage the driving force of sintering seems to be the excessive surface energy of fine particles.

Let us consider, for example, porous compact consisting of spherical particles with radius R and initial porosity θ_0. The excessive free energy is accumulated in its volume due to the existence of a developed free surface. The ratio of excessive free energy to the volume gives the value of pressure squeezing the compact to reduce its free surface. The value of this pressure, supposing the compression is uniform, can be evaluated from the relation

$$P \approx \frac{\alpha}{R}\theta(1 - \theta_0) \tag{2.85}$$

where α is the value of surface tension; θ is the current porosity value. Assuming that $R = 2{*}10^{-6}$ m, $\alpha = 1$ J/m^2, $\theta_0 = 0.5$ and $\theta = 0.4$, we obtain $P = 10^5$ N/m^2 = 1 atm. Therefore, even under such pressure, due to sliding of particles along one another, the shrinkage of bulk material can occur. The residual stresses also play a considerable role in compacting of the porous product. At boundaries, cohesion between particles and resistance to sliding are much lower, so in this case considerable shrinkage rates may be expected.

Sliding along the grain surfaces becomes impossible when the tight packing of particles is reached. Further shrinkage and decrease of porosity only proceed due to mass transfer in bulk.

2. Intermediate stage

At this stage the growth of connecting necks, formed at the initial stage, is observed. The contacts between particles expand and, eventually, the net of grain boundaries forms. Meanwhile, the pores transform into communicating channels; further they grow smaller, and by the end of the intermediate stage only isolated pores are observed. Their location depends mainly on the sintering mechanism – they can occur either at grain boundaries or inside the grain. If initial pore distribution is uniform, the uniform shrinkage of porous compact and an increase of its mechanical rigidity are usually observed.

The processes occurring at the intermediate stage are often modelled by sintering of spherical particles (supposing the real particles to be close to the sphere) to a flat surface. It can be seen from Table 2.42 that the value of the characteristic coefficient n in kinetic equation describing neck surface growth

$$\left(\frac{x}{R_0}\right)^n = A\tau \tag{2.86}$$

is specific for different sintering mechanisms and can be used to determine the mechanism of the process involved.

It should be noted that relations presented in Table 2.42 are true only if initial particles are spherical monocrystals and there is no boundary between grains in the neck. At the same time, if such a boundary exists it can considerably affect the sintering process controlled by volume diffusion. The grain boundaries are, in fact, drains for vacancies, i.e. excessive vacancies move from concave neck surface towards them; as the result of vacancy annihilation, the necks grow fast and shrinkage takes place.

Obviously, regarding the sintering of two polycrystalline spheres, the situation would be considerably complicated.

The polycrystalline grains have additional sources and drains of vacancies, which are boundaries of coherent domains and other extended defects. Thus, the sintering intensity is mainly determined by the type of dominant defects and their

TABLE 2.42 The basic equations, fitting the sintering kinetics of single crystalline spherical particles.

	$(x/R_0)^n = A\tau$	
Mechanism	*n*	*A*
Viscous flow	2	$\frac{3}{2}\frac{\alpha}{R_0\eta}$
Volume diffusion	5	$\frac{20\pi\alpha\Omega D_0}{R_0{}^3kT}$
Surface diffusion	7	$\frac{28\alpha\Omega D_s\delta_s}{R_0{}^4kT}$
Transfer of substance through the gaseous phase under low pressure of inert gas	3	$\left(\frac{18}{\pi}\right)^{1/2}\left(\frac{m}{kT}\right)^{3/2}\frac{\alpha P_0}{d^2R_0{}^2}$

Notes: R_0 is the radius of particles, x is the radius of the contact area between particles, α is the surface energy, η is the coefficient of viscosity; Ω is the volume of one vacancy; D_0 is the coefficient of volume diffusion of atoms (ions); D_s is the coefficient of surface diffusion of atoms (ions); δ_s is the effective thickness of surface layer; m is the mass of atom (ion); d is the density of substance in condense state; P_0 is the partial pressure of vapour in the volume of pore; k is the Boltzmann constant; T is the temperature.

distribution. For example, if defects close to the boundary of sintered particles act as vacancy drains, the flux of atoms moving to this boundary lessens and the rate of sintering decreases.

3. The final stage of sintering

This involves vanishing of isolated pores formed by the end of the intermediate stage. There are two basic approaches describing the behaviour of isolated pores in solids. The first one was developed by Frenkel (1946), who showed that 'spontaneous' shrinkage of solids occurs because of the tendency for the particles' surface energy to produce the lowest possible value of free energy. In this case, the kinetics of the process are determined by the rate of viscous flow (creep rate) (Figure 2.104a).

According to Frenkel, the process proceeds under the influence of capillary pressure applied to curved pore surface. The kinetics of isolated pore vanishing can be described by the equation

$$\frac{\mathrm{d}a}{\mathrm{d}\tau} = -\frac{4}{3}\frac{\sigma}{\eta} \tag{2.87}$$

its integration gives the expression for the time of complete vanishing of the pore

$$\tau_b = -\frac{3}{4}\frac{\eta a}{\sigma} \tag{2.88}$$

In Equations 2.87 and 2.88, a and a_0 are current and initial pore radii, respectively; τ is the process time, σ is the surface tension, η is the internal friction coefficient determining the viscosity of polycrystalline solid.

From Equation 2.87, it can be concluded that the rate of pore lessening is a constant during sintering. However, Frenkel has shown that the pressure of gas held by the pore and insoluble in sintered material would grow with a decrease of pore volume:

$$p = p_0\left(\frac{{a_0}^3}{a^3}\right) \tag{2.89}$$

where p and p_0 are current and initial gas pressures within the pore, respectively.

When the pressure in isolated pores reaches the capillary pressure, the sintering comes to an end.

Another approach was developed by Pines (1946) who gave an atomistic explanation of the process. He was the first one to point out the fact that the well-known Thomson formula describing excessive vapour pressure over concave

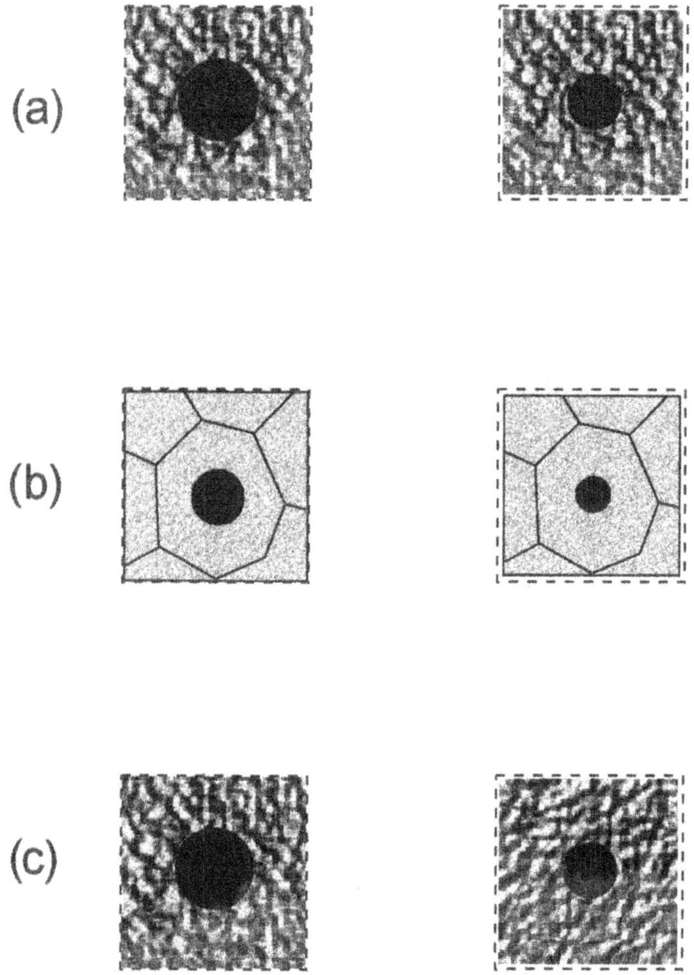

FIGURE 2.104 Pore filling by substance during sintering in different regimes: (a) diffusive viscous flow; (b) vacancy dissolution; (c) self-consistent motion of grains toward pore centre.

liquid surface does not involve the mass of the particles and can be applied to vacancies in crystal lattice:

$$\Delta p = \frac{\alpha\, \Omega}{a\, kT}\, p_0 \tag{2.90}$$

where α is the coefficient of surface tension for the drop, a is the radius of the drop, Ω is the volume of the molecule, p_0 is the vapour pressure over a flat liquid surface, k is the Boltzmann constant and T is the temperature.

Similarly with vapour pressure over a small liquid drop, near the surface of a closed pore with a small curvature radius, vacancy concentration considerably exceeds the equilibrium concentration and can be described by the equation

$$C_1 = C_0\left(1 + \frac{2\sigma\delta^3}{akT}\right) \tag{2.91}$$

where C_0 is the equilibrium concentration of vacancies on the flat surface of crystal, σ is the coefficient of surface tension, δ is the lattice constant and a is the radius of surface curvature.

The gradient of vacancy concentration results in a vacancy diffusion flux from the pore to the crystallite surface, equivalent to the counterflux of atoms. Eventually, this leads to the decrease of pore radius (Figure 2.104b).

The rate of this process can be described by the equation

$$\frac{\mathrm{d}a}{\mathrm{d}\tau} = \frac{2\sigma}{a^2}\frac{\sigma^3}{kT}D \tag{2.92}$$

and the time of pore vanishing by the relation

$$\tau_b = \frac{a^3}{\delta^3}\frac{kT}{6D\sigma} \tag{2.93}$$

The concepts developed by Pines improved the qualitative agreement between theory and experimental data. Nevertheless, the values calculated from Equation 2.93 sometimes exceed experimental results considerably. One reason for these discrepancies seems to be the internal vacancy drains occurring in real polycrystalline materials (Herring, 1950). The boundaries of coherent micro-domains, whose linear size determines the efficient creep of polycrystalline solids, would be such drains. Other sites of internal vacancy drains in real solids would be pores.

Theoretical analysis of pore ensembles (it should be emphasized that the Frenkel and Pines models describe the behaviour of isolated pores) has shown that pore distribution according to radius can conveniently be characterized not only by mean pore radius a, but also by the pore critical size a_c (Lifshitz and Slyozov, 1958, 1959). During sintering, pores $a_i < a_c$ would diminish and pores $a_i > a_c$ would grow, being the sources and drains of vacancies, respectively.

This conclusion is of great practical importance, because it suggests that with the pore surface decreasing their total volume remains the constant, and, thus, there should be an optimal value for the sintering time at constant temperature.

Exceeding this value leads to the emergence of large pores, which, consequently, will have a negative influence on structure-sensible properties.

Lifshitz (1963) has generalized the theories of Frenkel and Pines and provided strict theoretical elaboration of viscous flux mechanism for real polycrystalline materials. At present, this mechanism is experimentally proved to have a dominant role in the sintering of the most parts of metals and ceramics.

Finally, Geguzin and Lifshitz (1962) have shown that the mechanism of pore vanishing can be determined from parameter θ, which is equal to the ratio of pore mean size to mean linear size of coherent microdomains l:

$$\theta = \bar{a}/l \tag{2.94}$$

If $a \gg l$, the mechanism of diffusion viscous flux, proceeding at rather high rates, is observed. The condition $a \gg l$ is usually characteristic for initial powder compact. Hence, such a mechanism works at the initial stages of sintering.

In the progress of shrinkage, the a value decreases and the l value, in opposite, increases. This leads to essential changes in sintering mechanism. At $l \gg a$ ($\theta \ll l$) the vacancial dissolution of pores in a crystal matrix is observed; produced vacancies are absorbed either by internal drains (grain boundaries, coherent microdomain boundaries, dislocations, large pores) or by sample surface.

When $\theta = l$, the pores are located at the grain boundaries and the rate of shrinkage is determined by the self-consistent motion of grains towards pore centres.

It should be noted that the value of θ and, therefore, the mechanism of pore vanishing depends considerably on the chemical, thermal and mechanical prehistory of powders. More often, the prehistory is determined by the empirically chosen precursors and conditions, with the main objective of obtaining highly dense material. The high density is usually considered as one of the most important factors affecting the structure-sensible properties of the material. From the considerations of sintering mechanisms above, the most appropriate way to solve the problem seems to be the sintering under conditions providing the minimum quantity of isolated pores by the end of the intermediate stage. That is, the condition $a \gg l$ should be kept until sufficiently high density is reached. Taking into account that the value of a continuously decreases during sintering, the main requirements for the powders can be formulated as follows: the size of coherent microdomains for sintered material should be quite small and, meanwhile, stable at high temperatures (Figure 2.104c).

Numerous experimental data on the fine crystal structure of oxide powders with different chemical and thermal prehistory show (Fadeeva *et al.*, 1976; Voinov, 1976; Panchenko *et al.*, 1979) that only cryochemical powders keep a relatively small size of coherent microdomains up to the temperatures of their intense sintering.

TABLE 2.43 Coherence dispersion area size *D* (nm) of the $Ni_{0.3}Zn_{0.7}Fe_2O_4$ ferrite powder obtained from different hydrate forms of sulphate precursor at various calcination temperatures.

	t, °C		
Salt precursor	850 °C	1000 °C	1150 °C
AP	110	140	160
Monohydrate	–	85	120
VS	80	105	115
VS → AP	65	90	115
VS → AP → monohydrate	–	120	125

VS – vacancy structure; AP – amorphous phase.

For example, the data presented in Table 2.43 demonstrate that coherent microdomain sizes for ferrite powder $Ni_{0.3}Zn_{0.7}Fe_2O_4$ are relatively small and change only slightly at 1000–1150 °C. These features of the fine crystal structure also provide the specific behaviour of powders during sintering.

Thus, shrinkage of nickel–zinc ferrite, sintered for the short time, proceeds more intensely at 1240–1260 °C (Shlyakhtin, 1981). However, with an increase of sintering duration, ferrite density reaches a certain limit and does not change further. At the same time, the sintering at 1200–1220 °C (the rate is lower at initial stage) makes it possible to prepare the ceramics with higher density (Figure 2.105).

Similar irregular changes can be observed, when the final density is determined by at least two processes with opposite dependencies on the temperature:

1. the temperature growth leads to higher diffusion mobility of atoms and atomic defects, resulting both in lowering of viscosity and increase of coherent microdomain size (at 1200–1220 °C the first process dominates, yielding maximal relative values of density);
2. at $t > 1220$ °C the increase of coherent microdomain size prevails, resulting in lower relative density of the material.

Besides the high density, another objective of sintering is the microstructure with certain size distribution of crystallites. The crystallite growth and final size distribution is mainly determined by three consecutive processes, following each other with temperature increasing:

1. primary recrystallization;
2. assembling recrystallization;
3. secondary recrystallization.

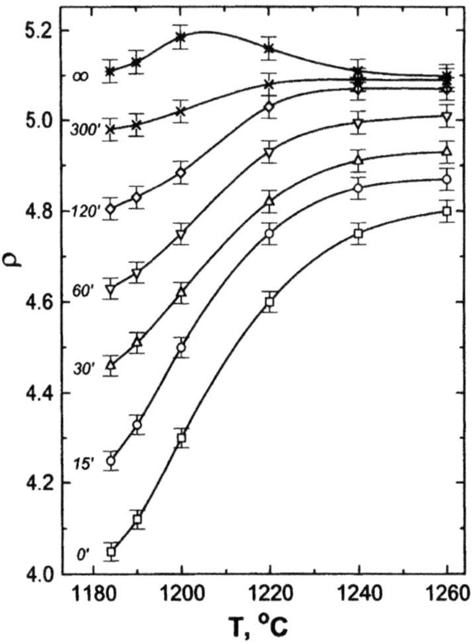

FIGURE 2.105 Isochrones of densification for $Ni_{0.3}Zn_{0.7}Fe_2O_4$ ferrite samples at different temperatures (top curve – estimated limit values).

The distribution of crystallites in sintered material by size D (or $\ln D$) usually obeys the normal distribution law (Levin *et al.*, 1978):

$$U(x) = \frac{1}{\sigma\sqrt{2\pi}} e^{-(x-\mu/2\sigma^2)} \tag{2.95}$$

where μ is the mathematical expectation of random value x; σ is the mean square deviation.

It was experimentally found that random value obeying the normal distribution law more often corresponds to $\ln \bar{D}$ (congregating recrystallization). Sometimes (if sintering starts from primary recrystallization), it may correspond to crystallite diameter. In this case, we can write:

$$\frac{N_i}{N} = \frac{1}{\sqrt{2\pi}} e^{-(D_i-\bar{D}/2\sigma^2)} \tag{2.96}$$

where N_i is the number of particles with diameter D_i; N is the total number of particles.

Therefore, the accumulation curve can be represented by the equation

$$\frac{\sum N_i}{N} = \frac{1}{\sqrt{2\pi}} \int_0^{D_i} e^{-(U^2/2)}\, \mathrm{d}U \tag{2.97}$$

where $\sum N_i$ is the number of particles with $D < D_i$; $U = (D_i - \bar{D})/\sigma$.

The right side of Equation 2.97 represents the probability integral, whose values are tabulated. Using proper tables towards experimentally obtained data of $\sum N_i/N$, the values $U_i = (D_i - D)/\sigma$ can be calculated. If normal distribution law is obeyed, the correlation between U and D is linear; the tangent of slope is equal to $1/\sigma$, and $D = D_i$ at $U = 0$. The last is true at $\sum N_i/N = 0.5$.

In those cases where the dependence between variables U and D is non-linear (i.e. curve $N_i/N = f(D_i)$ is strongly asymmetric), the coefficients of asymmetry A and of excess E can be calculated to characterize the degree of deviation from normal distribution law.

The collection and treatment of experimental data are usually performed using methods of stereometric metallography, most parts of which are now automated.

Let us consider recrystallization on the basis of quantitative information about regularities of crystallite growth.

Primary recrystallization

This involves the formation of crystallization centres in compacted fine powder and their subsequent growth. The driving force of this process is enhanced specific free energy ($2*10^3$–$4*10^3$ J/kg) of the fine powder in the state of plastic deformation. At this stage, the linear growth of mean grain size D at constant temperature is observed.

$$\bar{D} = \omega(\tau - \tau_0) \tag{2.98}$$

where ω is the rate of crystallite growth, cm/s; τ is the process duration: τ_0 is the induction period.

The most common size distribution of crystallites observed at primary recrystallization stage is shown in Figure 2.106a. It means that for any intermediate state of the process (curves 2, 3 and 4) the distribution $N_i/N = f(D)$ with two maxima corresponding to initial and formed crystallites is characteristic. In most cases, the position of the second maximum is constant at $\bar{D}_n$ during the whole process.

Assembling recrystallization

The driving force of this process is also an excessive (compared to large crystals) free energy of fine disperse phase; however, its value is considerably lower than observed for the primary recrystallization: $5*10^2$–$2*10^3$ J/kg.

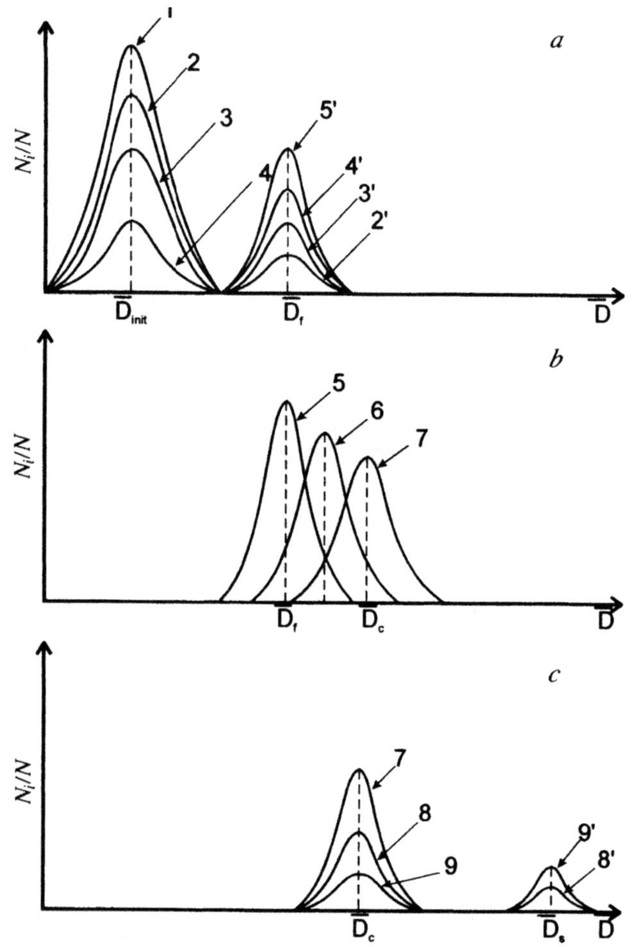

FIGURE 2.106 Particle size behaviour upon different kinds of crystallization: (a) primary crystallization; (b) assembling and (c) secondary recrystallization.

It was experimentally found that during crystallite growth the boundaries of the crystallite move to the centre of curvature so as to reach angles of 120° at contact of three crystallites (Kingery, 1976). From simple geometrical considerations, it can be shown that the angles can close to 120° only if crystallites grow in hexahedral form. Taking this into account, it can be stated that crystallites with less than six sides will diminish and those with more than six will grow. Feltham (1957) has shown that grain growth at this stage can be described by equation

$$\overline{D_{\tau_1}{}^n} - \overline{D_{\tau_0}{}^n} = \frac{k}{T}(\tau_1 - \tau_0)e^{-(Q/RT)} \tag{2.99}$$

where $(\tau_1 - \tau_0)$ is the time of isothermal annealing, $\bar{D}_{\tau_1}$ and $\bar{D}_{\tau_0}$ are the mean diameters of grains in the beginning and at the end of annealing, k is a constant, T is the absolute temperature, R is the universal gas constant, Q is the activation energy of crystallite growth and n is a coefficient depending on mechanism of crystallite growth.

Let us consider the changes in grain size distribution during sintering (Figure 2.106a). Their most specific feature is singular (after completion of primary recrystallization) distribution $N_i/N = f(D)$. During congregating recrystallization mean grain size grows monotonously and relative width of distribution considerably increases (Shershnev, 1973).

It should be noted that chemical and thermal prehistories also greatly affect the crystallite size distribution (Sokolov, 1978).

From Equation 2.99 it can be seen that mean grain size would approach a certain 'limit' value, which is quite steady during a long time.

Some processes occurring on grain boundaries may hamper grain growth; they are the following:

1. impurity concentration;
2. segregation of impurity phases;
3. pore concentration;
4. certain relative orientation of grain boundaries.

As an example of such influence, we can mention the kinetic data on mean grain size changes during sintering of hematite obtained by thermal decomposition of Mohr salt crystallized in equilibrium conditions, as well as prepared by cryochemical technique (Figure 2.107) (Sokolov, 1978).

Secondary recrystallization
This involves non-uniform grain growth appearing when one part of the grains grow more intensely, absorbing the others. More often, secondary recrystallization takes place if, due to fluctuations, the boundaries emerged are able to move with anomalous high velocity, leading to non-uniform crystallite growth.

As a result, non-uniform grain size distribution is observed, and two maxima (Figure 2.106c) relatively distant one from another are characteristic. The shift of the first maximum can be observed, because the areas with no secondary recrystallization still exist (Equation 2.99). At the stage of secondary recrystallization the increase of mean grain size $\bar{D}_b$ obeys linear correlation, as at the stage of primary recrystallization (Kingery, 1976).

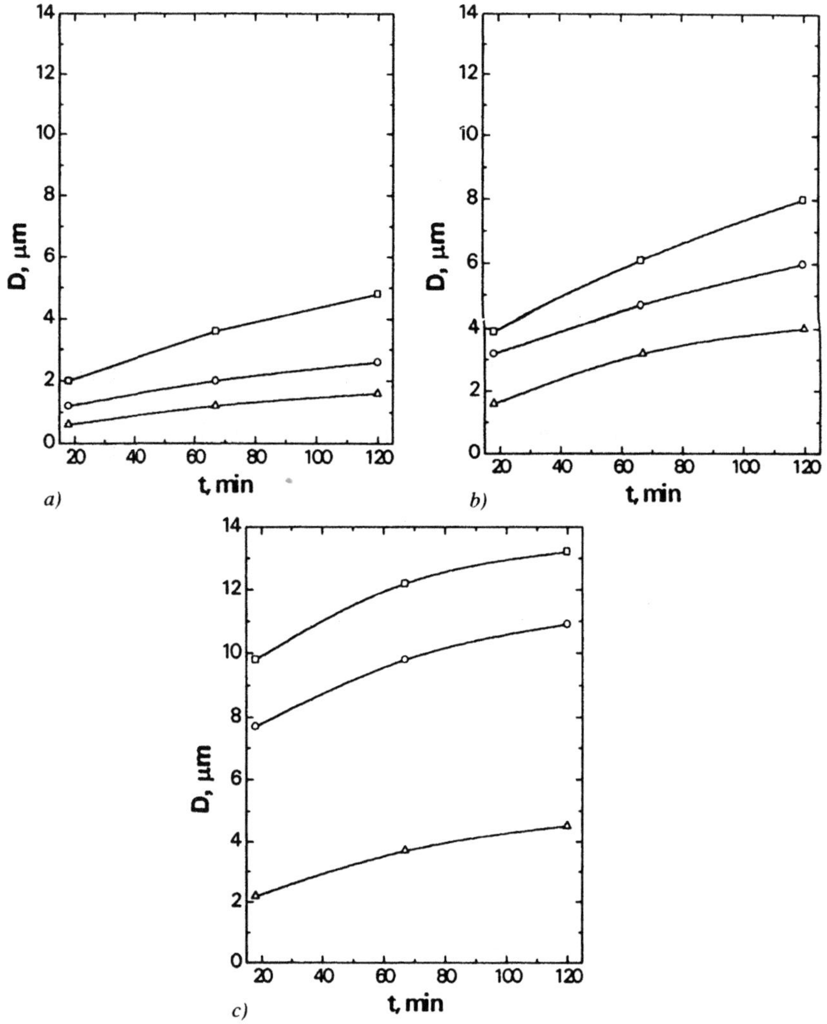

FIGURE 2.107 Growth kinetics of hematite crystals during sintering of different precursors at $T = 1100$–$1300\,°C$.

2.6 CRYODISPERSION PROCESSES

2.6.1 The nature of mechanical dispersion and activation of solids

Supplying mechanical energy into physico-chemical systems may result in various processes, which mainly depend on the methods of the energy input and on the nature of involved phases. They can be divided into three interrelated groups: mechanical, physical and chemical processes.

One of the most common techniques of mechanical treatment is pulverization. The pulverization of solids increases the surface of contacts between reacting substances and results in considerable acceleration of solid-state reactions. At the same time, due to cooperative interatomic interactions, the defect state of solids can change, i.e. pulverization is accompanied by the activation of solid.

According to particle size of initial substances and final products, the processes of pulverization fall into several classes (Table 2.44).

Let us consider the changes occurring in a solid bearing a load (Khodakov, 1972). A schematic state diagram in coordinates: relative elongation λ *vs.* applied stress *P*, which is correct for short-time stresses λ, is represented in Figure 2.108. The following main areas are shown in the diagram.

1. the area of elastic deformation *OA* (P_{el} is the elasticity limit);
2. the area of plastic deformation *AB* (P_{pl} is the plasticity limit);
3. the area of strengthening *BC*, where the increase of mechanical strength is observed (P_{st} is the limit of strength).

An elementary stage of plastic deformation is dislocation shift, which occurs at $P_{el} < P_{pl} < P_{st}$. Meanwhile, local or regular energetic barriers can obstruct the dislocation movement. The regular potential barriers are those separating neighbour equilibrium positions of dislocations and corresponding to the minima of crystal-free energy. The local ones are various defects. The local potential barriers can occur in crystal lattices with several planes of sliding. On countermovement in crossing planes, the so-called sedentary dislocations, tightly interconnected and quite immovable, are formed. The local barriers can also form as a result of point and impurity defect displacements towards dislocations, leading to formation of alien atom chains which complicate the dislocation shift considerably. Thus, the elasticity limit for a single crystal of pure aluminium is

TABLE 2.44 Classification of the milling processes.

Type of milling process	*Size of particle before milling* $\times 10^{-6}$, nm	*Size of particle after milling* $\times 10^{-6}$, nm
Crushing:		
big	1000	250
middle	250	20
small	20	1–5
Grinding		
coarse	1–5	0.1–0.04
medium	0.1–0.04	0.005–0.015
thin	0.1–0.04	0.001–0.005
colloidal	<0.1	<0.001

equal to $4*10^4$ kg/m^2, while for commercial aluminium it is about 10^7 kg/m^2. The moving dislocation nearing in the same sliding plane towards the dislocation held by regular or local barrier undergoes repulsion of its elastic stress field and will stop at a certain distance. Another dislocation moving in the same plane will be stopped by the first dislocation and so on. Finally, an ensemble of dislocations will form.

Under considerable loading $P > P_{st}$ (see Figure 2.108), fragile solids (some metals, oxides, sulphides etc.) can be destroyed; this occurs due to the formation and development of germ fissures induced by the dislocations congregated near the barrier. In other words, on free movement of dislocations the crystal 'flows'; but when an obstacle occurs, and dislocation movement is impeded, the fissures appear and grow larger with an increase in dislocation density.

At high temperatures the dislocations can bypass the obstacles, improving the plastic properties of the material. It should be noted that the plastic properties are determined by the ability of dislocations not only to move, but also to generate new dislocation defects during deformation. Emergence of new dislocations maintains the development of plastic deformation and can account for activation of solids during dispersion. In contrast, plastic deformation ceases after all dislocations initially present in the solid are 'used'. Actually, the absence of free sliding dislocations near an area with $P = P_{pl}$ results in the loss of plasticity. At the same time, the congregation of dislocations at potential barriers leads, on the one hand, to crystal strengthening and, on the other, to its fragility.

The energy supplied into the solid in the course of its mechanical treatment is thus spent to increase the surface energy (pulverization) and to increase the

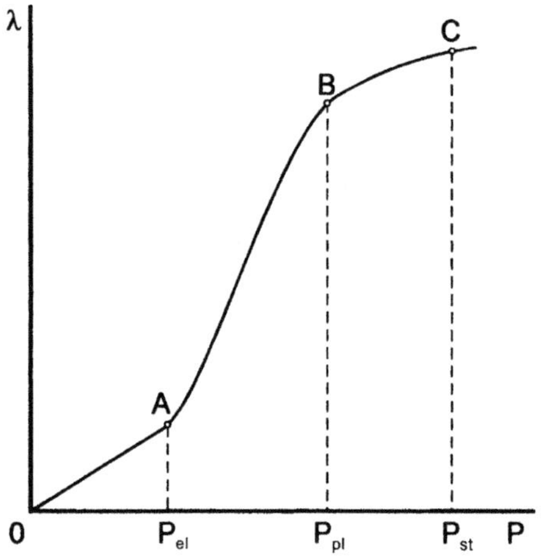

FIGURE 2.108 Relative elongation of solid upon loading.

volume energy (activation). Development of controlled processes, which would allow a change in ratio between surface and volume energies still remains the main problem of mechanochemistry (Avvakumov, 1986). This ratio, obviously, depends on many factors, including the design of devices, the nature of solids, temperature, environment and, of course, the technique of mechanical treatment. Actually, there are many kinds of mechanical actions: splitting, smashing, cutting, rubbing, pounding and their combinations.

For fine grinding, the pulverizers of pounding (disintegrators and dismembrators) or pound-abrading actions are usually used; in particular, they include vibration, planetary and gyroscopic mills. In the pound-abrading apparatus, dispersion is provided either by pounding with sliding onto pulverized material, or by pounding with rotation around any axis.

At present, there are a large variety of devices for material pulverization based on many kinds of mechanical action.

Regarding pulverization at low temperatures, there are rather few actual data on this problem. Interpretation of mechanical properties of solids in terms of a dislocation model is well developed only for the processes observed at room and enhanced temperatures, i.e. when dislocations are quite mobile. At lower ($T < 200$ K) and very low (about 50–100 K) temperatures, the potential energy of interactions between point and impurity defects and dislocations increases and, therefore, the number of high-energy potential barriers also increases. In other words, if at room temperature most crystal defects may be overcome by dislocations due to easy thermal activation of the migrations through defects, at low and extra-low temperatures practically all defects represent obstacles for dislocation motion. As a result, near each barrier (due to their great quantity) a limited number of dislocations are accumulated, and considerable impulse stresses should be applied (still increasing with temperature decrease) to cause their congregation and the formation of fissure germ (microfissure).

Making use of such features of material behaviour at cooling, the effective techniques of cryodispersion were proposed. They make it possible to carry out pulverization at low temperatures, down to the boiling point of liquid nitrogen (77 K).

First, this allows energy requirements for pulverization to be reduced; furthermore, it makes it possible to obtain a fine dispersed product of the materials which are rather difficult to pulverize at room temperature (Savitsky, 1954).

It should be noted that liquid nitrogen, used as a cooling agent, possesses high heat capacity and absorbs the heat of pulverization. Moreover, evaporated nitrogen (or argon) provides a dry and non-toxic atmosphere within the device and over the pulverized product.

At present, low-temperature pulverization is widely used to process polymers (polyethylene, polyvinylchloride), plastics, rubbers, caoutchoucs, bitumens, metals, foods and drugs. Cryochemical grinding can also be used for the utilization of industrial wastes, including metal scraps and chips.

Cryopulverization involves the following stages:

1. preliminary cooling of materials with vapours of cooling agents;
2. cooling by direct contact with cooling agents;
3. grinding in mills of various kinds.

The type of applied device depends on the nature of the pulverized material and on the method of pulverization (crushing, dispersion). For example, the devices for pulverization of metallic chips and scraps include the preliminary pressing of materials in compact bulks. Pulverization itself is carried out after cooling or at the moment of the material's contact with the cooling agent. It is usually performed using a hammer, rotor, bill, or vibration mill.

Depending on the required dispersion degree (the cryopulverization technique produces powders with a particle size from 1 to 40 microns) and ability of the material to break at low temperatures, different cooling agents can be used. The majority of them are liquid gases (nitrogen, air, argon); solid carbon dioxide can also be applied. Special attention should be given to the rational use of a cooling agent. For this purpose, closed circulation systems, incorporating counterflow heat exchangers both in outlet and inlet, are designed, and efficient heat isolating materials and hermetic devices are widely applied. All this makes it possible to reduce the amount of cooling agents to 2 kg per 1 kg of the product, i.e. to use 90% of their cooling ability.

2.6.2 Physico-chemical regularities of cryodispersion

Intensive industrial promotion of powder metallurgy techniques, whose beneficial features are the absence of wastes and the possibility of multiple usage of expensive materials, brought up actual problems concerning the dispersion of metals and alloys. The principal way to make the metal pulverization more efficient is to change treated metals from the plastic state to the fragile.

Well-known and widely used effect of plasticity reduction by using the small quantities of surfactants (the Rebinder effect) is hardly appropriate for metals with their high surface free energy ΔG_s. The reason for that is the low surfaceability of organic media; on absorption of such media, variations of ΔG_s are rather small (about 10^{-2} J/m^2). Considerably more active towards metals are the melts of other low-melting metals. They may reduce free surface energy of the matrix metal to several joules per square metre (J/m^2). However, this leads to considerable content of impurities in the treated substance, which are certainly undesirable for high-quality products.

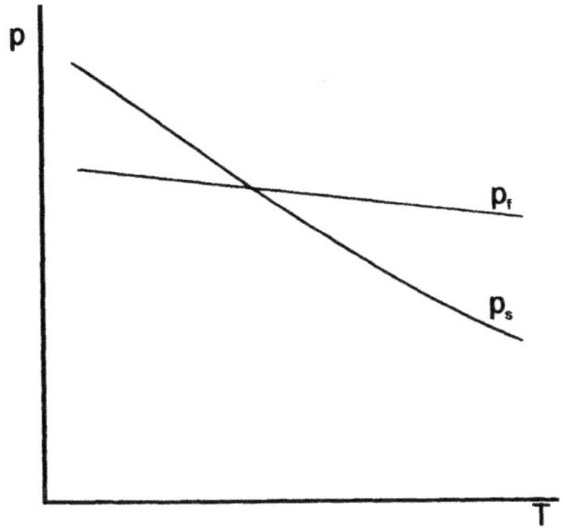

FIGURE 2.109 Temperature dependencies of flowability limit P_f and strength P_s for coldbreaking materials.

Another efficient way to reduce plasticity and, hence, to increase the efficiency of dispersion processes is cooling. It is well known that many materials at low temperatures are converted from the plastic state to the friable one (cold brittleness). For cold-brittle materials the yield stress usually depends on the temperature being much stronger than the tensile strength. From Figure 2.109 it can be seen that in cooled material the tensile strength is achieved prior to the yield stress, and, thus, the friable degradation of the material occurs without plastic deformation.

Changing the stressed state of the material, the range of transitions from plasticity to fragility can be shifted either to high temperatures or to low. The reason for this is the different metal structure. Among the majority of metal lattices the face-centred cubic (FCC), body-centred cubic (BCC) and hexagonal close-packed (HCP) structures are the most common. For FCC (Cu, Au, Ag, Al, Pt, etc.) and HCP (Mg, α-Be, Se, Zn, Cd, α-Co, etc.) the coordination numbers are equal to 12; for BCC (α-Fe, Na, K, Ba, β-Ti etc.) the coordination number of 8 is characteristic. The tendency to change plasticity with a lowering temperature is not the same for metals with different spatial structures.

Metals with FCC lattice do not change their plasticity down to very low temperatures. Upon a decrease in temperature, their tensile strength increases faster than the yield stress, because they have four sliding planes. At low temperatures, the number of sliding planes decreases, because it becomes impossible for dislocations to overcome the energy barriers; eventually, this leads to a remarkable increase in material strength. However, it is quite probable that at least one sliding plane may remain, and this fact may account for the low-temperature plasticity of metals with FCC lattice (Startsev, 1975).

For the metals with HCP lattice, possessing one sliding plane parallel to the prism's base, the possibility of plastic deformation by shift mechanism quickly disappears. At the same time, for some metals with HCP lattice due to the emergence of a great number of twinning defects with a structure similar to FCC lattice, plasticity can occur. This makes it impossible to foresee the changes in mechanical properties for metals with HCP lattice under various temperature conditions.

Metals with BCC lattice all manifest cool brittleness with a decrease in temperature. The reason for this seems to be lower packing density of atoms in the BCC lattice (68%) compared with FCC (74%). Sliding of one loose layer along another is obstructed, because of the deeper 'falling down' of one layer's atoms into cavities between the atoms of another layer.

For some metals (for example, Na and Cs), the increase of plasticity observed at low temperatures is conditioned by the transition of BCC lattice to FCC.

Taking into account the peculiarities of cryodispersion (the action of impulses with high mechanical energy and low frequency), the perceptible activation of treated material can hardly be expected. However, the studies of cryodispersion of metallic zinc (HCP) at the temperature of liquid nitrogen have shown that in addition to pulverization activation takes place. According to the data of sieve analysis, the cryopulverization of zinc proceeds quite intensely, and the average size of particles d decreases from $4*10^{-5}$ to $4*10^{-4}$ nm (at room temperature zinc can be deformed, but not pulverized).

In this case, the reactivity of zinc powders increases considerably; at least three reasons may account for this:

1. accumulation of electropositive impurities by contamination with mill materials;
2. the increase of dispersion degree;
3. changes in the state of the zinc surface.

The first factor can be excluded from consideration, because the impurity content changes very little in the course of cryogrinding. Besides, the dissolution rate of powder zinc 'monofraction' ($1.65*10^{-5}$ to $4*10^{-4}$ nm) increases considerably with the increase of cryodispersion time. Therefore, in addition to the growth of total surface the substructure of the metal also changes. By means of X-ray analysis it was found that the principal kind of structure imperfections, which undergo changes during low-temperature pulverization, are the areas of coherent dispersion (CDA), decreasing their size from $D > 200$ nm (the limit of sensibility for the technique used) to $D = 100$ nm (Pyagai, 1982).

There is a correlation between the rate of zinc dissolution in acid ω and the CDA size, which can be written as follows:

$$\ln \omega = A + B/D \tag{2.100}$$

where A and B are constant coefficients.

From comparison (2.100) with the Arrhenius equation

$$\ln k = \ln k_0 - E_a/RT \tag{2.101}$$

the expression for apparent activation energy of zinc dissolution process can be drawn:

$$E_a = RT[(\ln k_0 - A) - B/D] \tag{2.102}$$

If evaluation of E_a is carried out using samples with different values CDA $(\bar{D}_1, \bar{D}_2)$ but in the same solvent, then

$$E_{\overline{D_1}} - E_{\overline{D_2}} = RT\left[B\left(\frac{1}{\overline{D_1}} - \frac{1}{\overline{D_2}}\right) - \ln\frac{k_0''}{k_0'}\right] \tag{2.103}$$

It was experimentally found (Pyagai *et al.*, 1982) that $E_{\overline{D_1}} - E_{\overline{D_2}} \approx 0$ (within measurement error limits), while CDA size changes from 100 to 170 nm. Therefore, in Equation 2.103

$$E_{\overline{D_1}} - E_{\overline{D_2}} = RT\left[B\left(\frac{1}{\overline{D_1}} - \frac{1}{\overline{D_2}}\right) - \ln\frac{k_0''}{k_0'}\right] = 0 \tag{2.104}$$

from which it follows that

$$\ln\frac{k_0''}{k_0'} = B\left(\frac{1}{\overline{D_1}} - \frac{1}{\overline{D_2}}\right) \tag{2.105}$$

Thus, for metals subjected to cryodispersion, the reaction rate constant is determined only by the CDA size, and the boundaries of CDA are the sites of reaction localization.

Therefore, cryodispersion of metals is usually accompanied by a change in their surface energy. However, at low temperatures some increase of volume energy can also occur, and this has a considerable influence on the reactivity of metal powders obtained by cryodispersion.

One of the traditional branches of mechanochemistry is intensification of crystallographic transformations (phase transitions), which are difficult to cause by thermal treatment (Table 2.45).

At the same time, none of the transformations indicated in Table 2.45 can be performed completely, because of the so-called equilibrium between initial and final phases. In this case, low-temperature mechanochemical action could be especially favourable, because often only direct reactions are initiated and, thus, a better yield of the product is obtained.

Let us consider as an example the polymorphous transformation β-PbO–α-PbO. As is seen from Figure 2.110, the limit value of conversion degree α changes non-uniformly with temperature (Senna and Couny, 1971).

TABLE 2.45 Phase transformations induced by mechanical treatment.

Type of mechanical treatment	*Substance*	*Phase transformation*
	SiO_2	β Quartz–Amorphous phase
Dispergation	TiO_2	Anatase–Brookite–Rutile–Amorphous phase
or	$CaCO_3$	Calcite–Argonite
friction	ZnS	Wurtzite–Sphalerite
	FeS_2	Marcesite–Pyrite
	P	Red, yellow-black
	C	Graphite–Diamond
Mechanical	PbO	β-PbO, Litharge, (red)-α-PbO, Massicot (yellow)
shock	PbO_2	β-PbO_2 (ruthile)-α-PbO_2 (rhombohedral)
	SiO_2	β Quartz–Amorphous phase
	CdS	Wurtzite–Sphalerite

The complete transformation takes place at 178 K, while at room temperature maximum conversion degree $\alpha = 0.8$, and at the boiling point of liquid nitrogen only 0.25. Therefore, for the reaction β-PbO–α-PbO there exists the optimal temperature range for dispersing (which includes 178 K), where polymorphous transformation occurs completely.

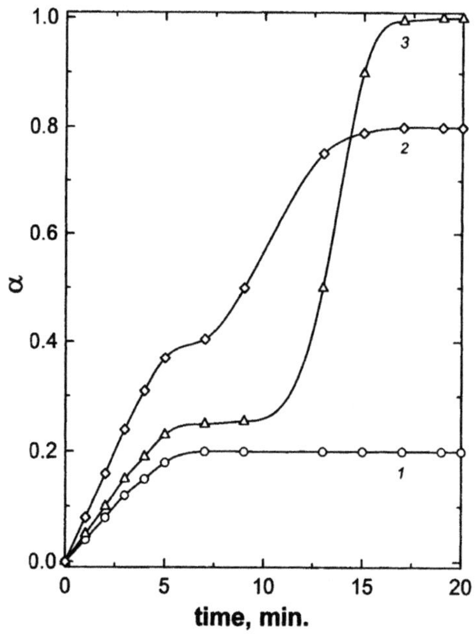

FIGURE 2.110 Degree of polymorph transformation β-PbO–α-PbO *vs.* grinding duration at different temperatures: (1) 77 K; (2) 300 K; (3)178 K.

The indicated fact is of great interest for technology with respect to the processing of lead-containing concentrates (in oxidation as well as dissolution in acids and alkalines the reactivity of α-PbO is much greater than that of β-PbO).

It could be expected that the mechanochemical action at low temperatures will also be useful for other selective processes of raw materials treatment.

References

Akselrud, G.A. (1970) *Mass Transfer in the Systems 'Solid State–Liquid'*, L'vov: LGU.

Akselrud, G.A. and Lysianskiy, V.M. (1974) *Extraction ('Solid State– Liquid' System)*, Leningrad: Chimiya.

Akselrud, G.A. and Molchanov, A.D. (1977) *Dissolution of Solid Substances*, Moscow: Chimiya.

Anastasyuk, N.V., Oleinikov, N.N. and Tretyakov, Yu.D. (1971) Synthesis of Li-containing ferrites by atomization of sulphate solutions in cooled hexane, *Synthesis of Inorganic Materials in Non-aqueous Media*, Moscow: Nauka, 2 (Russ.).

Anastasyuk, N.V. (1972) Studies of effectiveness of chemical methods of ferrites preparation, PhD thesis, Moscow: Moscow State University.

Angell, C.A. and Sare, E.J. (1970) Glass-forming composition region and transition temperatures for aqueous electrolyte solution, *J. Chem. Phys.*, **53**(3), 1058–68.

Angell, C.A. and Tucker, J.C. (1980) Heat capacity changes in glass-forming aqueous solutions and the glass transition in vitreous water, *J. Phys. Chem.*, **84**(3), 268–72.

Angell, C.A. (1995) Formation of glasses from liquids and biopolymers, *Science*, **267**(5206), 1924–35.

Aslanian, S. and Balarew, C. (1977) Uber die Kristallstrukturdifferenzen der $MeSO_4*nH_2O$-Salztypen, *Kristall und Technik*, **12**, 435.

Avvakumov, E.G. (1986) *Mechanical Methods of Activation of Chemical Processes*, Novosibirsk: Nauka.

Barret, P. (1970) Application des techniques de pulverisation electrochimique par etincelles et de lyophilisation à l'obtention de solides de grandes surfaces ou de depots, *Silicates industriels. Colloque Proprietes des petites particules*, Mons, Octobre.

Barret, P. (1973) *Cinetique heterogene*, Paris: Gautier-Villars.

Beck, G., Black, G., Cote, P. *et al.* (1983) *Metallic Glasses: Ionic Structure, Electron Transfer and Crystallization*, Moscow: Mir.

Belousov, V.P. and Panov, M.Yu. (1983) *Thermodynamics of Aqueous Solutions of Nonelectrolytes*, Leningrad: Chimiya.

Belousov, V.P. and Panov, M.Yu. (1973) Thermodynamic properties of diluted aqueous solutions of alcohols, *Chemistry and Thermodynamics of Solutions*, Leningrad: LGU, 3–27.

Bezborodov, M.A. (1981) *Self-initiated Crystallization of Silicate Glasses*, Minsk: Nauka I Technika.

Blachere, I.R. and Younge, I.E. (1972) Freezing point of water in porous glass, *J. Am. Ceram. Soc.*, **55**(6), 306–11.

Braverman, E.M. and Muchnik, I.B. (1983) *Structural Methods of Processing the Empiric Data*, Moscow: Nauka (Russ.).

Brazhnikov, S.M. (1983) Heat and mass transfer during evaporative freezing in vacuum, PhD thesis, Moscow: MIChM.

Brown, M.E., Dollimore, D. and Galwey, A.K. (1980) *Reactions in the Solid State*, Amsterdam–Oxford–New York: Elsevier Sci. Publ.

Cameron, J.A., Keszthelyi, L., Nagy, G. and Kacsoh, L. (1973) Moessbauer effect and thermal analysis of frozen ferrous perchlorate solution, I., *J. Chem. Phys.*, **58**(10), 4610–13.

Chaves, F.A.B. and Garg, V.K. (1975) Moessbauer frozen solution studies of iron (III) nitrate and iron (III) perchlorate, *J. Inorg. Nucl. Chem.*, **37**(11), 2283–5.

Chupakhin, A.P. (1983) The effect of the water on the kinetics of reaction boundary movement during dehydration of crystal hydrates, PhD thesis, Novosibirsk: Institute of Solid State Chemistry.

Churagulov, B.R., Oleinikov, N.N., Lyubimov, S.L. *et al.* (1995) Synthesis of $YBa_2Cu_3O_{7-x}$ by the rapid expansion of supercritical aqueous solutions of yttrium, barium and copper nitrates, *Russ. J. Inorg. Chem.*, **40**(2), 202–7.

Cohen, R.L. and West, K.W. (1972) Moessbauer spectroscopy of frozen solution, *Chem. Phys. Letters*, **13**, 482–4.

Collongues, R. (1971) *La non-stoechiometrie*, Paris, Masson et Cie.

Dalglish, J.McN. (1990) *Freeze-drying for the Food Industries*, London, New York: Elsevier Sci. Pub.

Delmon, B. (1969) *Introduction a la cinetique heterogene*, Paris: TECHNIP editions.

Delmon, B. and Boldyrev, V. (1979) *The Control of the Reactivity of Solids*, Amsterdam–New York: Elsevier Scientific Publ. Co.

Dorosh, A.K. (1981) *Structure of the Condensed Systems*, L'vov: Vysha shkola.

Dowell, I.G. and Rinfert, A.P. (1960) Low-temperature forms of ice as studied by X-ray diffraction, *Nature*, **188**, 1144–8.

Dupuy, J., Jal, J.F., Ferradon, C. *et al.* (1982) Controlled nucleation and disordered growth of ice crystals from low temperature electrolyte solutions, *Nature*, **296**(5853), 138–40.

Eisenberg, D. and Kauzmann, W. (1969) *The Structure and Properties of Water*, Oxford: Oxford University Press.

Fadeeva, V. I., Portnoi, V. K., Oleinikov, N.N. and Tretyakov, Yu.D. (1976) The influence of the dislocation defects on the sintering kinetics of the ferrite powders, *Sci. of Sintering*, **8**(2), 99–117.

Fedotov, G.N. (1984) Synthesis and properties of structurized piezocomposites PZT-polymer obtained by cryochemical methods, PhD thesis, Moscow: Moscow State University.

Feltham, P. (1957) Grain growth in metal, *Acta Metall.*, **5**(2), 97–105.

Fletcher, N.H. (1970) *The Chemical Physics of Ice*, Cambridge: Cambridge University Press.

Le Floc, L. (1986) *Principles of Freeze-drying*, Maurepas: Usifroid.

Fokin, V.A., Belov, V.V., Zhabotinskii, V.A. *et al.* (1979) The technique to prepare PZT-based polycrystalline oxide material, *Pat. USSR* No. 2824814/29-26 from 28.09.79.

Franks, F. (ed.) (1986) *Water and Aqueous Solutions at the Temperatures below 0 °C*, Kiev: Naukova dumka.

Frenkel, Ya.I. (1946) Viscous flow in crystalline bodies, *J. Exp. Theor. Phys.*, **16**(1), 29–52 (Russ.).

Frenkel, Ya.I. (1975) *Kinetic Theory of Liquids*, Leningrad: Nauka.

Frohlich, K. and Keszthelyi, L. (1973) Mössbauer effect and thermal analysis of frozen ferrous perchlorate solution II, *J. Chem. Phys.*, **58**(10), 4614–16.

Gavrilova, E.V. (1978) Studies of heat and mass transfer during freeze-drying in the case of granules formation in vacuum, PhD thesis, Moscow: Moscow Institute of Chemical Mashinery.

Geguzin, Ia.E. and Lifshitz, I.M. (1962) On the mechanism and kinetics of isolated pore elimination in the crystalline body, *Solid State Physics*, **29**(5), 1326–33.

Geguzin, Ia.E. (1984) *Physics of Sintering*, Moscow: Nauka.

Goldansky, V.I. (1968) *Chemical Applications of Mossbauer Spectroscopy*, New York: Academic Press.

Golovchansky, A.V. (1987) Physico-chemical processes during freezing of water–salt systems with high cooling rates, PhD thesis, Moscow: Moscow State University.

Gordon, G. (1979) *Organic Chemistry of Solutions of Electrolytes*, Moscow: Mir.

Gubanov, A.I. (1965) *Quantum Electron Theory of Amorphous Conductors*, New York: Consultants Bureau.

Guhman, A.A. and Ermakova, E.A. (1953) About peculiarities of heat transfer during ice sublimation in vacuo, *Russ. Eng. Phys. J.*, **8**, 114–29.

Guhman, A.A. and Volynets, A.Z. (1968) On the character of ice sublimation in vacuum, *Russ. Eng. Phys. J.*, **15**(5), 777–83.

Guthrie, A. and Wakerling, R.K. (eds) (1949) *Vacuum Equipment and Techniques*, New York: McGraw-Hill, 242.

Hello, M.O. (1995) Mass and heat transfer and structure formation during freezing of aqueous salt solutions in freeze-drying technology of fine grained powders, PhD thesis, Moscow: MIChM.

Herring, C. (1950) Diffusional viscosity of a polycrystalline solid, *J. Appl. Phys.*, **21**(5), 437–45.

Hibbert, D.B. and Tseung, A.C.C. (1979) A critical examination of a cryochemical method for the preparation of high surface area semiconducting powders. Part 2. Non-aqueous solvents. The behaviour of solid solutions in freeze-drying, *J. Mater. Sci.*, **14**, 2665.

Hodge, I.M. and Angell, C.A. (1978) Ionic hydration and secondary relaxation in vitrified solutions, *J. Phys. Chem.*, **82**(15), 1761–6.

Ischenko, V.V., Shlyakhtin, O.A. and Oleinikov, N.N. (1997) Specific features of formation of the microstructure of oxide powders obtained by thermal decomposition of salts, *Russ. J. Inorg. Materials*, **33**(9), 6.

Ivanov, V.I. (1993) Heat and mass transfer during solution freezing in the freeze-drying technology of fine powdered inorganic salts, PhD thesis, Moscow: MIChM.

Ivensen, V.A. (1971) *Kinetics of Densification of Metal Powders during Sintering*, Moscow: Metallurgia, 99–117.

Karabanov, A.V. (1987) Heat-, mass transfer and formation of structure upon freezing in the course of freeze-drying synthesis of fine powders, PhD thesis, Moscow: MIChM.

Karimova, A.F. (1981) ESR spectra and structure of aquacomplex $[Fe(H_2O)_6]^{3+}$ in frozen aqueous ferrous nitrate solutions, *J. Theor. Exp. Chem.*, **17**(2), 268–73.

Kelly, J., Hibbert, D.B. and Tseung, A.C.C. (1978) A critical examination of a cryochemical method for the preparation of high surface area semiconducting powders. Part 1. Rate of freeze-drying, *J. Mater. Sci.*, **13**, 1053.

Khodakov, G.S. (1972) *Physics of Comminution*, Moscow: Nauka (Russ.).

Kim, Y.S. and Monforte, F.R. (1971) Theoretically dense (99.9%) polycrystalline alumina prepared from cryochemically processed powders, *Ceramic Bulletin*, **50**(6), 532.

Kingery, W.D. (1976) *Introduction to Ceramics*, New York: Wiley & Sons.

Kolotyrkina, M.A. (1983) Influence of conditions of microstructure formation on the structure-sensitive properties of ferrites with square hysteresis loop, PhD thesis, Moscow: Moscow State University.

Komm, T.Z., Kaul, A.R., Kalinina, L.A. and Tretyakov, Yu.D. (1981) The technique to prepare beta-alumina doped by yttrium oxide, *Pat. USSR* No. 3232496/22-02 from 04.01.81.

Kramers, H. (1958) *Rate Controlling Factors in Freeze-drying*, London: Society of Chemical Industries, 57.

Krishnaraj, P., Lelovic, M., Eror, N.G. and Balachandran U. (1993) Synthesis and microstructure of $Bi_{1.8}Pb_{0.4}Sr_2Ca_{2.2}Cu_3O_x$ obtained from freeze-dried precursors, *Physica C*, **215**, 305–12.

Kulakov, A.B. (1989) Thermal decomposition of nitrates in the synthesis of Y–Ba cuprates, PhD thesis, Moscow: Moscow State University.

Kulakov, A.B., Mozhaev, A.P., Tesker, A.M. and Churagulov, B.R. (1992) Physico-chemical analysis of cryocrystallization processes of Y, Ba, Cu nitrate solutions and their mixtures, *Russ. J. Inorg. Chem.*, **37**(9), 2137–42 (Russ.), 1106–9 (Int. edn).

Kuleshov, I.V., Gordeev, I.V., Kesler, Ya.A. and Tretyakov, Yu.D. (1983) Dissolution enthalpy of ice in acetone, *Courier of Moscow State University, Chemistry*, **24**(1), 87–9.

Kuleshov, I.V., Gordeev, I.V., Mozhaev, A.P. and Tretyakov, Yu.D. (1986) Thermochemical studies of the system 'dimethylformamide–water' at 248 K, *Courier of Moscow State University, Chemistry*, **27**(4), 436–7.

Kurz, V. and Zahm, P.R. (1980) *Directed Crystallization of Eutectic Materials*, Moscow: Metallurgy.

Kutepov, A.M., Volynets, A.Z. and Brazhnikov, S.M. (1989) Sublimation processes and apparatus of cryochemical technology, *Ultrapure Substances*, **3**, 59–63 (Russ.).

Landsberg, A. and Campbell, T.T. (1965) Freeze-dry technique for making ultra-fine metal powder, *J. Metals*, 856.

Lebedev, D.P. and Perelman, T.A. (1973) *Heat and Mass Transfer in the Processes of Sublimation in Vacuum*, Moscow: Energiya.

Legasov, V.A., Oleinikov, N.N. and Tretyakov, Yu.D. (1986) The problem of flexibility in the technology of ceramic materials, *Russ. J. Inorg. Chem.*, **31**(7), 1637–43.

Levin, B.E., Tretyakov, Yu.D. and Letyuk, L.M. (1978) *Physico-chemical Fundamentals of Preparation, Properties and Application of Ferrites*, Moscow: Metallurgy.

Lifshitz, I.M. and Slyozov, V.V. (1958) On the kinetics of diffusional decay of supersaturated solid solutions, *J. Exp. Theor. Phys.*, **35**, 479–92.

Lifshitz, I.M. and Slyozov, V.V. (1959) On the theory of coalescence of solid solutions, *Solid State Physics*, **26**(9), 1401–10.

Lifshitz, I.M. (1963) On the theory of diffusional viscous flow of polycrystalline bodies, *J. Exp. Theor. Phys.*, **44**(4), 1349–67.

Lindenmeyer, G.S. and Chalmers, B. (1966) Growth rate of ice dendrites in aqueous solution, *J. Chem. Phys.*, **45**(8), 2807–8.

Livage, J. (1985) The gel route to transition metal oxides, *J. Solid State Chemistry*, **64**, 322–30.

Lloyd, I.K., Blachere, J.R., Beyer, L.A. *et al.* (1989) Characterization of freeze-dried alumina powders using electron microscopy, *Ultramicroscopy*, **29**, 153.

Lurie, Yu.Yu. (1979) *Handbook on Analytical Chemistry*, Moscow: Chimiya.

Luyet, B. (1960) On various phase transitions occurring in aqueous solutions at low temperatures, *Ann. N.Y. Acad. Sci.*, 549–69.

Luyet, B. and Rasmussen, D. (1968) Study by differential thermal analysis of the temperatures of instability of rapidly cooled solutions of glycerol, ethyleneglycol, sucrose and glucose, *Biodynamica*, **10**, 167–91.

Lyikov, A.V. (1968) *Theory of Drying*, Moscow: Energiya, 334–80.

Lyikov, A.V. and Vasil'ev, L.L. (1980) Heat and mass transfer of capillary porous bodies in the low pressure gas stream, *Heat and Mass Transfer at Low Temperatures*, Minsk: Nauka I Technika, 8.

McAdams, W.H. (1954) *Heat Transmission*, New York: McGraw-Hill.

Macklin, W.C. and Ryan, B.F. (1969) Growth velocities of ice in supercooled water and aqueous sucrose solution, *Phil. Mag.*, **17**(143), 83–7.

McKenzie, A.P. and Luyet, B.J. (1964) Apparatus for freeze-drying at very low, controlled temperatures, *Biodynamica*, **9**, 177–91.

Magneli, A. (1953) *Acta Crystallographica*, **6**, 495.

Malykh, O.A. (1985) Studies of liquid phase dehydration of crystallohydrates by organic solvents, Diploma Theses, Moscow: MSU.

Mishchenko, K.P. and Poltoratskiy, G.M. (1976) *Thermodynamics and Structure of Aqueous and Nonaqueous Solutions of Electrolytes*, Leningrad: Chimiya.

Moerup, S., Knudsen, Y.E., Mielson, M.K. and Trampy, G. (1976) Moessbauer spectroscopic studies of frozen aqueous solution of iron (3+) salts, *J. Chem. Phys.*, **65**(2), 536–43.

Moor, H. (1964) Die Gefrierfixation lebender Zellen und ihre Anwendung in der Gefrieraetzung, *Z. Zellforsch.*, **62**, 581–601.

Mott, N.F. (1987) *Conduction in Non-crystalline Materials*, New York: Oxford University Press.

Mozhaev, A.P. (1988) Low temperature processes during synthesis of solid state materials with special magnetic and electric properties, DSc. thesis, Moscow: MSU.

Mozhaev, A.P., Shabatin, V.P., Sheredeko, M.V. and Tretyakov, Yu.D. (1982) The technique to prepare polycrystalline oxide material, *Pat. USSR* No. 3445579/23-26 from 24.05.82.

Mozhaev, A.P., Shabatin, V.P. and Pershin, V.I. (1989) Methods of synthesis of high T_c superconductors, *J. Russ. Mendeleev Chem. Soc.*, **34**(4), 504–8.

Mutin, J.C., Watelle, G. and Dusausoy, Y.I. (1979) Study of a lacunary solid phase. I. Thermodynamic and crystallographic characteristics of its formation, *J.Solid State Chem.*, **27**, 407.

Mutin, J.C. and Watelle, G. (1979) Study of a lacunary solid phase. II. Morphological and kinetic characteristics of its formation, *J. Solid State Chem.*, **28**, 1.

Mylay, L.N. and Selwood, P.W. (1954) The hydrolysis of Fe^{3+} ion: magnetic and spectrophotometric studies on ferric perchlorate solutions, *J. Am. Chem. Soc.*, **76**, 6207–8.

Naumov, S.V. (1982) Physico-chemical processes during cryocrystallization of water–salt systems, PhD thesis, Moscow: Moscow State University.

Naumov, S.V., Mozhaev, A.P. and Tretyakov, Yu.D. (1982a) Separate crystallization during freezing of water solutions of sulphates, *Russ. J. Phys. Chem.*, **56**(2), 440–1.

Naumov, S.V., Mozhaev, A.P., Gordeev, I.V. *et al.* (1982b) Cryocalorimetry of dissolution of frozen droplets of ammonium trioxalatoferriate solution in ethyl alcohol, *Russ. J. Phys. Chem.*, **56**(3), 556–60.

Nicolis, G. and Prigogine, I. (1989) *Exploring Complexity: An Introduction*, New York: W.H. Freeman.

Ohya, T. and Ono, K. (1973) Paramagnetic relaxation effects in ferric perchlorate solution, *J. Phys. Soc. Japan.*, **34**(2), 376–8.

Oleinikov, N.N., Pershin, V.I., Tretyakov, Yu.D. *et al.* (1974) The technique for powder preparation, *Pat. USSR* No. 2053978/33 from 09.08.74.

Oleinikov, N.N., Tretyakov, Yu.D., Anastasyuk, N.V. *et al.* (1978) The technique to prepare ferrite powder, *Pat. USSR* No. 2634446/22-02 from 27.06.78.

Oleinikov, N.N., Tesker, A.M., Putlyaev, V.I. and Tretyakov, Yu.D. (1985a) The influence of rheological properties of ferrite powders on their behaviour during forming, *Russ. Chem. Ind.*, **7**, 412–15.

Oleinikov, N.N., Shlyakhtin, O.A. and Tretyakov, Yu.D. (1985b) Thermal analysis of freeze-dried frozen aqueous solutions, *Thermochimica Acta*, **93**, 309–11.

Oleinikov, N.N. (1988) Features of ferrites formation during synthesis of magnetic materials with controlled microstructure and properties, DSc thesis, Moscow: Moscow State University.

Oleinikov, N.N. (1995) The nature and the role of topochemical memory effect in the synthesis of solid state substances and materials, *J. Russ. Mendeleev Chem. Soc. (Mendeleev Chemistry Journal)*, **39**(2), 85–93.

Oswald, H.-R., Guenter, J.R. and Dubler, E. (1975) Topotactic decomposition and crystal structure of white molybdenum trioxide monohydrate: prediction of structure by topotaxy.

Panchenko, L.A., Mozhaev, A.P., Zver'kova, I.I. *et al.* (1979) Substructure and sintering ability of grinded ferrite powders, *Powder Metallurgy*, **7**, 32–7 (Russ.).

Pankov, V.V., Paulus, M. and Dubon, A. (1982) Powder production for rapid sintering of oxides suspended in liquid solution of other cations by the combined suspension-solution freeze-drying technique, *Ann. Chim. Fr.*, **7**, 377.

Pershin, V.I. (1977) Studies of physico-chemical processes during preparation of ferrites and ferrite forming oxides by cryochemical method, PhD thesis, Moscow: Moscow State University.

Petrenko, S.I., Lyashchenko, A.K., Karataeva, I.M. and Mozhaev, A.P. (1990a) Solubility diagrams in triple aqueous salt systems with yttrium, barium and copper nitrates at 25 °C, *Russ. J. Inorg. Chem.*, **35**(6), 1575–80.

Petrenko, S.I., Lyashchenko, A.K., Karataeva, I.M. and Mozhaev, A.P. (1990b) Solubility diagrams of $Y(NO_3)_3$–$Ba(NO_3)_2$–$Cu(NO_3)_2$–H_2O at 25 °C developed by calculation–experimental method, *Russ. J. Inorg. Chem.*, **35**(7), 1857–9.

Pines, B.Ya. (1946) About the viscous break, *J. Exp. Theor. Phys.*, **16**(6), 744–55 (Russ.).

Plachinda, A.S. and Makarov, E.F. (1972) Moessbauer investigation of the glass-to-crystal transformation in frozen aqueous iron(III) chloride solution, *Chem. Phys. Letters*, **15**(4), 627–30.

Pushkar, N.S., Belousov, A.M., Itkin, Yu.A. *et al.* (1977) *Low Temperature Crystallization in Biological Systems*, Kiev: Naukova dumka.

Pyagai, R.N. (1982) Influence of cryotreatment on the reactivity of solids, PhD thesis, Moscow: Moscow State University.

Pyagai, R.N., Oleinikov, N.N. and Tretyakov, Yu.D. (1982) Influence of cryomilling on acid dissolution of Zn metal, *Powder Metallurgy* (Russ.), **3**, 6–12.

de Quervain, M.R. (1975) Crystallization of water, a review. In S.A. Goldblith, L. Rey and W.W. Rothmayne (eds), *Freeze-drying and Advanced Food Technology*, London–New York–San Francisco: Academic Press, 3–16.

Rao, K.J., Helphrey, D.B. and Angell, C.A. (1973) Thermodynamic properties of M(I)–M(II) mixed nitrate glasses and supercooled liquids, *Phys. Chem. Glasses*, **14**(2), 26–32.

Rasmussen, H. and McKenzie, A.P. (1971) The glass transition in amorphous water. Application of the measurements to problems arising in cryobiology, *J. Phys. Chem.*, **75**(7), 967–73.

Read, R., Prausnitz, G. and Sherwood, T. (1982) *Properties of Gases and Liquids: Handbook*, Leningrad: Chimiya.

Riehle, U. (1968) Schnelgefrieren organische Praeparate fuer die Electronen-Mikroskopie. Die Vitrifizierung verduennter waessriger Loesungen, *Chem.-Ing.-Techn*, 40 Jahrgang, Heft 5, 213–18.

Rossini, F.D., Wagman, D.D., Evans, W.H. *et al.* (1950) *Chemical Thermodynamic Properties*, Nat. Bur. Stand., USA.

Rozenthal, O.M. (1972) Features of ice formation in water and solutions II. Kinetics of crystallization of aqueous solutions of electrolytes, *Russ. J. Phys. Chem.*, **46**(3), 657–9.

Rozenthal, O.M. (1973) Interconnections of structural properties of ionic solutions with ice formation, *Russ. J.Phys. Chem.*, **14**(5), 797–801.

Ruby, S.L., Zabransky, B.I., Stevens, J.G. (1971) Metastable phase in ferrous chloride hydrates, *J. Chem. Phys.*, **54**(11), 4559–62.

Rudobashta, S.P. (1979) *Kinetics of Mass Transfer in the Systems with Solid Phase*, Moscow, MIChM.

Rudobashta, S.P. (1980) *Mass Transfer in the System with Solid Phase*, Moscow: Chimiya.

Saltykov, S.A. (1970) *Stereometric metallography*, Moscow: Metallurgia.

Samokhvalov, Yu.P. (1975) On the mechanism of heat and mass transfer during sublimation in vacuo. In B.M. Smol'sky (ed.), *Phase and Chemical Transformation during Contact of Bodies with Gas Flow*, Minsk, 92–105.

Savchenko, L.A. (1982) Specific features of the freeze drying in the fast frozen water–salt systems, PhD thesis, Moscow: Moscow State University.

Savitsky, E.M. (1954) *Influence of the Temperature on the Mechanical Properties of Metals and Alloys*, Moscow, (Russ.).

Sauer, H.A. and Lewis, J.A. (1972) Freezing droplets of aqueous solutions for the cryochemical process, *AIChE J.*, **18**(2), 435.

Schaeffer, R.W., Macho, J., Salomon, R.E. *et al.* (1991) Synthesis and characterisation of YBaCuO superconductors from solutions in liquid ammonia, *J. of Superconductivity*, **4**(5), 365–74.

Schmidt, W.W. and Breitschwerdt, K.L. (1976) Electron spin resonance study on phase transition and structure of frozen ionic solution, *Chem. Phys. Letter*, **44**(3), 557–64.

Schnettler, F.J., Monforte, F.R. and Rhodes, W.W. (1968) A cryochemical method for preparing ceramic material, *Science of Ceramics*, **4**, 79–90.

Schnettler, F.J., Monforte, F.R. and Rhodes, W.W. (1974) Cryochemical method for preparing ceramic materials, *Chem. Eng. World*, **9**(9), 89–94.

Segal, D. (1991) *Chemical Synthesis of Advanced Ceramic Materials*, Cambridge: Cambridge University Press.

Senna, M. and Couny, H. (1971) Polymorphous transformation of PbO by isothermal wet ball-milling, *J. Amer. Ceram. Soc.*, **54**, 259.

Sergeev, G.B. (1981) *Cryochemistry*, Moscow: Mir Publ.

Shabatin, V.P. (1986) The development of solid state materials synthesis methods using cryocoprecipitation of hydroxides, oxalates and carbonates, PhD thesis, Moscow: Moscow State University.

Shabatin, V.P., Mozhaev, A.P., Skazin, V.E. *et al.* (1983a) The technique to prepare polycrystalline ferrite material with garnet structure, *Pat. USSR* No. 3654908/23-26 from 20.10.83.

Shabatin, V.P., Barkovskii, A.I., Mozhaev, A.P. *et al.* (1983b) The technique to prepare precursor for composite metal–ceramic electric contacts, *Pat. USSR* No. 3680469/22-02 from 23.12.83.

Shabatin, V.P., Kalinchenko, A.V., Mozhaev, A.P. and Tretyakov, Yu.D. (1984a) Determination of oxalic acid solubility in ethyl alcohol and water–ethanol solutions, *Courier of Moscow State University, Chemistry*, **25**(2), 204–5.

Shabatin, V.P., Mozhaev, A.P., Tretyakov, Yu.D. *et al.* (1984b) The techniquc to prepare doped nickel–zinc ferrites, *Pat. USSR* No. 3810877/22-02 from 02.11.84.

Shabatin, V.P., Tretyakov, Yu.D., Sokolovsky, V.R. *et al.* (1987) The technique to prepare powder compositions, *Pat. USSR* No. 4213515/31-02 from 26.01.87.

Shabatin, V.P., Sokolovsky, V.R., Mozhaev, A.P., *et al.* (1988) The technique to prepare glass powders, *Pat. USSR* No. 4366362/33 from 19.01.88.

Sharikov, F.Yu. (1991) Cryochemical synthesis of finely dispersed oxide powders using ion exchange, PhD thesis, Moscow: Moscow State University.

Shatnyi, V.I. (1988) Heat and mass transfer and structure formation during evaporation freezing process of freeze-drying technology of ultrafine powders, PhD thesis, Moscow: MIChM.

Shchukin, E.D., Pertsov, A.V. and Amelina, E.A. (1982) *Colloid Chemistry*, Moscow: MSU, 332 (Russ.).

Shershnev, N.G. (1973) Influence of chemical and thermal prehistory on the sintering and ferrite formation activity of ferric oxide, PhD thesis, Moscow: Moscow State University.

Shishov, V.V. (1978) Specific features of synthesis and properties of Li, K, Na metaniobates and their solid solutions, PhD thesis, Moscow: Moscow State University.

Shlyakhtin, O.A. (1981) The influence of doping and sintering conditions on some structure-sensitive properties of Ni–Zn ferrites prepared by cryochemical method, MSc thesis, Moscow: Moscow State University.

Shlyakhtin, O.A. (1985) The role of the water in topochemical processes during cryochemical synthesis of Ni–Zn ferrites, PhD thesis, Moscow: Moscow State University.

Shlyakhtin, O.A., Oleinikov, N.N. and Tretyakov, Yu.D. (1985) Specific features of interaction of freeze-drying salt products with atmospheric moisture, *Russ. J. Inorg. Materials*, **21**(11), 1940–4.

Shlyakhtin, O.A., Oleinikov, N.N., Pokholok, K.V. and Tretyakov, Yu.D. (1988) Freeze-drying of frozen aqueous sulphate solutions, *Russ. J. Inorg. Materials*, **24**(3), 368–71 (Engl.), 453–6 (Russ.).

Shlyakhtin, O.A., Badun, Yu.V., Shpanchenko, R.V. *et al.* (1991) Rare earth–copper coupling at the early stages of $RBa_2Cu_3O_x$ nitrate synthesis, *International Workshop on Chemistry and Technology of High-temperature Superconductors* (MSU-HTSC II), Moscow, 61.

Shlyakhtin, O.A., Shpanchenko, R.V., Antipov, E.V. *et al.* (1993a) RCu_6NO_{11} – new intermediate product of $RBa_2Cu_3O_x$ nitrate synthesis, *First International Conference on Materials Chemistry*, University of Aberdeen, Scotland, UK, 19–22 June, 119.

Shlyakhtin, O.A., Shpanchenko, R.V., Antipov, E.V. *et al.* (1993b) *New Family of Binary Oxynitrates – $RCu_6O_8NO_3$ (R=Ho–Lu), 3rd International Workshop on Chemistry and Technology of High Temperature Superconductors (MSU-HTSC-III)*, Moscow, 20–23 September, Programme and Abstracts, 37.

Sirotin, Iu.I. (1982) *Fundamentals of Crystal Physics*, Moscow: Mir Publ.

Skripov, V.P. and Koverda, V.P. (1984) *Spontaneous Crystallization of Supercooled Liquids*, Moscow: Nauka.

Sokolov, I.S. (1978) Influence of chemical and thermal prehistory on the properties of ferrite powders obtained by cryochemical method, PhD thesis, Moscow: Moscow State University.

Spiro, Th.G., Allerton, S.E., Renner, J. *et al.* (1966) The hydrolytic polymerization of iron (III), *J. Am. Chem. Soc.*, **88**, 2721–9.

Stairs, R.A. (1980) Viscosity of binary solutions of polar liquids, *Can. J. Chem.*, **58**(3), 296–301.

Startsev, V.I. *et al.* (1975) *Plasticity and Hardness of Metals and Alloys at Low Temperatures*, Moscow (Russ.).

Startseva, I.V., Levina, M.E., Tretyakov, Yu.D. and Fokin, V.A. (1977) The technique to prepare polycrystalline oxide material, *Pat. USSR* No. 2503211/29-33 from 23.06.77.
Stillinger, F.H. (1995) A topographic view of supercooled liquids and glass formation, *Science*, **267**(5206), 1935–9.
Stolyarov, E.A. and Orlova, N.G. (1976) *Calculation of Physico-chemical Properties of Liquids*, Leningrad: Chimiya [Mo, 190].
Tamman, G. (1935a) *Glassy State*, Moscow: ONTI, [Mo, 80].
Tamman, G. (1935b) *Guide Book on Heterogeneous Equilibria*, Leningrad: Chimteoret, 197–200 [Mo, 225].
Telegina, L.I., Mozhaev, A.P. and Tretyakov, Yu.D. (1983) Kinetics of granulated ice dissolution in organic solvents, *Russ. J. Phys. Chem.*, **57**(2), 386–90.
Tesker, A.M. (1986) Principles of diagnostics of multistage synthesis of Ni–Zn and Mn–Zn ferrite powders, PhD thesis, Moscow: Moscow State University.
Thierauf, A., Egger, Ch., Sporn, D. and Keck, K. (1992) Crystallization behaviour of freeze-dried $YBa_2Cu_3O_{7-x}$ powders at low temperatures, *Supercond. Sci. Technol.*, **5**, 123–9.
Timmermans (1960) *Physico-Chemical Constants of Binary Systems*, New York, 154.
Tretyakov, Yu.D. (1978) *Solid State Reactions*, Moscow: Chimiya (Russ.).
Tretyakov, Yu.D., Oleinikov, N.N., Anastasyuk, N.V. and Pershin, V.I. (1973a) The technique to prepare polycrystalline materials, *Pat. USSR* No. 1937660/29-33 from 14.06.73.
Tretyakov, Yu.D., Oleinikov, N.N. and Granik, V.A. (1973b) *Physico-chemical Fundamentals of Thermal Processing of Ferrites*, Moscow: MSU.
Tretyakov, Yu.D., Shabatin, V.P., Sokolovskii, V.R. (1987a) The technique to prepare silver powder, *Pat. USSR* No. 4186083/31-02 from 26.01.87.
Tretyakov, Yu.D., Oleinikov, N.N. and Mozhaev, A.P. (1987b) *Fundamentals of Cryochemical Technology*, Moscow: Vysshaya Shkola.
Tretyakov, Yu.D., Oleynikov, N.N. and Goodilin, E.A. (1995) Materials engineering through evolutional development of solid state systems, *Fabrication and Characterization of Advanced Materials. Proc. 3rd IUMRS Intern. Conf. in Asia (IUMRS-ICA-95)*, Seoul, Korea, 17–20 Oct., vol. 3, 1137–18.
Ubellode, A.R. (1982) *Molten State of the Matter*, Moscow: Metallurgy.
Vasil'ev, V.P. (1982) *Thermodynamic Properties of Solutions of Electrolytes*, Moscow: Vysshaya Shkola.
Verkin, B.I., Getmanets, V.F. and Mikhalchenko, R.S. (1980) *Thermophysics of Low Temperature Sublimation Cooling*, Kiev: Naukova Dumka.
Vertegel, A.A., Kalinin, S.V., Oleynikov, N.N. and Tretyakov Yu.D. (1995) The fractal particles of iron(III) hydroxonitrate: from solution to solid state, *J. Non-Crystalline Solids*, **181**, 146–50.
Vigdorchik, E.M. and Sheinin, A.B. (1971) *Mathematic Modelling of Continuous Dissolution Processes*, Leningrad: Chimiya.
Vigdorovich, V.N., Vol'pyan, A.E. and Kurdyumov, G.M. (1976) *Directional Crystallization and Physico-chemical Analysis*, Moscow: Chimiya.
Vishnevskaya, G.P., Gumerov, F.M. and Yagdfarov M.Sh. (1979) Studies of phase composition of frozen aqueous Mn(II) nitrate solutions by dynamic calorimetry and ESR methods, *Russ. J. Phys. Chem.*, **53**(3), 609–12.
Vishnevskaya, G.P., Gumerov, F.M. and Kozyrev, B.M. (1978) Features of glass formation in aqueous solutions of chromium nitrate according to ESR and paramagnetic relaxation data, *Reports of USSR Academy of Sciences*, **239**(5), 1121–4.

Voinov, I.D. (1976) Influence of chemical and thermal prehistory on the substructure and catalytic activity of cobalt ferrite, PhD thesis, Moscow: Moscow State University.
Volfkovitch, S.I., Mitichenko, M.G., Savchenko, L.A. *et al.* (1979) The technique to prepare porous alumophosphate, *Pat. USSR* No. 2815234/23-26 from 23.08.79.
Voss, H. (1970) Gelbfieberimpfstoff, Herstellung und Kontrolle. II. Mitteilung: Die Gefriertrocknung von Gelbfieberimpfstoff, *Zentbl. Bakt. I. Abt. Orig.*, **215**, 404–28.
Wertheim, G.K. (1964) *Moessbauer Effect: Principles and Applications*, New York: Academic Press.
West, A.R. (1984) *Solid State Chemistry and its Applications*, New York: Wiley & Sons.
Willemer, H. (1975) Low temperature freeze-drying of aqueous and non-aqueous solutions using liquid nitrogen. In S.A. Goldblith, L. Rey and W.W. Rothmayne (eds), *Freeze-drying and Advanced Food Technology*, London–New York–San Francisco: Academic Press, 461–77.
Young, D.A. (1966) *Decomposition of Solids*, New York: Pergamon Press.
Zalishchanskii, M.E., Shabatin, V.P., Kulakov, A.B. *et al.* (1988) The method for preparation of high Tc superconductors, *Pat. USSR* No. 4384238/31-33 from 25.02.88.
Zhiboedov, V.G., Makarov, V.M., Skotnikov, A.S. *et al.* (1976) The technique to prepare granular polycrystalline oxides of group III and IV elements of the Periodic Table, *Pat. USSR* No. 2434366 from 27.12.76.
Zhiboedov, V.G., Fokin, V.A., Tretyakov, Yu.D. and Skotnikov, A.S. (1977) The technique to prepare polycrystalline materials, *Pat. USSR* No. 2511848/29-33 from 21.07.77.

3

Application of cryochemical processes in advanced materials preparation

The development of new generations of advanced materials proved to be one of the most important branches of scientific and engineering activities. Being exclusively multiform in their composition, structure, properties and applications (Figure 3.1) these materials have brought about the development of completely specific technologies, one of which is cryochemical technology. The cryochemical technology made it possible to obtain a wide range of necessary products or their precursors. Some of them are presented in Figure 3.2. In each particular case specific features of the application of cryochemical processes are predetermined by the specific properties of the materials. Cryochemical technology has big advantages because it mixes components (at the stage of solutions preparation) with high precision, it reduces the losses during syntheses (especially using freeze-drying), and stabilizes the physico-chemical prehistory of materials because the nucleation of crystal phases in substance microvolumes occurs under identical conditions (the cryocrystallization stage). Because of all these features the reproducibility of properties of the products increases, the yield of useful products grows, their functional parameters improve, the need to correct the technological process disappears, and so on. The last is the most important because it opens the possibilities of automatization of individual technological stages and of the whole process of materials production, makes it flexible and adaptable for the production of a quickly varying range of products. Most parts of the production process ought to make different materials using the same equipment due to the rapidly changing demands of the materials market. The ability of different synthesis methods (and corresponding equipment) to be involved in such flexible materials technologies is rather different.

In the present chapter most of the principal publications known to the authors and concerning the experience of cryochemical technology applications for

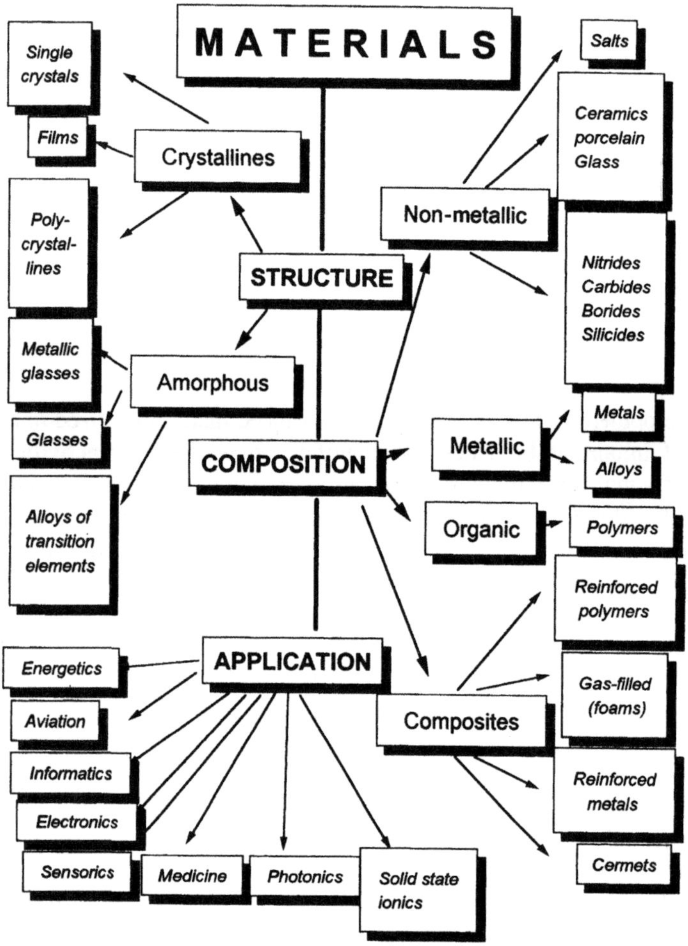

FIGURE 3.1 Classification of materials by different parameters.

preparation of advanced materials are discussed. The discussion is based on the classification of materials by their properties and applications, with an emphasis on those studied most intensively. Unfortunately, a significant part of the publications usually contain information only about applied cryochemical techniques and rheologic properties of powder products, and gives no idea of functional parameters of materials and their improvement using cryochemical processes.

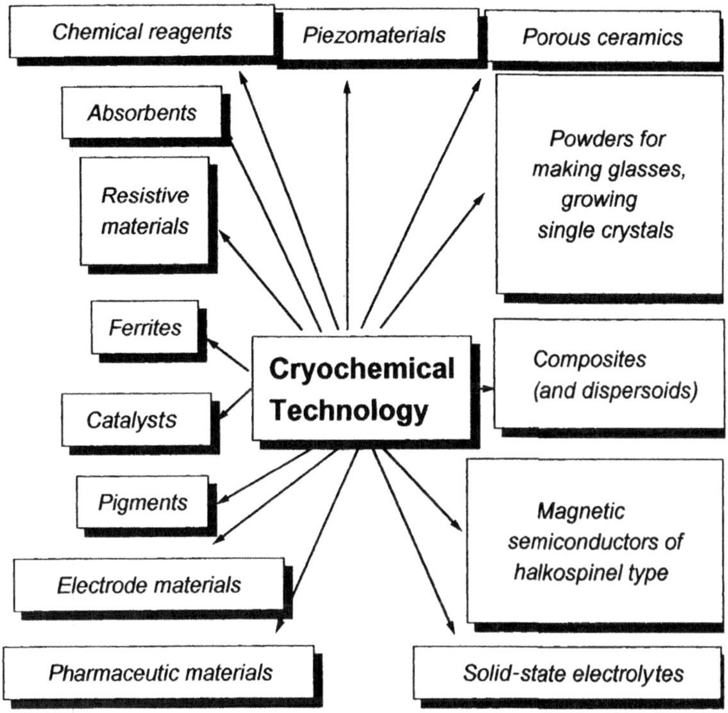

FIGURE 3.2 Main products of cryochemical technology.

3.1 PHOSPHOR MATERIALS

The rapid progress of video display terminals has generated significant interest in display technologies. Among existing displays, the cathode ray tube (CRT) ones are the most widespread, providing large-area, high-resolution, full-colour displays. In their production, phosphor powder is deposited on the faceplate of the CRT tube, and the visual image is induced by the fluorescence of phosphor powder excited by electron beams. The creation and development of projection displays have fostered a need for high-intensity phosphors.

Phosphor powders are also used in other display technologies. For example, in electroluminescent (EL) displays, phosphor powders are placed between two electrodes and an applied electric field leads to electroluminescence. Both powder and thin-film phosphors can be used in EL displays. However, the powdered phosphors have an advantage because of their better control of stoichiometry and crystal perfection, which are essential for good luminescent properties. Their use also avoids the problem of low optical output in thin-film EL displays resulting from the internal reflection between the thin film phosphor layer and the electrode, even though recent development has raised the optical output of certain thin-film devices above that of powdered phosphor displays. Furthermore, in powdered EL technology a large area ($\geq$100 cm^2) display can be produced by depositing a uniform layer of defect-free phosphor powder between electrodes, while such a large area structure is difficult to obtain by thin-film technology.

Most commercial phosphor powders, especially the garnet-based phosphors, are rather expensive because of the stringent requirements on high purity, precise stoichiometry, exact crystalline phases and controlled particle size and morphology. Furthermore, the performance of CRT and EL display systems essentially depends on the quality of the phosphor powder. Thus, in the development of current techniques it is of great importance to produce phosphor powders with novel compositions, crystalline phases, and particle morphologies, improve their quality and provide the competitive product for display technologies.

Cryochemical technology seems to be promising for the preparation of powder materials, but so far effects in this direction have been limited.

Cerium-doped $Y_3Al_5O_{12}$ powders with nominal compositions of $Ce_xY_{3-x}Al_5O_{12}$ (x=0.005, 0.01, 0.03, and 0.05) were prepared by freeze-drying of sulphate and nitrate solutions of the constituent cations (Yan *et al.*, 1987). Water-soluble salts of the constituent cations were chosen as the precursors. Reagent grade $(NH_4)_2Ce(NO_3)_6$ anhydrous $Y_2(SO_4)_3$ and hydrated $Al_2(SO_4)_3*18H_2O$ were used as cation sources of cerium, yttrium, and aluminium, respectively.

Both $(NH_4)_2Ce(NO_3)_6$ and $Al_2(SO_4)_3*18H_2O$ can be readily dissolved in water. However, $Y_2(SO_4)_3$ has a limited solubility in water. In fact, it was shown that $Y_2(SO_4)_3$ was about 10 times less soluble than the cerium and aluminium salts. Thus, the maximum cation concentration in precursor solution was limited by solubility of yttrium salt. In order to increase the solubility of $Y_2(SO_4)_3$, sulphuric acid was added.

In a typical precursor solution, appropriate quantities of $(NH_4)_2Ce(NO_3)_6$, $Y_2(SO_4)_3$ and $Al_2(SO_4)_3*18H_2O$ were added to deionized water so that the precursor solution had a total cation concentration of 0.33 mol/litre. About 0.30 mol H_2SO_4 per litre of solution was added to the precursor solution to promote the solubility of $Y_2(SO_4)_3$. The precursor solution was heated to 65 °C and stirred for 16 h. After the salts were completely dissolved and the solution was cooled to room temperature, 0.60 mol NH_4OH per litre of solution was added to the precursor solution to increase the pH. Just before freezing, prepared precursor solutions had pH values of 1.6–2.0.

The prepared precursor solution was sprayed through a glass nozzle with internal diameter of 670 μm into stirred hexane, which was cooled to −80 °C by a mixture of dry ice and acetone. The solution was frozen at the rate of about 20 g/min. The frozen solution was placed into stainless-steel trays and, then, in a vacuum drier. At the early stage, it was kept at −50 °C and a liquid-nitrogen trap was used to collect evaporated hexane. When the pressure in the vacuum drier had been below 0.4 mm Hg (53 Pa), the temperature was gradually increased to the maximum temperature of 70 °C, for 20 °C every hour, while the pressure was always kept below 0.5 mm Hg (67 Pa). After the precursor salts were dried for 16 h at 70 °C and 0.4 mm Hg (53 Pa), the dried powder was removed from the vacuum drier. The freeze-dried powder was placed in platinum boats and heat-treated at temperatures between 1350 °C and 1550 °C in oxygen and then cooled at different rates.

Table 3.1 represents phase compositions of heat-treated powder according to X-ray diffraction (XRD) analysis. These data show that for the freeze-dried powder garnet phase readily formed at a rather low temperature: 1300 °C. However, a relatively high temperature >1550 °C was necessary to obtain garnet phase in powders prepared by mixing the constituent oxide powders. Thus, the formation of the garnet phase was significantly enhanced by homogeneous mixing of the precursor powder in the freeze-dried sample (Yan *et al.*, 1987).

Table 3.2 shows the emission peak intensities for powders prepared by the freeze-drying and the conventional methods after annealing at different temperatures. The data show that the emission intensity generally increased with the annealing temperature. Furthermore, the powders prepared by freeze-drying had much higher emission intensities (2–4 times) than the conventionally

TABLE 3.1 Phase compositions of heat-treated powders.

	Powder preparation technique	
Temperature (°C)	*Freeze-drying*	*Conventional mixing*
1200	Garnet + spinel	Multiphase
1300	Garnet	Multiphase
1400	Garnet	Multiphase
1500	Garnet	Multiphase
1550	Garnet	Garnet + spinel

TABLE 3.2 Emission intensities of commercial phosphor powder (Derby P96) and YAG:0.03 Ce powders prepared by freeze-drying and conventional methods.

	Emission intensity	
Annealing temperature (°C)	*Freeze-drying synthesis*	*Conventional synthesis*
1300	42.4	11.2
1400	52.1	23.0
1550	67.0	39.3
P96 (Derby)		100

prepared powders. The reasons for this are higher homogeneity of the freeze-dried powder and, thus, more ready formation of garnet phase at low temperatures. Conventionally prepared powders were multiphase due to the non-homogeneous mixture of the constituent components.

3.2 MAGNETIC MATERIALS

The application of cryochemical technology for preparation of magnetic materials is limited to ferrites – compounds of iron oxide with oxides of other metals. Ferrites, as well as magnetic alloys, possess domain structure, high magnetic permeability and typical hysteresis magnetization curve. At the same time, in contrast to magnetic alloys ferrites usually have high electric resistance and, hence, low-energy losses in demagnetization which makes them indispensable in the high- and ultra-high-frequencies engineering.

Many of the magnetic properties of ferrites (for example, coercive force, saturation magnetization, magnetic permeability) are the structure-sensitive ones, i.e. they depend considerably on ceramic structure of materials, including size and shape of crystallites, size, shape and distribution of pores. For this reason, the problem of preparation of ferrites with predetermined and well-reproducible properties becomes reduced to the problem of preparation of materials with determined chemical composition and desirable ceramic structure (Table 3.3).

3.3 FERROELECTRIC MATERIALS

In contrast with usual dielectrics, ferroelectric materials have exclusively high dielectric permeability – the ability to keep residual electric polarization after the removal of the external electric field and to manifest a hysteresis dependence between polarization and potential. Spontaneous polarization of ferroelectrics is caused by asymmetric disposition of ions composing the crystal lattice. The most famous among ceramic ferroelectrics is barium titanate with a perovskite structure. Below its Curie point (120 °C) barium titanate becomes spontaneously polarized due to the displacement of gravity centres of titanium ions relative to oxygen ions composing their octahedral surrounding. It should be noted that in its polarized state every ferroelectric possesses piezoelectric activity, i.e. the ability to be polarized by elastic deformation and, inversely, to be deformed in the external electric field.

In ferroelectrics similar to $BaTiO_3$ the domain structures are formed due to spontaneous parallel orientation of TiO_6 distorted octahedra. The size of domains differs from each other and usually reaches several tens or hundreds of angstroms. Within the limits of an individual domain the dipoles are polarized along one crystallographic axis.

TABLE 3.3 Application of the cryochemical methods to ferrite preparation.

Ferrite composition	*Precursor*	*Applied techniques*	*Heat treatment*	*Main results*	*Reference*
$Ni_{0.34}Zn_{0.66}Fe_2O_4$	Sulphates	Freeze-drying of aqueous solutions	Controlled humidity treatment at room temperature. Decomposition at 850 °C, sintering at 1200–1230 °C	The strong influence of hydrate composition of the dried product on the frequency dependence of magnetic permeability of ferrite was observed. The optimum storage and processing conditions are proposed	Shlyakhtin, 1985
$Ni_{0.5}Zn_{0.5}Fe_2O_4$ $NiFe_2O_4$	Oxalates	Freeze-drying of precipitate of complex oxalates $MFe_2(C_2O_4)_3 * nH_2O$ (M = Ni or Ni + Zn)	Heating at 300 °C, 3 h + 500 °C, 3 h, in air	Single phase target product (spinel) is obtained at rather low temperatures	Langbein and Fischer, 1991
$Ni_{1-x}Zn_xFe_2O_4$ $Mn_{1-x-y}Zn_xFe_{2+y}O_4$	μ-oxo-acetate-complexes	Freeze-drying of complex acetates solutions of $[Fe_2MO(CH_3COO)_6(H_2O)_3]*nH_2O$ (M = Ni + Zn or Mn + Zn + Fe)	Annealing at 400 °C for 1–5 h	Single phase target product (spinel) is obtained at reduced temperatures	Langbein *et al.*, 1991
$Co_{0.003}Zn_{0.567}Ni_{0.430}Fe_2O_4$	Sulphates	Freeze-drying of aqueous solutions	Decomposition: 850 °C, 2 h; sintering: 1100–1150 °C, 2 h	Significant lowering of H_c values and enhancement of magnetic induction B_s at high B_r/B_s ratio due to high homogeneity of microcomponent distribution	Pershin, 1977
$Co_{0.003}Zn_{0.567}Ni_{0.430}Fe_2O_4$	Double sulphates $(NH_4)_2SO_4 * MSO_4 * 6H_2O$ (M = Co + Zn + Ni + Fe)	Freeze-drying of aqueous solutions	Fluidized bed decomposition at 700–850 °C, sintering at 1100–1150 °C, 2 h	Increased initial magnetic permeability resulted from improved homogeneity of microcomponent distribution	Sokolov, 1978; Tretyakov and Oleinikov, 1980
$(Co_{0.003}Zn_{0.567}Ni_{0.430})Fe_2O_4$	Double sulphates $(NH_4)_2SO_4 * MSO_4 * 6H_2O$ (M = Co + Zn + Ni + Fe)	Freeze-drying in vibrofluidized layer (amplitude 0.1 mm and frequency 20 Hz)	Fluidized bed decomposition at 850 °C; sintering at 1150 °C, 2 h	Significant cutting of freeze-drying duration without damage of the product homogeneity	Oleinikov *et al.*, 1978
$Ni_{0.5}Zn_{0.5}Fe_2O_4$	Sulphates	Freeze-drying of aqueous solutions	Decomposition at 850–950 °C for 2 h; powder disintegration by cryogrinding, sintering at 1200 °C for 2 h	Cryomilling is proved to be effective for density increase	Tesker, 1986
$Ni_{0.36}Zn_{0.64}Fe_2O_4$	Sulphates	(a) Cryoextraction from aqueous solutions by acetone or propanol at 248–263 K (b) Freeze-drying of aqueous solutions (c) Cryoprecipitation from aqueous solutions by alcohol solution of oxalic acid at 263 K	(a, c) Atmospheric drying at room temperature, decomposition: 850 °C, 40 min; sintering: 1150 °C, 4 h (b) Dehydration at 80 °C, 2 h, N_2; decomposition: 900 °C, 40 min, sintering: 1150 °C, 4 h	Properties of ferrites obtained using cryoextraction and freeze-drying are comparable with each other. Application of cryoprecipitation promotes enhancement of initial magnetic permeability at $f = 0.1–5.0$ MHz	Mozhaev, 1988; Mozhaev *et al.*, 1982

TABLE 3.3 *(Continued)*

Ferrite composition	*Precursor*	*Applied techniques*	*Heat treatment*	*Main results*	*Reference*
(a) $Ni_{0.5}Zn_{0.5}Fe_2O_4$ (b) $Li_{0.5}Fe_{2.5}O_4$	Hydroxonitrates $[Fe(OH)_{3-x}(NO_3)_x]/$ $x = 0.02–0.1/$ (a) Zn + Ni nitrates or (b) Li nitrate	Freeze-drying of colloidal solutions, obtained by anion exchange treatment of nitrate solution	(a) Decomposition at 600 °C for 1 h; (b) decomposition at 550 °C for 1 h	Gradual temperature decrease of spinel formation	Sharikov, 1991
$Ni_{0.36}Zn_{0.64}Fe_2O_4$ +0.005–0.02 CaO, $Ni_{0.36}Zn_{0.64}Fe_2O_4$ +0.002Al_2O_3, $Ni_{0.36}Zn_{0.64}Fe_2O_4$ +0.005–0.02TiO_2	Ferrite powder; solutions of $Ca(NO_3)_2$, $Al_2(SO_4)_3$ or $TiOC_2O_4$	Cryoimpregnation of ferrite powders	Decomposition at 900 °C, 1 h; sintering at 1200 °C, 1 h	Improvement of $\mu_0(f)$ dependence for sintered ceramics	Shabatin *et al.*, 1984; Shabatin, 1986
(a) $Ni_{0.3}Zn_{0.7}Fe_2O_4$ +0.01–0.02Fe_2O_3 (b) $Ni_{0.3}Zn_{0.7}Fe_2O_4$+ 0.01–0.03$CoFe_2O_4$ (c) $Ni_{0.3}Zn_{0.7}Fe_2O_4$+ 0.01–0.02WO_3	(a, b) Sulphates (c) ferrite powder + solution of $(NH_4)_6W_7O_{24}$	(a, b) Freeze-drying of aqueous solution (c) cryoimpregnation of ferrite powders	Decomposition at 800 °C, 4 h; sintering at 1200 °C, 1–6 h	Better $\mu_0(f)$ characteristics for sintered ferrite, decrease of optimum dopant concentration	Shlyakhtin, 1981
$Ni_{0.5}Zn_{0.5}Fe_2O_4$+ different dopants, including MnO, Fe_2O_3, CoO, NiO, ZnO, CuO, $CuFe_2O_4$, WO_3, TiO_2	Ferrite powder; corresponding solutions of sulphates, $TiOC_2O_4$ and $(NH_4)_6W_7O_{24}$	Cryoimpregnation of ferrite powders	Decomposition at 850 °C, 1 h; sintering at 1250 °C, 1 h	Cryoimpregnation resulted in better control of grain growth during sintering and final magnetic properties enhancement due to better dopant distribution	Bashkuev, 1985; Tesker, 1986
$Zn_{0.03}Mg_{0.27}Mn_{0.9}$ $Fe_{1.8}O_{4+\gamma}$	Individual and double sulphates $(NH_4)_2SO_4{}^*$ $MSO_4{}^*$ $6H_2O$ (M = Zn + Mg + Mn + Fe)	Freeze-drying of aqueous solutions	Decomposition at 850–950 °C, 1–5 h; sintering at 1100–1350 °C, 2–5 h	Possibility of precise controlling coercive force in the wide range of values at $B_r/B_s > 0.95$	Oleinikov, 1988
$CoFe_2O_4$	Double sulphates	Freeze-drying of aqueous solutions	Decomposition at 850–1000 °C, 1.5 h	Lower stacking faults concentration and smaller size of coherent X-ray scattering areas in spinel structure of the product in comparison with other synthesis techniques	Fadeeva *et al.*, 1978; Voinov, 1976
$Ni_{0.90}Cu_{0.10}Fe_{1.75}$ $Al_{0.15}Mn_{0.02}O_4$	Sulphates	Freeze-drying of aqueous solutions	Decomposition at 900 °C, 8 h; sintering at 1000–1250 °C, 4–20 h	Accurate control of the grain size between 4 and 40 μm	Rigterink, 1972
$Ni_{0.8}Mg_{0.15}Co_{0.03}$ $Mn_{0.04}Fe_{1.98}O_4$	Sulphates	Freeze-drying of aqueous solutions frozen by hexane at 195 K	Decomposition at 1000 °C, 3 h; sintering at 1240 °C, 2 h or hot pressing at 1025–1075 °C	Low porosity and uniform grain size distribution in sintered ceramics	de Lau *et al.*, 1975
$Li_{0.5}Fe_{2.5}O_4$	Aqueous solution of $Fe(NO_3)_3$ and Li_2CO_3	Freeze-drying of aqueous solutions	Decomposition at 450–640 °C, sintering at 1000 °C	Lowering of sintering temperature without dopants	Bermejo *et al.*, 1994
$Li_{0.3}Zn_{0.4}Fe_{2.3}O_4$	Aqueous solution of $Fe(NO_3)_3$, $Zn(NO_3)_2$ and Li_2CO_3	Freeze-drying of aqueous solutions	Decomposition at 300–700 °C	Temperature decrease of spinel formation and preparation of ultrafine ferrite particles	Bermejo *et al.*, 1995

TABLE 3.3 *(Continued)*

Ferrite composition	*Precursor*	*Applied techniques*	*Heat treatment*	*Main results*	*Reference*
$Li_{0.5}Fe_{2.5}O_4$	Formates	Freeze-drying of solution	Decomposition at 250–700 °C	Spinel formation at 250 °C and superstructure formation at 700 °C	Bonsdorf *et al.*, 1995
$Li_{0.5}Fe_{2.5}O_4$	Lithium oxalate $+[(NH_4)_3Fe(C_2O_4)_3]$	(a) Freeze-drying of aqueous solution (b) ice cryoextraction by saturated solution of $Li_2C_2O_4$ in acetone	Decomposition at 500 °C, 2 h	The cryoextraction products are similar to freeze-drying products	Mozhaev, 1988
(a) $Li_{0.5}Fe_{2.5}O_4$ (b) $Li_{0.5}Mn_{0.15}Fe_{2.35}O_4$	(a) Oxalates (b) Citrates	Freeze-drying of aqueous solutions	Decomposition at 250 °C, 16 h followed by calcination at 400–1035 °C for 1 h	Single phase formation at 400 °C and possibility to control the grain growth by annealing at preselected temperatures	Johnson *et al.*, 1974
$LiFeO_2$	Hydroxonitrate $[Fe(OH)_{3-x}(NO_3)_x]$, $x = 0.02$–0.1	Freeze-drying of colloidal solutions obtained by anion exchange treatment of nitrate solution	Decomposition of freeze-dried product in mixture with Li_2CO_3 at 400–600 °C	Low temperature formation of metastable α-$LiFeO_2$ modification usually formed at temperature above 672 °C	Sharikov, 1991
$Li_{0.470}Cu_{0.030}Co_{0.004}Fe_{2.496}O_4 + 0.01Bi_2O_3$	Double sulphates $(NH_4)_2SO_4$* MSO_4*$6H_2O$ (M = Cu + Co + Fe)	Freeze-drying of aqueous solution	Decomposition of freeze-dried product mixed with Li_2CO_3 at 750 °C, 1 h + 800 °C, 1 h (CO_2) with intermediate grinding, sintering with Bi_2O_3 at 960–1000 °C	High density of magnetic ceramics achieved at the temperatures as low as 1000 °C to avoid phase decomposition upon sintering	Bich *et al.*, 1983
(a) $Li_{0.8}Fe_{2.5}O_4$ +0.01NiO (b) $Na_{0.036}Li_{0.464}Fe_{2.5}O_4 + 0.01NiO$	Double sulphates $[(Li, Na, NH_4)_2SO_4]$* MSO_4* $6H_2O$ (M = Fe + Ni)	Freeze-drying of aqueous solutions	Decomposition of freeze-dried product in fluidized layer at 800–1000 °C, 20 min	Considerable lowering of synthesis temperature	Anastasyuk, 1972; Tretyakov *et al.*, 1973
$Y_3Fe_5O_{12}$	Nitrates	Cryoprecipitation by aqueous ammonia at 238 K followed by freeze-drying	Decomposition of freeze-dried product at 1100 °C, 1 h + hot pressing 1150–1200 °C, 15 min	High density of ceramics (99.5%) accompanied by the low grain size (<2 μm) and considerable increase of threshold power ΔH_k	Shabatin *et al.*, 1983; Shabatin, 1986; Mozhaev, 1988
$Bi_{3-2x}Ca_xFe_{5-x}V_xO_{12}$	Nitrates + $VOSO_4$	Freeze-drying followed by cryoprecipitation by aqueous ammonia at 243 K	Decomposition of freeze-dried product at 1100–1200 °C, 15 min	95% yield of garnet-structure product relative to 75% yield for mixed oxides technique	Shabatin *et al.*, 1983; Shabatin, 1986; Mozhaev, 1988
$Y_3Fe_5O_{12}$	Hydroxonitrate $[Fe(OH)_{3-x}(NO_3)_x]$, ($x = 0.02$–0.1); $Y(NO_3)_3$	Freeze-drying of colloidal solutions obtained by anion exchange treatment of nitrate solution	(a) Decomposition of freeze-dried product at 700–900 °C, 1 h (b) Decomposition of freeze-dried product at 500 °C, 1 h + cold pressing at $P > 6$ GPa	Considerable lowering of the temperature of garnet-structure formation: (a) up to 700 °C (b) up to 500 °C	Sharikov, 1991; Vertegel *et al.*, 1996
$BaFe_{12}O_{19}$	(a) Nitrates (b) Nitrates + complexing agents like citric acid	Freeze-drying of aqueous solutions	Decomposition of freeze-dried products at 700–1200 °C, 5 min	Preparation of single-domain magnetic particles of pure barium hexaferrite in case (b)	Borisova *et al.*, 1992

TABLE 3.3 *(Continued)*

Ferrite composition	*Precursor*	*Applied techniques*	*Heat treatment*	*Main results*	*Reference*
$BaFe_{12}O_{19}$	Hydroxonitrate $[Fe(OH)_{3-x}(NO_3)_x]$ $(x = 0.02–0.1)$; $Ba(NO_3)_2$	Freeze-drying of colloidal solutions obtained by anion exchange treatment of nitrate solution	Decomposition of freeze-dried products at 550–800 °C, 1 h	Formation of single-phase barium hexaferrite at temperature ~750 °C	Sharikov, 1991
$Ba_3Co_2Fe_{24}O_{41}$ ('Co_2Z'-phase)	Hydroxonitrate $[(Fe(OH)_{3-x}(NO_3)_x]$, $(x = 0.02–0.1)$; $Ba(NO_3)_2$; $Co(NO_3)_2$	Freeze-drying of colloidal solutions obtained by anion exchange treatment of nitrate solution	Decomposition of freeze-dried products at 550–1000 °C, 1 h	Formation of pure 'Co_2Z'-phase at temperature below 900 °C	Sharikov, 1991
Fe_3O_4	Ferrous and ferric sulphates coprecipitated by NaOH at pH = 11	Freeze-drying of hydroxides mixture	Decomposition conditions are not described	Non-aggregated ultrafine particles for magnetic recording	Sugano and Karita, 1985a
Fe_3O_4	Ferrous and ferric sulphates coprecipitated by NaOH at pH = 11	Freeze-drying of hydroxides mixture	Decomposition conditions are not described	Non-aggregated ultrafine particles for magnetic fluids	Sugano and Karita, 1985b
γ-Fe_2O_3	Hydroxychlorides $Fe(OH)_{3-x}Cl_x$ $(x = 0.05–0.15)$	Freeze-drying of colloidal solutions obtained by anion exchange treatment of chloride solution	Decomposition of freeze-dried product at 300–350 °C, 1 h + 400–450 °C, 1 h	Formation of non-aggregated ultrafine particles stable up to 550 °C	Sharikov, 1991
Co-doped γ-Fe_2O_3	γ-Fe_2O_3; Na stearate	Freeze-drying of suspension	No heat treatment	Homogenized distribution of stearate for preparation of magnetic recording medium	Sato, 1985

At present, several hundreds of ferroelectric materials are known and among them there are a wide group of oxide compounds with the structure of deformed (non-cubic) perovskite.

Ferroelectric oxides are applied in capacitors due to their extremely high dielectric permeability at temperatures close to T_c (Figure 3.3). For this reason, to enhance dielectric permeability it is necessary to develop the materials with Curie point close to the temperature of the environment. Thus, the Curie point of $BaTiO_3$ (Figure 3.3) can be considerably decreased, and transition temperature range can be extended by partial exchange of Ba^{2+} or Ti^{4+} for other cations: the exchange of Ba^{2+} for Ca^{2+} results in compression of unit cell and decrease of T_c; the exchange of 'active' Ti^{4+} ions for other 'inactive' four-charged ions, for example Zr^{4+} and Sb^{4+}, leads to a sharp decrease of T_c.

Spontaneous polarization is also observed for antiferroelectrics and its nature is similar to that in ferroelectrics. Individual dipoles in antiferroelectrics are ordered in such a way that each dipole is antiparallel to neighbouring ones. As a result, the spontaneous polarization of the material itself is equal to zero. Above the antiferroelectric Curie point the materials are normal para-electrics. Lead zirconate $PbZrO_3$ (233 °C), sodium niobate $NaNbO_3$ (638 °C) and ammonium dihydrophosphate $NH_4H_2PO_4$ (−125 °C) are examples of substances with antiferroelectric properties (the values in parentheses correspond to Curie points).

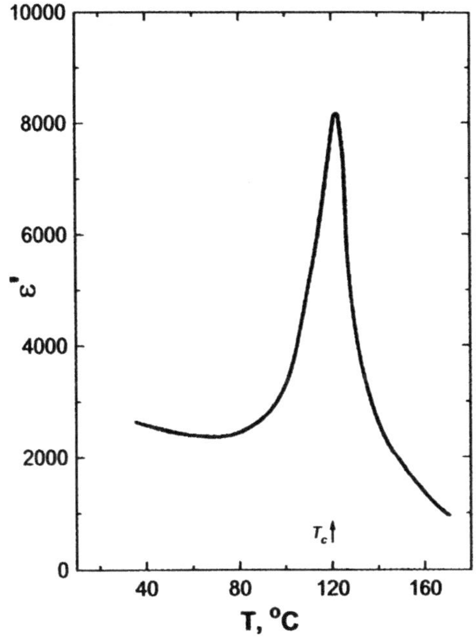

FIGURE 3.3 Temperature dependence of dielectric permeability for ceramic $BaTiO_3$.

The electrical properties of ferroelectrics and antiferroelectrics differ considerably. Because in antiferroelectric state there is no spontaneous polarization, the hysteresis phenomena in alternative electric field are not observed despite the great increase of dielectric permeability ε near T_c for these materials (for $PbZrO_3$ ε is about 100 at 200 °C and about 3000 at 230 °C). Sometimes, the antiparallel dipoles ordering in antiferroelectrics and the parallel one in ferroelectrics have similar stability. In these cases, little variations in external conditions can lead to phase transition, for example under external electric field $PbZrO_3$ passes from antiferroelectric state to ferroelectric one (Figure 3.4a); the value of electrical potential determining the transition depends on the temperature. The behaviour of polarization in the external field corresponding to this case is represented in Figure 3.4b: at low temperatures there is no hysteresis because under these conditions $PbZrO_3$ is antiferroelectric, while in stronger fields (of both directions) $PbZrO_3$ passes into the ferroelectric state and a hysteresis loop appears.

The most commonly used ferroelectric materials are those based on the system $PbZrO_3$–$PbTiO_3$ (PZT) within the composition range close to the so-called morphotropic boundary, i.e. the phase boundary separating the areas corresponding to solid solutions with rhombohedral and tetragonal structures.

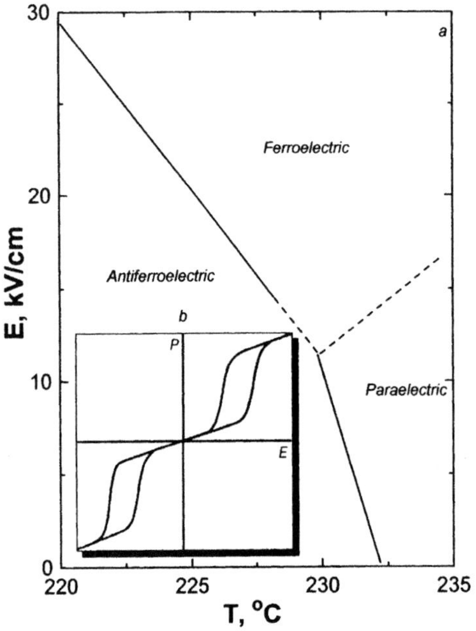

FIGURE 3.4 Voltage dependence of antiferroelectric–ferroelectric transition for $PbZrO_3$ and its polarization behaviour upon this transition.

They are characterized by high Curie point (300–400 °C) and high coefficients of electromechanical interplay, as well as by good polarizability (up to 50 mcC/ cm^2). The properties of PZT ceramics can be varied in a wide range by doping. For example, substitution of some Pb^{2+} ions by La^{3+} ions suppresses the formation of lead defects and facilitates the manufacturing of materials transparent within the whole visible spectral range.

In all the cases, the properties of ferroelectric ceramics are very sensitive towards preparation techniques, and this has been the main reason for the wide application of cryochemical techniques in producing ferroelectric materials (Table 3.4).

3.4 SOLID-STATE IONICS

Ceramic electrolytes comprise a universal group of ceramic materials with high ionic mobility and corresponding high ionic conductivity. In contrast with the classic liquid electrolytes, the conductivity of ceramic electrolytes is usually unipolar and caused by disorder of one of the crystal sublattices. The most

TABLE 3.4 Application of cryochemical methods to ferroelectric materials preparation.

Ferroelectric composition	*Precursor*	*Applied technique*	Heat treatment	*Main results*	*References*
$BaTiO_3$	$BaCl_2$; $TiCl_4$	Co-precipitation by aqueous NH_3 and H_2O_2 followed by (a) freeze-drying of aqueous suspension; (b) ethanol drying and freeze-drying of suspension in tert-butyl alcohol	Decomposition at 150°, 30 min, 560°, 20 min and 680°, 60 min. Sintering at 1045–1275° for 25–600 min	Density over 90% th.d. is achieved at the temperature as low as 1050 °C. Reduced aggregation of $BaTiO_3$ powder obtained by (b) procedure resulted in significant acceleration of sintering	Xue *et al.*, 1986
$PbZr_{0.53}Ti_{0.47}O_3$	$Pb(NO_3)_2$ $ZrO(NO_3)_2$ $TiO(NO_3)_2$	Freeze-drying of solutions frozen by hexane at 179 K after adding NH_4OH up to pH = 1.5	Decomposition at 450–600 °C, 10–60 min	Considerable increase of interaction degree (~90%) at moderate temperatures resulted in lowering Pb losses upon subsequent sintering	Tretyakov *et al.*, 1977; Levina *et al.*, 1978; Tretyakov and Fokin, 1982
$PbZr_{0.53}Ti_{0.47}O_3$	$Pb(NO_3)_2$ $ZrO(NO_3)_2$ $TiO(NO_3)_2$	Freeze-drying of solutions frozen by hexane at 179 K after adding NH_4OH up to pH = 1.5	Decomposition at 550 °C, 10 h and 700 °C, 2 h; sintering at 1220 °C, 2 h in $PbZrO_3$ powder	Low general porosity (<4%) and no open porosity in sintered ceramics, better values of tgδ, ε, K_p and d_{31}	Fokin, 1977
$PbZr_{0.53}Ti_{0.47}O_3$	$Pb(NO_3)_2$ $ZrO(NO_3)_2$ $TiO(NO_3)_2$	Anion exchange treatment of nitrate solution followed by freeze-drying	Decomposition of freeze-dried products at 450–550 °C, 10–60 min	Single-phase perovskite is obtained at the temperature as low as 450 °C	Fedotov, 1984
$Pb_{1-x}Bi_{2x/3}Zr_{0.53}Ti_{0.47}O_3$ ($x = 0.0025$–0.011)	$Pb(NO_3)_2$ $Bi(NO_3)_3$ $ZrO(NO_3)_2$ $TiO(NO_3)_2$	Cryoprecipitation by aqueous ammonia at 243 K followed by freeze-drying	Decomposition of freeze-dried product at 350 °C, 3 h, sintering at 1100 °C	Reduced temperature of sintering and enhanced grain growth in ceramics due to homogeneous distribution of the dopant	Levina *et al.*, 1980
$Pb_{0.94}La_{0.09}Zr_{0.65}Ti_{0.35}O_3$	$Pb(NO_3)_2$ $La(NO_3)_3{*}6H_2O$ $[ZrO(NO_3)_2{*}nH_2O]$ $TiO(OC_3H_7)_4$	Freeze-drying of true solutions	Decomposition at 550 °C, 16 h; annealing at 950 °C, 16 h in sealed container, ball milling and hot pressing in O_2 at 1250 °C, 16 h	Optically transparent ceramics with improved optical transmission value at $\lambda = 0.50\ \mu m$ (65%)	Thomson, 1974
$Pb_{0.92}La_{0.08}Zr_{0.59}Ti_{0.41}O_3$	$Pb(NO_3)_2$ $La(NO_3)_3{*}6H_2O$ $[ZrO(N_0)_2{*}nH_2O]$	Freeze-drying of H_2O_2 containing solution treated by aqueous ammonia up to pH = 3.0	Decomposition at 550 °C, 10 h, sintering at 1220 °C, 2 h in the media of $PbZrO_3$ powder	Higher density of ceramics and optical transmission value at $\lambda = 0.63\ \mu m$ (55% *vs.* 45% for ceramics obtained by ordinary technology)	Fokin *et al.*, 1979
$Pb_{0.92}La_{0.08}Zr_{0.59}Ti_{0.41}O_3$	Nitrates	Cryoprecipitation by aqueous ammonia at 256–270 K followed by freeze-drying	Atmospheric drying at 100–110 °C, decomposition at 530–550 °C, sintering at 1000 °C	Enhanced sintering ability at relatively low temperatures	Startseva *et al.*, 1977; Startseva, 1981; Tretyakov, and Fokin, 1982
$PbZr_{0.53}Ti_{0.47}O_3$	Zirconyl and titanyl hydroxo-nitrates; $Pb(NO_3)_2$	Freeze-drying of bulk frozen colloidal solution obtained by anionite treatment of nitrate solution	Decomposition at 450–550 °C, 30 min, sintering of products framework at 1000–1100 °C, 0.5–2 h	Extremely porous (90–96% porosity) strong ceramic framework. Polymerization of monomer, filled this framework, results in flexible PZT-based composite with outstanding piezosensitivity	Fedotov, 1984; Fedotov *et al.*, 1993

TABLE 3.4 (*Continued*)

Ferroelectric composition	*Precursor*	*Applied technique*	*Heat treatment*	*Main results*	*References*
$LiNbO_3$	$Li_2C_2O_4$, $H_3[NbO(C_2O_4)_3]$	Freeze-drying of aqueous solution	Decomposition at 500–1000 °C, 0.5–4 h	Reduction of $LiNbO_3$ formation temperature to 500 °C compared to 750 °C for ceramic method	Gallagher and Schrey, 1970
$MNbO_3$ (M = Li, Na, K); $Na_{0.95}Li_{0.05}NbO_3$; $Na_{0.5}Li_{0.5}NbO_3$	$Li_2C_2O_4$; $Na_2C_2O_4$; $K_2C_2O_4$; $H_3[NbO(C_2O_4)_3]$	Freeze-drying of aqueous solution	Decomposition of freeze-dried product at 400–600 °C, 1 h, sintering at 1100 °C, 4 h, pure oxygen	Considerable lowering of the temperature of perovskite formation; high density (>98.5%) and zero water adsorption of ceramics; outstanding threshold voltage value	Shishov, 1978
$PbMg_{0.33}Nb_{0.67}O_3$	$Pb(OAc)_2$ $Mg(OEt)_2$ $Nb(OEt)_5$	Freeze-drying		Very high homogeneity of the final product due to avoiding intermediate pyrochlore phase formation	Ho *et al.*, 1995

common oxide materials with cationic conductivity are compounds with the β-alumina structure.

High cationic conductivity combined with chemical and thermal stability of the polycrystalline ceramics based on β-Al_2O_3 makes them exclusively promising as solid electrolytes in various electrochemical systems, including solid-phase batteries and accumulators with high-energy capacity, gas analysers and memory elements.

β-Alumina is one of the sodium polyaluminates $Na_2O*11Al_2O_3$ ($NaAl_{11}O_{17}$); its hexagonal layer structure consists of two-dimensional spinel blocks, each of them being formed by four layers of closely packed oxygen ions; the octahedral and tetrahedral lattice sites are occupied by aluminium ions as in spinels. The parallel spinel blocks are separated by a large layer where bonding oxygen and sodium ions are located. Both kinds of ions do not form a close packing and fill only 82 and 35% of available sites, respectively (Figure 3.5). For this reason, the mobility of sodium ions along the layer between spinel blocks is enhanced. It is noticeable that the diffusion coefficient of sodium ions in single crystals of β-Al_2O_3 (10^{-5} cm^2/s at 300 °C) is practically the same as the diffusion coefficient of sodium in salt melts ($2*10^{-5}$ cm^2/s at 300 °C in sodium nitrate).

It was established that the formula usually attributed to β-alumina actually is hypothetical. In fact, the component combination according to the ratio 1:11 never corresponds to a single phase and under equilibrium conditions represents a mechanical mixture of α-Al_2O_3 and β-Al_2O_3. This means that β-alumina is a typical non-stoichiometric phase, stable within wide concentration range, with excess of sodium oxide relative to $NaAl_{11}O_{17}$ composition.

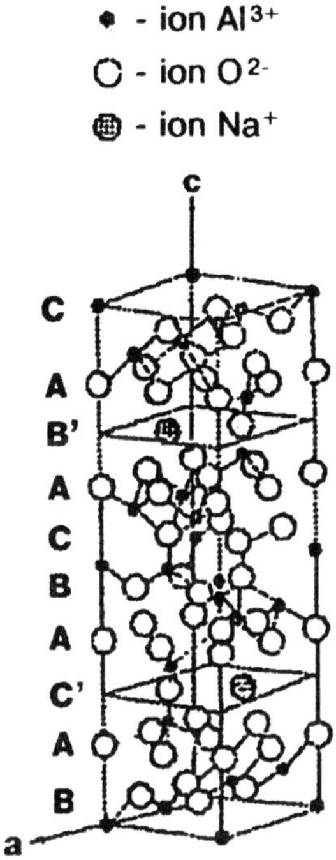

FIGURE 3.5 Crystal lattice of β-alumina.

In β-alumina sodium ions possessing high mobility can be completely or partially replaced by other monovalent or bivalent cations. Using the technique of ionic exchange in salt melts, the single-phase β-alumina samples $MO_x*nAl_2O_3$ were obtained, with lithium, potassium, rubidium, cesium, silver, thallium, gallium, zinc, cadmium, iron and ammonium as M.

In most cases the application of β-alumina as a solid electrolyte demands highly consolidated polycrystalline ceramics possessing maximum conductivity, mechanical resistance, chemical and thermal stability. The synthesis of such ceramics is difficult from a technological point of view, because the temperatures of formation (1200–1500 °C) and sintering (1700–2000 °C) of single-phase β-alumina are very high and uncontrolled losses of sodium oxide are possible under these conditions.

All these circumstances on the whole stimulate a wide application of cryochemical technologies for the preparation of ceramic electrolytes. The attempts to synthesize β-alumina as well as NASICON and other similar materials (see West, 1984) characterized by high values of cationic conductivity for Li^+ and Na^+, respectively, are given in Table 3.5.

TABLE 3.5 Application of the cryochemical methods to solid-state ionics preparation.

Material composition	*Precursor*	*Applied techniques*	*Heat treatment*	*Main results*	*References*
$Na_2O*6Al_2O_3$ ($Na_2Al_{12}O_{19}$)	Na_2CO_3; $Al_2(SO_4)_3*16H_2O$	Freeze-drying of aqueous solution	Decomposition at 1200–1400 °C, 1 h, hot pressing at $T = 350$–$1500\,°C$, $P = 28$–41 MPa, in vacuum	Lowering of β-alumina phase formation temperature compared to spray-dried samples, synthesis of 98% density β-alumina ceramics	Green and Hutchison, 1980
(a) $Na_2O*6Al_2O_3$ (b) $Na_2O*6Al_2O_3 + 0.03MgO$ (c) $Na_2O*6Al_2O_3 + 0.015Li_2O$	Na_2CO_3; $Al_2(SO_4)_3*16H_2O$; $MgSO_4*7H_2O$; Li_2CO_3	Freeze-drying of aqueous solutions	Decomposition at 1250 °C, 5 h, sintering at 1600 °C, 5–30 min	Dense (98% th.d) ceramics with ≥90% β''-Al_2O_3 content (samples b and c)	Pekarsky and Nicholson, 1980
$1.02Na_2O*6Al_2O_3 + 0.03MgO$	Na_2CO_3; $Al_2(SO_4)_3*16H_2O$; $MgSO_4*7H_2O$	Freeze-drying of aqueous solutions	Decomposition at 1330 °C, 5 h, sintering at 1600 °C, 8–15 min	Very dense (~100% th.d) ceramics with 85–89% β''-Al_2O_3 content, grain size 0.1–0.3 μm, ionic resistivity $R_{300\,°C} = 8.1$ Ohm*cm, the activation energy for Na^+ conduction 5.0 kcal/mol	Kuribayashi and Nicholson, 1980
$Na_2O*6Al_2O_3 + 0.02Y_2O_3$	$Na_2C_2O_4$; $Al_2(C_2O_4)_3*9.5H_2O$; $Y_2(SO_4)_3*8H_2O$	Freeze-drying of suspension, obtained by mixing oxalates and sulphate solution	Atmospheric drying at 140 °C under stirring, decomposition at 800 °C, sintering at 1650 °C in sealed container	Dense (94–95% th.d) ceramics with ionic resistivity $R_{300\,°C} = 32$ Ohm*cm. Homogeneous distribution of Y_2O_3 prevents grain growth	Komm *et al.*, 1981
$K_{1+x}Mg_xAl_{11-x}O_{17}$	K_2CO_3; $MgSO_4*7H_2O$; $Al_2(SO_4)_3*18H_2O$	Freeze-drying of aqueous solution	Decomposition at 1200–1600 °C, 2 h	Powder product containing up to 88% β''-Al_2O_3 phase due to homogeneously distributed magnesia	Nagai and Nicholson, 1982
$(K_{1-x}Na_x)_{1+y}Mg_yAl_{11-y}O_{17}$	K_2CO_3; Na_2CO_3; $MgSO_4$; $Al_2(SO_4)_3*18H_2O$	Freeze-drying of aqueous solution	Decomposition at 1300 °C, 4 h, sintering at 1610 °C, 1–3 min	K, Na-β''-Al_2O_3, ceramics with β'' phase content up to 86% and density up to 99% th.d	Nagai *et al.*, 1983
$Na_2O*6Al_2O_3$	$Al(NO_3)_3$; CH_3COONa	Freeze-drying of aqueous solution	Decomposition at 600–1300 °C, 8–12 h	Comparison of phase formation sequence for different synthesis methods	Hodge, 1983
(a) $Na_2O*6Al_2O_3$ (b) $Na_2O*0.175Li_2O*6.19Al_2O_3$	$Na_2C_2O_4$; $Al_2(C_2O_4)_3*9.5H_2O$	Freeze-drying of aqueous solution	Atmospheric drying at 140 °C under stirring, decomposition at 800 °C, sintering at 1650 °C	Reduced temperature of β-Al_2O_3 phase formation due to the better homogeneity of precursor	Komm *et al.*, 1981; Plokhih, 1985
(a) $Na_2O*7Al_2O_3$ (b) $Na_2O*0.175Li_2O*6.19Al_2O_3$ (c) $Na_2O*0.175Li_2O*6.19Al_2O_3$	(a) $Na_2O*8.5Al_2O_3$ (solid), $NaAlO_2$ (in solution) (b) $Na_2O*5Al_2O_3$, $Li_2C_2O_4$, $Al_2(C_2O_4)_3*9.5H_2O$ (c) $[Na_2O*0.264Li_2O*8.5Al_2O_3]$ (solid), $NaAlO_2$ (in solution)	Freeze-drying of suspension, obtained by (a, c) mixing of oxide phase and basic aluminate solution or (b) obtained by mixing of oxide phase and aqueous solution of oxalates	Atmospheric drying at 140 °C, decomposition at 800 °C, sintering at 1620 ± 10 °C, 7 min	(b) Dense (90%) ceramics with $\beta''/\beta' Al_2O_3$ ratio close to 1 (c) very dense (96%) ceramics containing 85% β''-Al_2O_3 with small grain size (0.1 μm). 10 days' boiling of (c) ceramics in conc. H_2SO_4 results in 80% Na for H substitution while mechanic strength of ceramics remains unchanged	Plokhih, 1985

TABLE 3.5 *(Continued)*

Material composition	*Precursor*	*Applied techniques*	*Heat treatment*	*Main results*	*References*
Na_2O*$6Al_2O_3$	$Al(OC_3H_7)_3$	Hydrolysis of $Al(OPr)_3$ solution in CH_3COOH followed by freeze-drying	Decomposition at 1000 °C, soaking in NaOH solution, atmospheric drying at 100 °C (24 h), sintering at 1400 °C (2 h)	Porous Na $\beta'\beta''$-Al_2O_3 ceramics for studies of influence of water vapour on the conductivity of β-alumina	Nagai *et al.*, 1989, 1990
$Li_{2.4}Ti_{3.4}O_8$	$(NH_4)_2[TiO(C_2O_4)_2]$; $Li_2C_2O_4$	Freeze-drying of aqueous solution	Decomposition at 1050–1100 °C; 2 h sintering at 1030 °C + 1 h at 1080 °C with intermediate grinding	Higher density (87% th.d) of single-phase ramsdellite ceramics compared to ceramics from mixed oxides precursor (70% th.d)	Mozhaev *et al.*, 1980
$Li_2Ti_3O_7$	$(NH_4)_2[TiO(C_2O_4)_2]$, $Li_2C_2O_4$	Freeze-drying of aqueous solution	Decomposition 1100 °C, 2 h; hot pressing	Single-phase dense (92% th.d) ramsdellite ceramics with improved reproducibility and uniform bulk distribution of electrophysical parameters	Kolotyrkin *et al.*, 1978; Kolotyrkin, 1980
$Na_5GdSi_4O_{12}$ (NASICON)	Na_2CO_3; Na_2SiO_3; Gd acetate	Freeze-drying of Gd hydroxide suspension in Na silicate/carbonate solution	Thermal decomposition at 530 °C, 24 h; sintering at 975–1050 °C, 5 min–67 h	50 times reduction of sintering time compared to ceramic method. Preparation of dense (99.5%) single phase NASICON ceramics	Bentzen and Nicholson, 1980
$Na_5GdSi_4O_{12}$ (NASICON), $Na_5Gd_{0.9}Ce_{0.1}Si_4O_{12}$, $Na_5Gd_{0.9}Eu_{0.1}Si_4O_{12}$	Na_2CO_3; Na_2SiO_3; Gd, Eu, Ce acetates	Freeze-drying of suspension prepared by mixing of aqueous solutions	Decomposition at 535 °C, 24 h; sintering at 1050–1060 °C, 30 min	Single-phase dense (94% th.d) Ce-substituted NASICON ceramics. High Na^+ conductivity of Eu-doped ceramics	Visco and Kennedy, 1983

3.5 SUPERCONDUCTING CERAMIC MATERIALS

The discovery of high-temperature superconductivity (HTSC) has spurred enormous scientific and practical interest to elucidate the nature of the phenomenon, as well as to find practical application for it (Larbalestier, 1988; Simon, 1989). It soon became evident that success in both of these directions was determined by the techniques applied for HTSC materials preparation and, especially, by the precursor powders for HTSC ceramics, coatings and composites.

At first, the so-called 'ceramic technique' was used. The process involved thorough mechanical mixing of oxides (for the 123 phase – yttrium and copper oxides with barium carbonate) and repeated 'beat and heat' cycles, to achieve complete solid-phase interaction of the reagents. In some cases, oxides were replaced by carbonates, nitrates or other salts (Metlin and Tretyakov, 1994).

From 1987 to the present time many investigations on HTSC technologies dealt with the development and application of different chemical procedures for powder preparation (Horowitz *et al.*, 1989; Rao *et al.*, 1993).

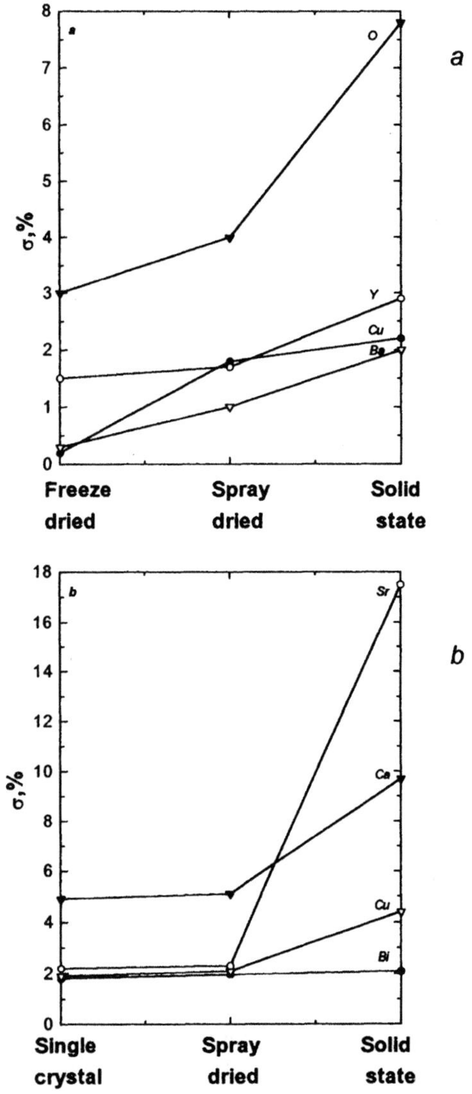

FIGURE 3.6 Standard deviations of SNMS analyses for $YBa_2Cu_3O_x$ and $Bi_2Sr_2CaCu_2O_{8+y}$ ceramics synthesized by different methods.

The efficiency of chemical techniques in increasing of ceramic homogeneity is illustrated in Figure 3.6 using values of standard deviation in the analysis of composition of Y- and Bi-containing ceramics by sputtered neutral mass spectrometry (SNMS) (Strelkov *et al.*, 1993).

Lower standard deviation implies greater reproducibility of layer-by-layer analysis of bulk samples. In turn, the reproducibility depends on chemical homogeneity. From the figure it can be seen that the chemical routes are preferable.

The use of chemical techniques for the preparation of powders may be useful even for the most popular melting techniques which are used to produce HTSC powders despite the considerable smoothing of morphological differences of powders with different backgrounds due to complete or partial melting. In particular, while investigating the nature and influence of various pinning centres on superconducting properties, much attention should be paid to the purity of the sample or to controlled content (size and distribution) of impurities; most easily this can be ensured by chemical techniques of synthesis.

At present several families of oxide HTSC phases are known, including

$La_{2-x}M_xCuO_4$ (M = Ba, Sr)
$Y_2Ba_4Cu_{6+n}O_{14+n}$ (n = 0–2)
$RE_2Ba_4Cu_{6+n}O_{14+n}$ (n = 0–2)
$Bi_2(Ca, Sr)_{n+1}Cu_nO_{2n+4}$ (n = 1–3)
$TlBa_2Ca_{n-1}Cu_nO_{2n+3}$ or $Tl_2Ba_2Ca_{n-1}Cu_nO_{2n+4}$ (n = 1–3)
$PbSr_2ACu_2O_6$ or $Pb_2Sr_2ACu_3O_8$ (A = Y, RE, RE + Sr, RE + Ca)
$RE_{2-x}Ce_xCuO_4$, $Sr_{1-x}M_xCuO_2$ (M = Sr, RE)

and

$HgBa_2Ca_{n-1}Cu_nO_{2n+3}$

They are all complex oxides and include oxides of copper, alkaline–earth elements, yttrium, rare-earth elements, bismuth, thallium, lead or mercury.

At the same time, the presence of such components as copper or alkaline–earth materials in all HTSC phases allows us to anticipate the applicability of various techniques. It is evident that co-precipitation using ammonia can disturb copper stoichiometry, while co-precipitation of carbon-containing salts is always accompanied by the formation of alkaline–earth–metal carbonates requiring prolonged times of thermal treatment. To prevent the losses of some components (thallium, mercury, lead) due to their high volatility, the synthesis is carried out by treating complex oxide precursors, prepared by chemical techniques, in vapours of volatile component. This technique was developed for thallium systems (Tang *et al.*, 1992) and was shown to be highly efficient for the synthesis of mercury-containing HTSC (Putilin *et al.*, 1993) materials.

Because the values of critical current density J_c ($J_c = f(T, B)$) were proved to be dramatically dependent on the microstructure peculiarities of HTSC-ceramics, many attempts have been made to apply different processes of cryochemical technology for production of different kinds of superconducting ceramics (Table 3.6).

3.6 ELECTRODE MATERIALS

Electrodes are indispensable elements of every electrochemical system designed to generate electric energy or to consume it in order to obtain certain chemical products by electrolysis. The common requirements of electrode materials are good electrical conductivity, high catalytic activity and selectivity, sufficient mechanical rigidity, as well as stability of these parameters in the course of time. The electrode materials used as components of electrochemical sensors, in addition, should possess high sensibility towards variations of activities (concentrations) of the particles (molecules, ions) to be detected. Taking into account these requirements it is reasonable to expect that the behaviour of electrode materials will be sensitive towards the technique of their preparation; this is the reason for the attempts to use the processes of cryochemical synthesis. From Table 3.7 it follows that some positive effects have been achieved for oxide, halogenide and metallic electrodes.

3.7 MATERIALS FOR CATALYSTS AND SORBENTS

Materials with high catalytic activity and sorbents are extremely variable in their composition, structure and properties. Metals and alloys, clays and salts, oxides, sulphides, nitrides and carbides of metals may act as catalysts. They may be single-phase substances and multiphase mixtures, crystalline, nanocrystalline and amorphous materials. They can have various dielectric, semiconductive and metallic properties. The most important parameters of catalysts are their activity and selectivity which are extremely sensitive to the method of the preparation, and in this sense the advantages of cryochemical technology attract the attention of many scientists. From Table 3.8 one can see that application of cryochemical processes for the preparation of catalysts and absorbents is usually connected with the use of complex oxide systems based on manganites, cobaltites and cuprates, showing synergism of concomitant cations. Nowadays there is some experience in the production of catalytic active borophosphates and alumophosphates, metals and alloys, immobilized enzymes. The cited materials are catalysts in oxidation of CO, methane and methanol, in hydrogenation of aromatic hydrocarbons, in reversible sorption of CO and other processes. Using cryochemical techniques, the temperatures of syntheses of catalysts were

TABLE 3.6 Cryochemical methods for HTSC ceramics preparation.

HTSC composition	*Precursor*	*Applied techniques*	*Heat treatment*	*Main results*	*Reference*
$La_{1.85}Sr_{0.15}CuO_{4-\delta}$, $La_{1.85}Sr_{0.15}Cu_{0.95}M_{0.05}O_{4-\delta}$ (M = Ti, V, Cr, Mn, Fe, Co, Ni, Zn)	Nitrates	Freeze-drying of aqueous solutions	Decomposition at 600 °C, 20 h, vacuum + 640 °C, 2 h, air + 6 h with intermediate grinding. Sintering at 1050 °C, 6 h + 1150 °C, 8 h + cooling 2 K/min up to room temperature	M-substituted cuprate ceramics with homogeneous dopant distribution is synthesized for electrophysical measurements. Density of sintered La–Sr cuprate ceramics is improved compared to mixed oxides method	Payzant and King, 1990
$La_{1.85}Sr_{0.15}CuO_{4-\delta}$	Acetates	Freeze-drying of aqueous solutions	Decomposition at 980 °C, 8 h. Sintering at 700–980 °C	Single phase ceramics with reduced grain size (0.5 μm)	Alconchel *et al.*, 1994
$La_{1.85}Sr_{0.15}CuO_{4-\delta}$	Acetates	Freeze-drying of aqueous solutions	Thermal decomposition at 980 °C, 8 h	Improved homogeneity of ceramics is achieved when heating rates ≥50 K/min was applied	Alconchel *et al.*, 1995a
$La_{2-x}Sr_xCuO_4$ (x = 0, 0.15, 0.33)	Acetates	Freeze-drying of aqueous solutions	Thermal treatment at 980 °C, 12 h, oxygen annealing at 500 °C, 6 h	Single phase ceramics	Alconchel *et al.*, 1995b
$YBa_2Cu_3O_{7-x}$	Nitrates, acetates	Freeze-drying of aqueous solutions	Thermal decomposition at 750–900 °C, O_2, 50 h. Sintering at 890–925 °C, 12–24 h. Annealing in O_2 at 700 °C and 400 °C, 8–12 h	Single phase 123 powder and dense (89% th.d.) ceramics with very sharp resistivity drop and large Meissner effect are obtained	Johnson *et al.*, 1987
$YBa_2Cu_3O_{7-x}$	Nitrates	Freeze-drying of aqueous solutions	Thermal decomposition at 850 °C, 15 min with heating rate over 500 K/min	Superconducting 123 powder with strong Meissner effect is obtained directly by fast thermal decomposition	Shabatin *et al.*, 1987
(a) $YBa_4Cu_5O_z$ (b) $La_{1.8}Sr_{0.15}Ca_{0.05}Cu_{0.95}Ag_{0.05}O_{4-\gamma}$ (c) $YBa_2Cu_3O_{7-x}$	Acetates	Freeze-drying of aqueous solutions	Thermal decomposition at 800 °C, 1 min with high heating rate. Sintering: (a) 1000 °C, 1.5 h, O_2 (b) 1000 °C, 1 h, air (c) 950 °C, 6 h, O_2	High-T_c ceramics with low width of superconducting transition ($\Delta T \leq 3$ K) due to high chemical homogeneity of sintered sample	Kulakov *et al.*, 1987
$RBa_2Cu_3O_{7-x}$ (R = Y, Lu, Eu)	$Cu(NO_3)_2$* $3H_2O$, $R(NO_3)_3$* xH_2O, $Ba(NO_2)_2$	Freeze-drying of aqueous solutions	Thermal decomposition at 825–900 °C, 5–15 min	R-123 high-T_c powders with strong Meissner effect obtained directly by thermal decomposition	Zalishchansky *et al.*, 1988
$YBa_2Cu_3O_{7-x}$	Nitrates	Freeze-drying of aqueous solutions	Thermal decomposition at 900 °C, 18 h, O_2. Sintering at 950 °C, 8–72 h, O_2	High T_c ceramics with improved chemical homogeneity and greater Meissner effect compared to mixed oxides method	Strasik *et al.*, 1988
$YBa_2Cu_3O_{7-x}$	$Y(NO_3)_3$, $Cu(CH_3COO)_2$, $Ba(CH_3COO)_2$	Freeze-drying of colloidal solution obtained by adding of aqueous NH_3 up to pH = 6.8	Thermal decomposition by slow heating to 500 °C. Firing at 750–900 °C in air. Annealing in O_2, 950 °C, 24 h	Grain size of high-T_c ceramics can be varied from 0.5 to 3 μm by changing ageing time to colloidal solution before freeze-drying	Barboux *et al.*, 1988
$YBa_2Cu_3O_{7-x}$	Nitrates	Freeze-drying of aqueous solutions	Thermal decomposition at 500–700 °C. Sintering 900 °C, 20 h + 950 °C, 12 h, O_2. Cooling from 650 °C to room temperature in O_2 for 24 h	123-phase formation at temperature as low as 700 °C	Clayton *et al.*, 1988

TABLE 3.6 (*Continued*)

HTSC composition	*Precursor*	*Applied techniques*	*Heat treatment*	*Main results*	*Reference*
$Yba_2Cu_3O_{7-x}$	$Cu(NO_3)_2*3H_2O$ $Y(NO_3)_3*5H_2O$ $Ba(NO_3)_2$ or $Ba(CH_3COO)_2$	Freeze-drying of aqueous solutions	Thermal decomposition: slow heating up to 600 °C + 825 °C, 12 h. Sintering at 925 °C, 12 h	Improved homogeneity of 123 powder resulted from adjustment of pH of starting solution. Dense ($\rho \geq 90\%$ th.d.) high T_c ceramics with grain size 2–5 μm	Medelius and Rowcliffe, 1989
$YBa_2Cu_3O_{7-x}$	Acetates	Freeze-drying of aqueous solutions	Thermal decomposition at 700 °C, 2 h + 900 °C, 0.5 h	Resistivity of 123 ceramics to atmospheric degradation is higher for cryochemical samples compared to $BaCO_3$-based ceramic method	Wang *et al.*, 1989
$YBa_2Cu_3O_{7-x}$	Nitrates	Freeze-drying of aqueous solutions	Thermal decomposition at 750–900 °C, O_2. Sintering at 900–1000 °C, 24 h	Dense (up to 90% th.d.) 123 ceramics with J_c up to 250 A/cm^2 (77 K, 0T) is obtained by optimization of sintering conditions	Pham Thi *et al.*, 1989
$YBa_2Cu_3O_{7-x}$	Nitrates	Freeze-drying of aqueous solutions	Air dehydration at 250 °C, 3 h. Decomposition at 950 °C, 15 h, O_2. Sintering: 900 °C, 15 h + annealing in O_2, 400 °C, 4 h	Improved reproducibility of electrical parameters of high T_c ceramics from homogeneous, freeze-dried precursors	Beltran-Porter *et al.*, 1989
$RBa_2Cu_3O_{7-x}$ (R = Y, Er)	Nitrates	Freeze-drying of aqueous solutions	Slow heating up to 600 °C. Decomposition at 600 °C, 20 h, vacuum + 600 °C, 2 h, air + 940 °C, 8 h. Sintering at 940 °C, 24 h. Cooling 2 K/min	Single phase Y- and Er-123 ceramics was prepared for precise thermal expansion coefficient measurements	King *et al.*, 1989
$YBa_2Cu_3O_{7-x}$	Nitrates	Freeze-drying of aqueous solutions	(a) Air dehydration at 110 °C, 3 h + 130 °C; thermal decomposition at 800 °C, 2 h (b) Thermal decomposition at 800 °C, 5 h + O_2-annealing, 400 °C, 40 h	Better homogeneity of 123 ceramics achieved due to absence of local melting for (a) samples leads to sharpening of high T_c transition. Additional air dehydration stage (regime a) leads to absence of Cu losses during thermal decomposition	Kulakov, 1988; Kulakov *et al.*, 1989
$YBa_2Cu_3O_{7-x}$	Nitrates of Y, Ba, Cu (1/4/3.5)	Precipitation by ammonium carbonate solution followed by freeze-drying of suspension of stoichiometric (1/2/3) composition	Thermal decomposition at 900–930 °C, air. Sintering at 930 °C, 1–24 h	Fine grained (300 nm) single phase 123 powders with improved sinterability. High T_c ceramics with 100% volume fraction of superconducting phase	Kimura *et al.*, 1990
$YBa_2Cu_3O_{7-x}$	(a) Nitrates (b) Acetates	Freeze-drying of aqueous solutions	Thermal decomposition: (a) 600–700 °C, 1–4 h, N_2 (b) 700–800 °C, 4 h, air	Significant diminution of 123-phase formation temperature	Schlenkrich *et al.*, 1990
$YBa_2Cu_3O_{7-x}$	Acetates	Freeze-drying of aqueous solutions	Thermal decomposition at 650 °C or 900 °C. Sintering at 975 °C. Annealing in O_2 at 650 °C	Improved homogeneity of dense (95% th.d.) 123 ceramics compared to samples prepared from mixed oxides is demonstrated by EDX measurements	Lacour *et al.*, 1990
$YBa_2Cu_3O_{7-x}$	Nitrates, acetates, perchlorates	Freeze-drying of frozen liquid ammonia solutions	Thermal decomposition at 900 °C, 2–4 h	Single-phase 123 powder with submicron particle size	Schaeffer *et al.*, 1991a, b

TABLE 3.6 (*Continued*)

HTSC composition	*Precursor*	*Applied techniques*	*Heat treatment*	*Main results*	*Reference*
$YBa_2Cu_3O_{7-x}$	Nitrates	Freeze-drying of aqueous solutions	Thermal decomposition at 600–900 °C in the controlled flow of powder through vertical furnace. Sintering at 900 °C, 12 h	Spherical 20–30 μm particles of 123. Improved homogeneity and higher rate of 123 phase formation for high (≥500 K/min) heating rates during thermal decomposition	Kimura *et al.*, 1991
$YBa_2Cu_3O_{7-x}$	Nitrates	Freeze-drying of aqueous solutions	Thermal decomposition in specially designed flat furnace at 750–850 °C. Sintering at 960 °C, 12 h	Stoichiometric composition of 123 powder resulted from 10% excess of Cu in precursor. Aggregation of 123 powders is diminished due to high heating rates at thermal decomposition	Ito *et al.*, 1991
$YBa_2Cu_3O_x$ $Y_{1.2}Ba_{1.8}Cu_3O_x$	Acetates	Freeze-drying of aqueous solutions	Thermal decomposition for 1 h at 650 °C (heating rate 200 K/min or 600 K/min) or at 900 °C (600 K/min). Sintering at 990 °C, 8 h, O_2 + 650 °C, 10 h, O_2	Improved grain size uniformity of 123 powders resulted in establishing correlations of J_c *vs.* grain size for isotropic and anisotropic grain growth	Lacour *et al.*, 1991
$YBa_2Cu_3O_{7-x}$	Nitrates	Precipitation by ammonium carbonate solution followed by freeze-drying of suspension	Thermal decomposition at 900 °C, 2 h. Sintering at 900 °C, 2 h	Homogeneity of 123 powders is improved due to higher porosity of precursor powder leading to better gas transport during thermal decomposition. Reproducibility of superconducting parameters of 123 ceramics is markedly improved	Tachivaki *et al.*, 1991a–c, 1993
$YBa_2Cu_3O_{7-x}$	Nitrates	Freeze-drying of aqueous solutions	Vacuum dehydration at 98–100 °C, heating to 150 °C. Thermal decomposition: fast heating to 600 °C, calcination 600 °C, 2 h, O_2, +925 °C, 2 h, air. Sintering at 940 °C, 1 h	Single-phase 123 powders with 3 μm grain size and 90% of Meissner response	Coppa *et al.*, 1992
$YBa_2Cu_3O_{7-x}$	Nitrates	Freeze-drying of aqueous solutions	Air drying at 110 °C, 5 h. Thermal decomposition at 600 °C, 2 h + 850 °C, 6 h, air or Ar. Air cooling to 400 °C, annealing at 400 °C, 5 h	Single-phase 123 powders with identical grain size and oxygen stoichiometry demonstrated different high T_c properties due to different size of magnetic domains. The last one can be controlled by thermal decomposition conditions	Oleinikov *et al.*, 1992; Kazin *et al.*, 1992
$YBa_2Cu_3O_{7-x}$	Nitrates	Freeze-drying of aqueous solutions	Thermal decomposition at 850 °C, 5 min, O_2. Annealing at 650–900 °C, 1 h with subsequent quenching	Single-phase 123 powders with identical grain size and oxygen stoichiometry demonstrated different high T_c properties due to different size of magnetic domains. The last one can be controlled by thermal decomposition conditions	Shabatin *et al.*, 1991a; Oleinikov *et al.*, 1992; Kazin *et al.*, 1992

TABLE 3.6 (*Continued*)

HTSC composition	*Precursor*	*Applied techniques*	*Heat treatment*	*Main results*	*Reference*
$YBa_2Cu_3O_{7-x}$	Acetates	Freeze-drying of aqueous solutions	Thermal decomposition at 460 °C, 30 min, O_2 + 750 °C, 20 h, O_2. Annealing in O_2, 500 °C, 10 h	Thick (100 μm) films obtained by fast annealing of precursors at MgO substrate. Single-phase powders of 123 without remarkable variation of composition by EDX data (10 nm sized microprobe)	Thierauf *et al.*, 1992
$YBa_2Cu_3O_{7-x}$	$Ba(NO_3)_2$/ $Ba(CH_3COO)_2$ = 1/1 $Cu(NO_3)_2$*$3H_2O$ $Y(NO_3)_3$*$5H_2O$	Freeze-drying of aqueous solutions	Thermal decomposition at 850 °C, 30 min. Sintering at 930 °C, 5 h	Improved homogeneity of 123 powder due to fast decomposition of nitrate–acetate mixture without local melting and $BaCO_3$ formation	Pershin *et al.*, 1993
$YBa_2Cu_3O_{7-x}$	Nitrates	Freeze-drying of aqueous solutions	Thermal decomposition by heating up to 850 °C at 7 K/min	Single-phase 123 powder with particle size 0.5 μm	Ezrahni *et al.*, 1993
$YBa_2Cu_3O_{7-x}$	Nitrates + urea	Freeze-drying of aqueous solutions	Thermal decomposition at 850 °C, 30 min	Precise control of oxide powder stoichiometry due to absence of Cu losses during thermal decomposition	Pershin *et al.*, 1994
$YBa_2Cu_3O_{7-x}$	Nitrates	Freeze-drying of aqueous solutions	Dehydration: 90–100 °C, 8–12 h. Decomposition at 800–850 °C, 30 min in air	High rate of 123 formation while stable intermediates appearance is avoided. A scheme of chemical transformations throughout synthesis chain is described	Shlyakhtin *et al.*, 1994
(a) $CaCuO_2$ (b) $Y_2Cu_2O_5$ (c) $Y_2Cu_2O_5$, Y_2BaCuO_5, $BaCuO_2$ (d) $Ba_{2+y}Cu_3O_6$ ($0 \leq 0.25$)	(a) Acetates (b) Nitrates (c) Nitrates (d) Nitrates	Freeze-drying of aqueous solutions	Thermal decomposition: 800 °C, a few hours, O_2, or 750 °C, 48 h, O_2, or 685 °C, a few weeks	Highly reactive powders of different phases for phase equilibrium studies of high-T_c systems	Sanchis *et al.*, 1993; Brosha *et al.*, 1994, 1995, 1996
$Y_4Ba_3O_9$	Nitrates	Freeze-drying of aqueous solutions	Thermal decomposition at 950 °C, 30 min, air	Metastable at 550 °C $YBa_2Cu_3O_{7-\delta}$ phase, obtained by reaction of $Y_4Ba_3O_9$ with CuO	Goodilin *et al.*, 1995
$YBa_2Cu_3O_{7-x}$	Nitrates	Freeze-drying of aqueous solutions	Thermal decomposition at 850 °C, 30 min. Sintering at 940 °C, 24 h	Direct comparative analysis of homogeneity of 123 ceramics, obtained from freeze-dried, mixed oxides and molten nitrates precursors	Strelkov *et al.*, 1993
$YBa_2Cu_3O_{7-x}$ $RBa_2Cu_3O_{7-x}$ (R = Y, Ho) $RBa_2Cu_3O_{7-x}$ (R = Ho, Sm, Eu, Y)	Nitrates	Freeze-drying of aqueous solutions	Thermal decomposition at 850 °C, 30 min. Sintering at 930 °C, 24 h	Dense 123 ceramics for the studies of: thermal expansion coefficient, magnetic properties, microgravimetric properties, microcontact properties, electric properties, temperature and field dependencies of the magnetization, ESR and magnetic properties, critical current studies	Basargin *et al.*, 1988; Avdeev *et al.*, 1987a, b; Rudnitskii *et al.*, 1988; Yanson *et al.*, 1988; Moschalkov *et al.*, 1988; Brandt *et al.*, 1988; Zhukov *et al.*, 1989, 1990, 1991

TABLE 3.6 *(Continued)*

HTSC composition	*Precursor*	*Applied techniques*	*Heat treatment*	*Main results*	*Reference*
$YBa_2(Cu_{1-x}Fe_x^{57})_3O_{7-\delta}$	Nitrates	Freeze-drying of aqueous solutions	Thermal decomposition at 850 °C, 30 min. Sintering at 930 °C, 10 h with subsequent annealing in O_2, 420 °C, ≥20 h	Fe^{57}-doped 123-ceramics for studies of 123 structure and its transformations by Mössbauer spectroscopy (absorption mode)	Zubov *et al.*, 1988; Shlyakhtin *et al.*, 1989
$YBa_2Cu_3O_{7-x}$	Nitrates	Freeze-drying of aqueous solutions	Thermal decomposition at 850 °C, 30 min. Sintering at 930 °C, 10 h, followed by annealing in O_2 at 420 °C	Single phase 123 ceramics for oxygen stoichiometry studies by Mössbauer spectroscopy (emission mode) after surface doping by ^{57}Co	Andrianov *et al.*, 1991, 1992
$YBa_2Cu_3O_{7-x}$	Nitrates	Freeze-drying of aqueous solutions	Thermal decomposition at 850 °C, 1 h. Hot pressing at 950 °C, $P = 100$ MPa, with subsequent temperature programmed oxygen annealing	High-density (94–99% th.d.) 123 ceramics for water vapour degradation studies	Os'kina *et al.*, 1990
$YBa_2Cu_3O_{7-x}$	Nitrates	Freeze-drying of aqueous solutions	Thermal decomposition at 850 °C, 30 min following the rapid heating (≥500 K/min). Hot pressing at 950 °C, $P = 50$–70 MPa, for 5–60 min	High-density 123 ceramics for studies of relations of T_c *vs.* grain size	Tesker *et al.*, 1989
$YBa_2Cu_3O_{7-x}$	Nitrates	Freeze-drying of aqueous solutions	Thermal decomposition at 850 °C, 30 min with subsequent planetary milling (*i*-PrOH, 30 min). Sintering on MgO substrate at 950 °C, 30 min, O_2	Thick (20 μm) films of 123 with 1–2 μm grain size obtained from highly reactive sinterable 123 powder	Lee *et al.*, 1992
$Y_{1+2x}Ba_{2+x}Cu_{3+x}O_{7-y}$	Nitrates	Freeze-drying of aqueous solutions	Thermal decomposition at 700 °C, O_2, 4 h, with subsequent melt treatment	Homogeneous fine precursor phases and phase mixtures for melt-grown 123 ceramics with finely dispersed Y_2BaCuO_5 inclusions	Dierickx *et al.*, 1994a, b; 1995a, b
$Y_{1.2}Ba_{2.1}Cu_{3.1}O_{7-x}$	Nitrates	Freeze-drying of aqueous solutions	Thermal decomposition at 850 °C, 30 min, with subsequent planetary milling (*i*-PrOH, 30 min) and QMG-processing	Enhanced (30–35*10^3 A/cm^2, 77 K) J_c for QMG-processed 123 ceramics resulted from uniform distribution of fine (2–3 μm) Y_2O_3 particles in quenched samples from freeze-dried precursors	Oleinikov *et al.*, 1993; Lee, 1992
$YBa_2Cu_3O_{7-x}$–Ag (0–60 vol.%Ag)	Nitrates	Freeze-drying of aqueous solutions	Thermal decomposition at 850 °C, 30 min. Sintering in dry O_2 at 910 °C, 30 min + 935 °C, 1 h	Improved mechanical parameters for composites with finely dispersed Ag (20–25%)	Elashkin *et al.*, 1989, 1990; Goldstein *et al.*, 1990
(a) $YBa_2Cu_3O_{7-x}$+ $0.2BaZrO_3$ (b) $YBa_{2.2}Cu_3Zr_{0.2}O_{7-x}$	Nitrates	Freeze-drying of aqueous solutions	Thermal decomposition at 850 °C, 30 min. Mixing of separately prepared fine powders of 123 and $BaZrO_3$ [for (a)]. Sintering: 930 °C, 10 h, MTG processing of sintered pellet	Enhancement of J_c (1.3–1.7 times) for (b) sample compared with undoped 123 MTG-ceramics due to finely distributed 1 μm $BaZrO_3$ particles	Shlyakhtin *et al.*, 1992; Terentiev *et al.*, 1993
$YBa_2Cu_4O_8$	Nitrates	Freeze-drying of aqueous solutions	Thermal decomposition: (a) 850 °C, O_2, 9 h, (heating 5 K/min); (b) 720 °C, 26 h, O_2, +sintering the mixture with Na_2CO_3 at 750–800 °C	124 ceramics with trace amounts of impurity phases obtained at normal pressures	Horn *et al.*, 1991

TABLE 3.6 (*Continued*)

HTSC composition	*Precursor*	*Applied techniques*	*Heat treatment*	*Main results*	*Reference*
$YBa_2Cu_4O_8$ $Y_{1-x}Ca_xBa_2Cu_4O_8$ $(0.1 \leq x \leq 0.2)$ $Y(Ba_{1-x}Sr_x)_2Cu_4O_8\,(x \leq 4)$	Nitrates	Freeze-drying of aqueous solutions	Vacuum dehydration by heating up to 150 °C. Thermal decomposition at 800 °C, O_2, 72 h	Single phase 124 ceramics with extended range of Sr and Ca doping obtained at normal pressure	Brosha *et al.*, 1991
$Y_{1-x}Ca_xBa_2Cu_4O_8$ $(0 \leq x \leq 0.1)$	Nitrates, nitrites	Freeze-drying of aqueous solutions	Thermal decomposition at 450 °C, 2 h. Sintering at 815 °C, 170 h, O_2	124 phase synthesized at normal pressure through intermediate 123 phase formation	Olesch *et al.*, 1992
$YBa_2Cu_4O_8$	Nitrates	Freeze-drying of aqueous solutions	Flash decomposition at 500–800 °C. Sintering at 830 °C in O_2, 24 h	Fast formation of 124 during sintering at normal pressure	Takahashi *et al.*, 1993
$RBa_2Cu_4O_8$ (R = Y, Sm)	Acetates	Freeze-drying of aqueous solutions	Thermal decomposition at 800 °C, 24 h. Sintering: 800 °C, 24 h + 60 h, O_2	Fast formation of Sm- and Y-124 phases in O_2 at $P = 1$ atm	Matsuda *et al.*, 1993
$Bi_2Sr_2CaCu_2O_{8+x}$	Nitrates	Freeze-drying of aqueous solutions	Thermal decomposition by flash heating to 850 °C followed by isothermal annealing	Single phase uniform sized 2212 powder with developed surface area for studies of oxygen stoichiometry	Krishnaraj *et al.*, 1994, 1995
$Bi_2Sr_{1.5}Ca_{1.5}Cu_2O_{8+x}$ $Bi_2Sr_2CaCu_2O_{8+x}$	Nitrates	Cryocoprecipitation by alcohol solution of oxalic acid followed by freeze-drying of suspension	Thermal decomposition at 800 °C, 30 min	Fast synthesis of high-T_c powders of $Bi_2Sr_{2-x}Ca_{1+x}Cu_2O_{8+y}$ composition	Shabatin *et al.*, 1989
$(Bi,Pb)_2Sr_2Ca_2Cu_3O_{10+x}$	Nitrates	Freeze-drying of aqueous solutions	Air drying at 200 °C, thermal decomposition at 830 °C, 10 h. Sintering at 840 °C, 30 h	Remarkable acceleration of 2223 phase formation compared to sol–gel and mixed oxides precursors due to better chemical homogeneity and absence of carbonates	Song *et al.*, 1990, 1991a
$Bi_2Sr_2Ca_2Cu_3O_{10+x}$	Nitrates	Freeze-drying of aqueous solutions	Air dehydration at 200 °C. Thermal decomposition at 800 °C, 12 h. Sintering: at 805–860 °C, 40–240 h, $P_{O_2} = 53$–20%	Pb-free 2223 ceramics synthesis and studies	M'Hamdi and Lacour, 1992, 1993
$Bi_{1.6}Pb_{0.4}Sr_2Ca_3Cu_4O_x$	Acetates	Freeze-drying of aqueous solutions	Thermal decomposition at 865 °C, 10 h. Annealing in pellet: 870 °C, 20 h	Remarkable acceleration of 2223 formation due to combination of high homogeneity of precursor and excess of $CaCuO_2$	Primo *et al.*, 1992
$Bi_{1.88}Pb_{0.4}Sr_2Ca_{2+x}Cu_{3+x}O_y$	Acetates	Freeze-drying of aqueous solutions	Thermal decomposition at 850 °C, 2 h. Annealing in pellet: 865 °C, 10 h + 870 °C, 10 h with intermediate grinding	Intermediate (Bi,Pb)-2212 phase formation	Primo *et al.*, 1993
$Bi_{1.8}Pb_{0.4}Sr_2Ca_{2.2}Cu_3O_x$	Nitrates	Freezing of aqueous solutions at cooled metal surface followed by freeze-drying	Air dehydration by heating (5 K/h) up to 125 °C. Thermal decomposition at 600–730 °C, 1 h. Annealing in pellet at 845 °C ($P_{O_2} = 7\%$) for 12 h with two intermediate grindings	2223 phase formation for the time as low as 12 h is achieved due to further improvement of precursor homogeneity and careful choice of annealing parameters. The presence of coherent 2201 particles within 2223 grains results in better pinning of magnetic field	Krishnaraj *et al.*, 1993; Eror, 1993

TABLE 3.6 (*Continued*)

HTSC composition	*Precursor*	*Applied techniques*	*Heat treatment*	*Main results*	*Reference*
$Bi_{1.74}Pb_{0.34}Sr_{1.94}Ca_{2.01}Cu_{3.07}O_x$	Nitrates	Freeze-drying of aqueous solutions	Vacuum dehydration by heating at 5 K/h up to 150 °C. Thermal decomposition at 625–790 °C, 2 min–40 h. Annealing in pellets: 865 °C, 37 h	Acceleration of 2223 phase formation by optimization of annealing temperature results in preventing the losses of volatile elements during annealing	Coppa *et al.*, 1994
(a) $Bi_{1.6}Pb_{0.4}Sr_{1.6}Ca_{2.2}Cu_3O_x$ (b) $Bi_{1.85}Pb_{0.35}Sr_{1.9}Ca_{2.03}Cu_{3.05}O_x$	Nitrates	Freeze-drying of aqueous solutions	Air dehydration at 200 °C (flash heating). Decomposition at 830 °C, 10 h. Annealing in Ag sheath: 820–845 °C, 60–120 h, at different P_{O_2} with intermediate mechanical treatments	Critical currents values up to 20 000 A/cm^2 (77 K, 0T) accompanied by high flexibility of Ag-2223 tapes have been achieved due to chemical and grain size uniformity of core 2223 powder and advanced optimization of processing conditions	Dou *et al.*, 1990, 1991a
(a) $Bi_{1.6}Pb_{0.4}Sr_{1.6}Ca_{2.2}Cu_3O_x$ (b) $Bi_{1.8}Pb_{0.4}Sr_2Ca_{2.2}Cu_3O_x$	Nitrates	Freeze-drying of aqueous solutions	Air dehydration at 200 °C. Decomposition at 830 °C, 10 h. Sintering: 840–850 °C, 100 h with intermediate grindings. Hot isostatic pressing: 650–840 °C, $P = 50$–200 MPa. Post annealing: 840–845 °C, 50 h	HIP processed 2223 samples showed high degree of grain alignment and density up to 90% th.d. Significant improvements of J_c stability in high magnetic fields and tenfold J_c increase compared to sintered sample has been achieved	Dou *et al.*, 1991b; Song *et al.*, 1991b
$Bi_{1.6}Pb_{0.4}Sr_2Ca_2Cu_{3.2}O_x$	Nitrates	Freeze-drying of aqueous solutions	Thermal decomposition at 750–800 °C	Cold drawing and rolling of Ag-sheathed oxide core obtained by thermal decomposition as well as optimized heat treatment lead to J_c enlargement up to 10^4 A/cm^2 (77 K, 0T)	Nikulin *et al.*, 1993, 1994
$Bi_{1.8}Pb_{0.4}Sr_2Ca_{2.2}Cu_3O_x$	Nitrates	Freeze-drying of splat frozen aqueous solutions	Thermal decomposition: 845 °C, 2 h at reduced P_{O_2}. Thermomechanical processing of Ag-sheathed oxide core: 810 °C, $P(O_2) = 0.07$ atm	$J_c = 1.1*10^5$ A/cm^2 (77 K, self-field) is achieved in 10 μm thick layer of 2223 superconductor adjacent to Ag sheath	Lelovic *et al.*, 1995a, b, 1996
$Bi_2Sr_2CaCu_2O_x + 0.2SrZrO_3$	Sr, Ca, Cu nitrates, $ZrO(NO_3)_2$, Bi acetate	Freeze-drying of aqueous solutions	Thermal decomposition at 820 °C, 30 min. Powder annealing: 800–805 °C, 2–10 h, $P_{O_2} = 0.01$–0.2 atm. Melt casting into preheated (600–850 °C) copper moulds with subsequent annealing at 850 °C	2212 ceramic matrix with finely distributed 0.1–0.3 μm $SrZrO_3$ inclusions is characterized by improved J_c and flux pinning values	Shlyakhtin *et al.*, 1995; Kazin *et al.*, 1995; Pupysheva *et al.*, 1995, 1996
$Bi_{1.85}Pb_{0.35}Sr_{1.91}Ca_{2.2}Cu_3O_9F_x$	Pb, Sr, Ca, Cu nitrates, Bi acetate, NH_4F	Freeze-drying of aqueous solutions	Thermal decomposition: 730 °C, 12 h. Sintering at 750–850 °C for 140 h	Doping by homogeneously distributed fluorine ($x \leq 0.4$) leads to acceleration of 2223 grain growth and to Meissner response enhancement during sintering at 840–850 °C	Scheurell *et al.*, 1993; Kemnitz *et al.*, 1993
$Tl_{2-x}M_xBa_2Ca_3Cu_4O_y$ $Tl_{1.5-z}M_zBa_2Ca_3Cu_4O_y$ (M = In, Ga; $x \leq 1.3$; $z \leq 1$)	Ba and Ca nitrites, Cu, In, Ga nitrates	Freeze-drying of aqueous solutions	Thermal decomposition at 850–870 °C, 5–10 min. Sintering of mixture with Tl_2O_3 at 890–920 °C, 15–90 min	Fast synthesis of Tl-based high T_c superconductors due to high reactivity of precursors	Zalishchansky *et al.*, 1989; Kopelev *et al.*, 1990

TABLE 3.6 *(Continued)*

HTSC composition	*Precursor*	*Applied techniques*	*Heat treatment*	*Main results*	*Reference*
$HgBa_2Ca_2Cu_3O_{8+x}$	Ba, Ca, Cu nitrates	Freeze-drying of aqueous solutions	Air dehydration at 110 °C, 10 h. Decomposition at 800 °C, 1 h. Sintering: 910 °C, 24 h. Annealing with HgO in evacuated quartz tube at 900 °C for 5 h	Directed formation of Hg-1223 phase and enhanced Meissner response are achieved due to advanced homogeneity of cuprate precursor	Lee *et al.*, 1995

TABLE 3.7 Application of cryochemical processes for the preparation of electrode materials.

Materials composition	*Precursors*	*Applied techniques*	*Heat treatment*	*Main results*	*References*
(a) $Ni_{1-x}Li_xO_{1-\delta}$ $(0 < x < 0.10)$ (b) $Ag_{1-2x}Cd_xCl$ $(x < 0.1)$	(a) Acetates (b) Chlorides	(a) Freeze-drying of aqueous solutions (b) Freeze-drying of aqueous ammonia solutions	Thermal decomposition in vacuo at 250 °C. Annealing at 300–1000 °C	Formation of Li-doped NiO with homogeneously distributed Li at the temperatures as low as 400 °C results in marked increase of surface area (12 m^2/g) compared to product of ceramic synthesis (1 m^2/g)	Tseung and Bevan, 1970; Bevan and Tseung, 1974
$NiCo_2O_4$	Nitrates	Freeze-drying of aqueous solutions or freeze-drying of quenched nitrate melt	Vacuum dehydration at 250–300 °C. Heat treatment at 300 °C, 24 h, air	Outstanding increase of electrochemical activity of spinel powder has been achieved by cryochemical method compared to co-precipitation and mixing of nitrates methods	King and Tseung, 1974a, b
$NiCo_2O_4$	Nitrates	Freeze-drying of aqueous solution frozen in droplets or on nickel screen	Thermal decomposition 400 °C, air	Significant improvement of graphite electrode performance is achieved by 10% doping by $NiCo_2O_4$ obtained from cryochemical precursor	Tseung and Yeung, 1978
$NiCo_2O_4$	Nitrates	Freeze-drying of frozen solution in glacial acetic acid	Thermal decomposition in vacuo, 250 °C, 3 h. Heat treatment at 400 °C, 10 h, air	20 times cutting of freeze-drying duration compared to frozen aqueous solution is accompanied by insignificant reduction of surface area and electrochemical activity of $NiCo_2O_4$	Hibbert and Tseung, 1979
$NiCo_2O_4$	Nitrates	Freeze-drying of aqueous solutions	Thermal decomposition at 300–400 °C, air	Improved corrosion resistance in alkaline solutions and extended up to 500 h mechanical stability of Teflon-bonded $NiCo_2O_4$ electrodes on nickel screens	Davidson *et al.*, 1982
$Ni_xFe_{3-x}O_4$ $(0.49 < x < 1.68)$	Nitrates	Freeze-drying of aqueous solutions	Heating up to 250 °C in vacuo. Air annealing at 350 °C, 1 h + 850 °C, 1 h	Synthesis of homogeneous finely dispersed solid solutions for Teflon-bonded electrode parameter measurements	Orehotsky *et al.*, 1979
Co_3O_4 $NiCo_2O_4$	Nitrates	Freeze-drying of aqueous solutions	Not discussed	Examples of application of freeze-drying synthesis in electrocatalyst's preparation	Poillerat, 1994
$SrTiO_3$	Chlorides	Slow (5 days) freeze-drying of aqueous solution	Thermal decomposition: 400 °C, 5 h + 600 °C, 24 h	2 times enlargement of H_2 production in photochemical cell compared to commercial $SrTiO_3$	Thampi *et al.*, 1988
Pt on WO_3 substrate	H_2PtCl_6	Freeze-drying of frozen suspension of WO_3 in aqueous H_2PtCl_6 solution	Reduction in H_2 at 50 °C	Better homogeneity of Pt distribution compared to fine mechanical mixture resulted in greater hydrogen uptake into WO_3	Hobbs and Tseung, 1972
Pt on WO_3 substrate	H_2PtCl_6	Freeze-drying of frozen suspension of WO_3 in aqueous H_2PtCl_6 solution	Reduction in H_2 at 50 °C, 30 min + 90 °C, 3 h	Homogeneous distribution of finely dispersed Pt-particles on the WO_3 surface leads to the appearance of synergetic effect in catalysis of H_2 evolution	Abbaro *et al.*, 1980

TABLE 3.8 Application of freeze-drying processes for the preparation of the catalysts and absorbents.

Material composition	*Precursors*	*Applied techniques*	*Heat treatment*	*Main results*	*References*
$La_{0.7}Sr_{0.3}MnO_3$ $La_{0.7}Pb_{0.3}MnO_3 + xPt$ $(0 \leq x \leq 0.0022)$	$La(NO_3)_3$, $Sr(NO_3)_2$, $Mn(NO_3)_2$, $Pb(NO_3)_2$, H_2PtCl_6	Freeze-drying of aqueous solutions	Thermal decomposition in vacuo by heating to 500 °C (50 K/h) followed by 4 h annealing. Reduction: 700 °C, 16 h in $(He + 0.02CO + 0.02O_2)$	Doping of $La_{0.7}Pb_{0.3}MnO_3$ (with developed surface area by Pt ≥1600 ppm) leads to synergetic enhancement of the overall catalytic activity of the material	Gallagher *et al.*, 1975
$La_{1-x}M(I)_xMn_{1-y}M(II)_yO_3$ (M(I) = Sr, Pb K, Ce; M(II) = Ni, Co, Li, Mg)	Nitrates	Freeze-drying of aqueous solutions	Thermal decomposition: 500 °C, reduced pressure + 700 °C, 2 h, O_2	Application of freeze-drying results in significant lowering of perovskite formation temperature. Considerable enhancement of catalytic activity has been achieved by Sr doping in A position and by Ni- in B position	Johnson *et al.*, 1976
$La_{0.5}Pb_{0.5}MnO_3$ $La_{0.5}Sr_{0.5}MnO_3$	Nitrates	Freeze-drying of frozen support impregnated by starting solution (cryoimpregnation technique)	Thermal decomposition at 500 °C, 1 h	Lower mobility of components during cryoimpregnation leads to better homogeneity of catalyst distribution on the support and higher activity for the CO oxidation. Annealing of catalyst on γ-Al_2O_3 support leads to its determination due to chemical reaction with substrate	Johnson *et al.*, 1977
$LaMn_{1-x}Cu_xO_{3-y}$ $(0 \leq x \leq 0.8)$	Nitrates	Freeze-drying of aqueous solutions	Thermal decomposition: 700 °C, 2 h, vacuum + 1100 °C, 4 h, O_2	Catalytic activity comparable to that of Pt and better thermal stability than ones of Co_3O_4 is achieved for $0.4 \leq x \leq 0.7$	Gallagher *et al.*, 1977; Vogel *et al.*, 1977
$La_{1-x}Sr_xCoO_{3-y}$	Nitrates	Freeze-drying of aqueous solutions	Thermal decomposition: vacuum, 500 °C, 1 h, +air, 1060 °C, 1 h	Due to application of freeze-drying technique the range of obtained single-phase solid solutions is extended up to $0.1 \leq x \leq 0.9$	Anderton and Sale, 1979
$LaCoO_3$	Nitrates	Freeze-drying of aqueous solutions	Thermal decomposition in vacuum, (250 °C, 2 h) and annealing in O_2 (500 °C, 10 h)	Single phase $LaCoO_3$ with extended surface area is obtained at temperatures as low as 500 °C	Wachowski *et al.*, 1980
$NiCo_2O_4$ $La_{0.5}Sr_{0.5}CoO_3$	$Co(NO_3)_2$, $Co(CH_3COO)_2$ La $(NO_3)_3$, $La(CH_3COO)_3$ $Sr(NO_3)_2$ $Sr(CH_3COO)_2$	Freeze-drying of aqueous solutions	Thermal decomposition: vacuum 300 °C. Annealing: at 400–1000 °C	Formation of solid solutions between salt components of precursor promotes achievement of higher surface area of perovskite powders	Hibbert *et al.*, 1987
$LaMn_{1-x}Cu_xO_{3-y}$	(a) Nitrates (b) Acetates	Freeze-drying of aqueous solutions	Thermal decomposition at 300 °C, 3 h + 850 °C, 5 h	Application of acetate precursors results in 4–20 times enhancement of catalytic activity of perovskite. Local inhomogeneities of cation distribution have been found	Mizuno *et al.*, 1989; Yasuda *et al.*, 1994

TABLE 3.8 *(Continued)*

Material composition	*Precursors*	*Applied techniques*	*Heat treatment*	*Main results*	*References*
$La_{1-x}Sr_xCoO_3$ ($x = 0.2, 0.5, 0.8$)	Nitrates	Freeze-drying of aqueous solutions	Vacuum annealing up to 300 °C. Air annealing at 570–1000 °C for 10 min, 32 h	Reduced temperature (570 °C) of phase formation allowed to obtain a series of vacancy-ordered metastable phases	Kirchnerova *et al.*, 1990, 1993
$La_{1-x}Sr_xCo_{1-y}Me_yO_3$ (Me = Mo, Fe, Mn, Ni)	Nitrates	Freeze-drying of aqueous solutions	Thermal decomposition: slow heating in vacuum up to 350 °C, air annealing 585 °C, 15 h	Freeze-drying synthesis combined with kinetic studies of methane combustion is used for the synthesis of new catalyst and optimization of its composition	Kirchnerova and Hibbert, 1993; Klvana *et al.*, 1994
$La_{0.8}Sr_{0.2}CoO_3$ $La_{0.4}Sr_{0.6}Fe_{0.4}Co_{0.6}O_3$ $La_{0.7}Sr_{0.3}CoO_3$ $La_{0.5}Sr_{0.5}Fe_{0.2}Co_{0.8}O_3$	Nitrates	Freeze-drying of aqueous solutions	Thermal decomposition: slow heating in vacuo up to 380 °C. Air calcination: 580–1050 °C, several hours	Optimization of freeze-drying and drying conditions allowed to obtain powder electrode with improved electrocatalytic activity	Kirchnerova and Klvana 1994; Kirchnerova *et al.*, 1995
$CoFe_2O_4$	$(NH_4)_2Fe(SO_4)_2$ *$6H_2O$, $(NH_4)_2Co(SO_4)_2$ *$6H_2O$	Freeze-drying of aqueous solutions	Thermal decomposition at 850–1000 °C in air	Synthesis of spinel catalyst with improved thermal stability of specific surface area	Voinov, 1976
$Zn_{0.89}Fe_{2.04}O_{4+\gamma}$ $Co_{0.89}Fe_{2.04}O_{4+\gamma}$ $Ni_{0.89}Fe_{2.04}O_{4+\gamma}$	$(NH_4)_2Fe(SO_4)_2$ *$6H_2O$, $(NH_4)_2Co(SO_4)_2$ *$6H_2O$, $(NH_4)_2Ni(SO_4)_2$ *$6H_2O$, $(NH_4)_2Zn(SO_4)_2$ *$6H_2O$	Freeze-drying of aqueous solutions	Thermal decomposition at 900 °C, 2 h	Spinel catalyst with high and selective catalytic activity in the reaction of *n*-butene dehydrogenation into butadiene	Savchenko, 1982
(a) BPO_4 (b) $AlPO_4$	NH_4OH, H_3PO_4, H_3BO_3, $Al(NO_3)_3$	(a) Freeze-drying of aqueous solutions (b) Precipitation by mixing of solutions followed by freeze-drying of suspensions	Vacuum dehydration at 50 °C. Thermal treatment at 600–800 °C, air	Outstanding improvement of specific surface area (480–770 m^2/g) leading to high catalytic activity in dehydration reactions	Savchenko 1982; Volfkovitch *et al.*, 1979a
$V_2Mo_9O_{32}$ $V_2Mo_{14}O_{47}$	NH_4VO_3, $(NH_4)_6Mo_7O_{24}$ *$4H_2O$	Freeze-drying of aqueous solutions	Thermal treatment at 150–500 °C	Improved selectivity of catalyst during oxidation of acrolein into acrylic acid	Savchenko, 1982
xNa_2O*yCo_3O_4*zFe_2O_3	$Na_3Co(NO_2)_6$, $FeSO_4$	Freeze-drying of aqueous solutions	Thermal decomposition at 700 °C after heating 10 K/min	Two times enhancement of NO_x capacity of sorbent compared to precipitation method	Gorst *et al.*, 1989
(a) $Ni(OH)_2$ $Mg(OH)_2$ (b) $Cu(OH)_2$ $Ti(OH)_4$	Nitrates	Precipitation of suspensions followed by freeze-drying	No heat treatment	Enlargements of specific surface 3 to 10 times compared to air drying of hydroxides	Trambouze, 1970; Kawakami *et al.*, 1990
$Ti(OH)_4$	H_2TiCl_6	Co-precipitation with enzyme by aqueous ammonia followed by freeze-drying of suspensions	No heat treatment	Optimization of pre-freezing treatment leads to simultaneous preparation of sorbent together with immobilized enzyme with retention of its enzymic activity	Kennedy and Kay, 1976

TABLE 3.8 *(Continued)*

Material composition	*Precursors*	*Applied techniques*	*Heat treatment*	*Main results*	*References*
SiO_2	$Na_2SiO_3*9H_2O$	Freeze-drying of gel monolites obtained by HCl hydrolysis of starting solution with subsequent ageing	No heat treatment	Specific surface area of obtained SiO_2 cryogels (600–750 m^2/g) attains that one of aerogels	Pajonk *et al.*, 1990
Ni/Al_2O_3	(a) Nickel acetate; secondary butylate of aluminium (b) $Al(NO_3)_3*9H_2O$; $Ni(NO_3)_2*6H_2O$	(a) Mixing of solutions of nickel acetate in ethylene glycol and aluminium butylate in 2-methyl-2-propanol followed by aqueous ammonia hydrolysis and freeze-drying of gel. (b) Ammonia hydrolysis of solution followed by freeze-drying of gel	Thermal decomposition at 400 °C, 12–16 h, air. Catalyst activation: 500 °C, 3 h, H_2	Ni/Al_2O_3 catalyst with specific surface area ~370 m^2/g and improved fluidizability is obtained. Its parameters are comparable with those of aerogel-based catalyst while preparation scheme is more economical	Klvana *et al.*, 1989, 1992

remarkably lowered and, thus, their high specific surface and homogeneity of microcomponents distribution were kept, ensuring high activity and selectivity.

3.8 REFRACTORY OXIDES

Individual and complex metal oxides considered in the present section are traditionally classed among refractory materials, i.e. materials with high durability, mechanical stability and chemical resistance at temperatures >1400 °C. However, actually most parts of these oxides are multifunctional. For example, aluminium oxide is widely used in the production of highly efficient internal combustion engines, high-temperature turbines, crucibles, acid-resistant coatings for chemical reactors, ceramic substrates of integrated circuits, for heat isolation and neutron shields in nuclear reactors, and for manufacturing transparent shells of high-pressure sodium lamps.

Similar multifunctional usage is also characteristic for ceramics based on zirconium dioxide doped with oxides of alkaline-earth or rare-earth metals, the concentration and distribution of which determine the possibility of manufacturing solid electrolytes with oxygen ionic conductivity, superplastic ceramics and high-resistant cutting tools. In the latter case, the effect is reached due to the formation of cubic modification of ZrO_2 followed by annealing at lower temperatures, eventually resulting in a great number (20 to 50 volume %) of uniformly distributed small crystallites of zirconium dioxide with tetragonal structure. The contact of the growing crack (the principal carrier of destruction energy) with such crystallite leads to its transformation in the monoclinic phase. This transformation is followed by the increase of crystallite volume (by about 4%) and by the decrease of crack energy; so, the crack growth stops. Since not every material has the ability to change its structure under action of temperature or pressure, additions of zirconium oxide are introduced to make materials more durable.

Multifunctional properties of metal oxides are closely related with their microstructure: with the size and shape of microcrystallites, and with the size, shape and distribution of pores. The factors determine the sensitivity of these properties to the prehistory of powders and ceramics preparation. The known attempts of cryochemical technique application for preparation of individual and complex refractory oxides are presented in Table 3.9.

3.9 ULTRAFINE METAL AND METAL ALLOY POWDERS

An intensive development of powder metallurgy a long time ago created an interest in the improvement of efficient techniques for the preparation of fine metal powders. Thirty years ago, in projects carried out by the US Federal Bureau of Mines, the techniques of cryochemical technology were firstly applied in the production of highly dispersed tungsten and molybdenum (Landsberg and Campbell, 1965). The freeze-drying technique usually yielding the oxide products was completed with the processes of hydrogen reduction of such products. Later, the cryochemical technology was successfully used in preparation of dispersed precursors of Ni–Fe, Cu–Ni–Fe and Cu–Ni–Co alloys, as well as for preparing active forms of Cu, Ag and Ni, used for metallization of dielectrics and diffusion welding, respectively. Finally, it was shown (Pyagai, 1982) that for disaggregation of metal powders the processes of cryodispersion can be successfully applied. The attempts of cryochemical process application to obtain ultra-fine metal and alloy powders are described in Table 3.10.

3.10 INORGANIC COMPOSITES

Cryochemical processes were firstly used for the preparation of inorganic composites, especially dispersion-strengthened materials, by scientists at the US Bureau of Mines (Landsberg and Campbell, 1965). The experiments concerned dispersion of zirconium dioxide in Mo, W and W–Re alloys. Later freeze-drying processes were applied for preparing other inorganic composites, including Cu–ThO_2, Ni–ThO_2, Pb–MgO, SiO_2–W, Fe–ThO_2 and Cu–Ni–ThO_2 (Roehrig and Wright, 1972). Cryochemical processes were found to be efficient in obtaining highly homogeneous precursors for metal–ceramic contacts based on

TABLE 3.9 Application of cryochemical methods for the preparation of refractory materials.

Material composition	*Precursors*	*Applied technique*	*Heat treatment*	*Main results*	*References*
MgO	$Mg(OCH_3)_2$	Controlled hydrolysis of alcoholic solution by aqueous ammonia followed by freeze-drying of suspension	Thermal decomposition: at 400–800 °C, 2 h. Sintering at 1300 °C for 6 h after heating 10 K/min	Sintered density is improved up to 98–99% th.d. compared to sol–gel synthesis	Tsai and Shih, 1993a, b
Al_2O_3	Sulphates	Freeze-drying of aqueous solution	Thermal decomposition: 800–900 °C, air	Finely dispersed γ-Al_2O_3 powders with size of crystallites ~150 Å are obtained. At 1150 °C formation of α-Al_2O_3 accompanied by crystallite growth up to 400 Å is observed	Johnson and Schnettler, 1970
$Al_2O_3 + x$MgO ($x < 0.015$)	Sulphates	Freeze-drying of aqueous solutions	Thermal decomposition: 1100 °C, 2 h. Presintering: 1350 °C, 2 h, air. Sintering 1700 °C, 1 h, H_2	Density of 99.9% at $x = 0.006$ accompanied by reduced grain size (2–10 mm) is attained due to homogeneous distribution of MgO dopant	Kim and Monforte, 1971
$Al_2O_3 + x$MgO ($x < 0.015$)	α-Al_2O_3 or γ-Al_2O_3, $MgSO_4$	Freeze-drying of alumina suspensions in aqueous solutions of $MgSO_4$	Thermal decomposition: 1200 °C, 2–35 h. Sintering: 1700 °C, 2 h + 1900 °C. 2–24 h vacuum	Increased up to 70% light transmittance values of theoretically dense sintered alumina bodies	Amato *et al.*, 1976
Al_2O_3 + 0.002MgO	Al_2O_3, $Mg(NO_3)_2$	Freeze-drying of suspension of alumina in aqueous solutions of $Mg(NO_3)_2$	Not described	Dense alumina ceramics with 5–6 μm grain size and outstanding resistance to high-temperature degradation	Ryabov *et al.*, 1978
$Al_2O_3 + x$MgO ($x < 0.015$)	Sulphates	Freeze-drying of aqueous solutions	Thermal decomposition at 1050–1300 °C	High surface area of obtained alumina (~150 m^2/g) can be controlled by concentration of initial solution	Alexeev *et al.*, 1991
Al_2O_3	Sulphates	Freeze-drying of aqueous solution	Air drying at 350 °C, 2 h. Thermal decomposition: 1120 °C, 6 h, or 1220, 6 h. Sintering or hot pressing at 1650 °C	Fine grained ($d < 35$ nm) γ-Al_2O_3 sintering aid, obtained by thermal decomposition, substantially increases the strength of Si_3N_4 ceramics. Strength of dense Al_2O_3 ceramics improved 1.5 times in comparison with ceramics from commercial Al_2O_3 precursor	Mikhailenko and Fomina, 1990
Al_2O_3	Sulphates	Freeze-drying of aqueous solutions	Thermal decomposition: 1000–1200 °C, 45 min	High surface area of obtained Al_2O_3 powders changes rather slowly during annealing at 1100 °C (170 m^2/g for 15 min and 130 m^2/g for 45 min)	Sarkar *et al.*, 1984
Al_2O_3	Sulphates	Freeze-drying of aqueous solutions	Air drying at 500 °C. Thermal decomposition at 900 °C, 3 h	Intensive heat supply during freeze-drying leads to marked decrease of surface area and enlargement of grain size of Al_2O_3, obtained by thermal decomposition	Beyer *et al.*, 1987

TABLE 3.9 (*Continued*)

Material composition	*Precursors*	*Applied technique*	*Heat treatment*	*Main results*	*References*
Al_2O_3	Sulphates	Freeze-drying of aqueous solutions	Thermal decomposition 1100 °C, 24 h. Sintering: 1500–1600 °C, O_2	Specific features of cryocrystallization can lead to bimodal distribution of pore size in Al_2O_3 particles with maxima at ~20 nm and 1 μm	Lloyd and Kovel, 1988; Lloyd *et al.*, 1989
Al_2O_3	(a) Sulphates	Freeze-drying of aqueous solutions	(a) Thermal decomposition at 800 °C, 3–6 h. Sintering at 1200–1400 °C.	Freeze-drying and thermal decomposition conditions have a serious influence on the aggregate structure of Al_2O_3 powders. Even moderate changes in this structure lead to dramatic changes of sintered density (40 to 99% th.d. for samples with equal surface area of powders)	Wang and Lloyd, 1991
	(b) Isopropoxides	Freeze-drying of aqueous solutions	(b) Air drying at 200–300 °C, 1 h; decomposition at 600–1200 °C, 1 h. Sintering at 1400 °C		Chang and Lloyd, 1993
ZrO_2	$ZrO(NO_3)_2$	Freeze-drying of aqueous solutions	Thermal decomposition at 495 °C after heating at 5 K/min	Synthesis of spherical granules of ZrO_2 (~200 μm) having enhanced mechanical strength is realized by starch addition to initial solution	Zhiboedov *et al.*, 1976
$ZrO_2 + 0.13CaO$	$ZrOCl_2$; $Ca(NO_3)_2$	Precipitation by ammonia and ammonium carbaminate followed by freeze-drying of suspension	Thermal decomposition at 700 °C for 30 min followed by milling in ethanol and air drying at 120 °C. Sintering at 1250–1600 °C for 3 h	High density of ceramics at reduced sintering temperatures can be achieved only from green compact with homogeneous microstructure. The last one is resulted from the absence of significant aggregation of ZrO_2 powder	Hausner and Roosen, 1983; Roosen and Hausner, 1983
$ZrO_2 + 0.02Y_2O_3$	$ZrOCl_2$; $Y(NO_3)_3$	Freeze-drying of suspension followed by precipitation with ammonia (pH = 8.5)	Thermal decomposition at 700 °C, 30 min. Wet milling in ethyl alcohol, air drying at 120°C. Sintering at 1100–1600 °C	Application of freeze-drying allowed to reduce sintering temperature for 200 K and to obtain T–ZrO_2 ceramics with low grain size (0.17 μm), 97% density and enhanced flexural strength	Ramme and Hausner, 1987
$ZrO_2 + 0.1CeO_2$	$ZrOCl_2$	Urea hydrolysis of starting solution followed by freeze-drying of suspension	Thermal decomposition 650 °C, 30 min	Continuous low-temperature freeze-drying is used to remove water and CO_2 from urea hydrolysis products	Luccini *et al.*, 1989
$ZrO_2 + xY_2O_3$	$ZrO(NO_3)_2$ $Y(NO_3)_3$	Freeze-drying of aqueous solution	Thermal decomposition, sintering at 1300–1500 °C	Dense (97% th.d) fine-grained Y-doped tetragonal zirconia ceramics	Iwaida *et al.*, 1994; Reetz and Moritz, 1991
ZrO_2	$ZrOCl_2$ or $ZrO(NO_3)_2$	Freeze-drying of gel obtained by potentiostatic reduction of starting solution at Pt electrode	Thermal decomposition by heating up to 600 °C	Single-phase fine-grained tetragonal zirconia powder	Nakajima *et al.*, 1995
$Ce_{1-x}M_xO_2$ (M = Ca, Sr, Sm, Gd, Y; $x = 0.2$)	Nitrates	Freeze-drying of hydrothermal treatment products	Sintering at 1400–1500 °C	Remarkable lowering of sintering temperature compared to conventional ceramic technique	Yamashita *et al.*, 1995

TABLE 3.9 (*Continued*)

Material composition	*Precursors*	*Applied technique*	*Heat treatment*	*Main results*	*References*
$ZrO_2 + 0.04Y_2O_3$ $ZrO_2 + 0.7MgO$ $ZrO_2 + 0.11MgO$	$ZrO(NO_3)_2$ $Y(NO_3)_3$	Freeze-drying of suspension followed by precipitation by aqueous ammonia	Not described	Aggregate structure of ZrO_2 powder markedly depends on the chemical nature and concentration of dopant	Rudolph and Reetz, 1988
Y_2O_3	$Y(NO_3)_3$	Freeze-drying of suspension obtained by precipitation (a) by aqueous ammonia (b) by aqueous ethylamine	Thermal decomposition at 1100 °C, 2 h after heating at 8 K/min. Sintering in vacuo: (a) 1800 °C, 1 h; (b) 1600–2000 °C, 1 h. Heating and cooling rate 20 K/min	Better sinterability of freeze-dried oxides compared to oven-dried leads to the preparation of Y_2O_3 ceramics of 95–98% density	Akinc and Rasmussen, 1983; Rasmussen *et al*., 1984
Y_2O_3	$Y(NO_3)_3$	Freeze-drying of suspension obtained by precipitation by aqueous ammonia and ageing of precipitates (80 °C, 4 days)	Thermal decomposition at 1000 °C, 1 h after heating 5 K/min. Sintering at 1250–1600 °C, 3 h, at heating and cooling rate 5 K/min	Sintered density up to 95–98% can be achieved at rather low temperatures (1200–1450 °C) if powders are processed in such a manner that green compact with small pores with narrow pore size distribution is formed	Dogan *et al*., 1987
$MgO–Al_2O_3$	Sulphates	Freeze-drying of aqueous solutions	Not described	The influence of different factors during cryocrystallization and freeze-drying on the properties of final oxide powder is considered	Lacour and Paulus, 1979
$MgAl_2O_4$	Sulphates	Freeze-drying of aqueous solutions	Not described	The role of different factors influencing the optical transmittance of $MgAl_2O_4$ ceramics is discussed with special attention being paid to uniformity and continuity of microstructure	Lacour, 1980
$MgAl_2O_4$	(a) Sulphates (b) $Al_2(SO_4)_3$ (c) $MgSO_4$	(a) Freeze-drying of aqueous solutions; (b) Freeze-drying of MgO suspension in $Al_2(SO_4)_3$ aqueous solutions (c) Freeze-drying of Al_2O_3 suspension in $MgSO_4$ aqueous solutions	Air dehydration at 150 °C, 24 h. Thermal decomposition at 1100 °C, 2 h. Sintering at 1540 °C, 1 h	Density of sintered pellets obtained using suspension freeze-drying (90–93%) is comparable with that one for pellet obtained using solution freeze-drying (96%) and much higher than for mixed oxides precursor (70–75%)	Pankov *et al*., 1983
$MgAl_2O_4$	Sulphates	Freeze-drying of aqueous solutions	Thermal decomposition at 950–1200 °C, 2 h. Sintering at 1700 °C, 3 h, vacuum	The controlled change of green compact preparation conditions can lead to 96–98% dense sintered pellet or porous permeable material with uniform distribution of pore size	Dudnik *et al*., 1988

TABLE 3.9 (*Continued*)

Material composition	*Precursors*	*Applied technique*	*Heat treatment*	*Main results*	*References*
$MgAl_2O_4$	Mg methoxide Al isopropoxide	Freeze-drying of sol obtained by hydrolysis and peptization of Al isopropoxide, followed by ageing (85 °C, 4 days) alumina sol in methanol solution of Mg methoxide	Thermal decomposition 1100 °C, 12 h. Sintering at 1500 °C, 200 mPa, 2 h, in argon	Recrystallization of single phase $MgAl_2O_4$ appeared at 400 °C proceeds up to 1100–1200 °C. Enclosed pores elimination by HIPing of dense (98–99%) ceramics results in remarkable improvement of optical transparency of material	Wang *et al.*, 1992
$LiAl_5O_8$	Li_2CO_3 $Al(NO_3)_3$*$9H_2O$	Freeze-drying of aqueous solutions	Thermal decomposition at 1000–1100 °C, 2 h. Hot pressing at temperature deviated from 1450 to 1550 °C	Dense ceramics transparent at 1 = 3–5 mm with bimodal grain size distribution (maxima at 1–2 and 20–25 μm)	Gazza, 1972
Starting mixtures for mineral synthesis	(a) Mg citrate, aqueous sol of SiO_2 (b) Fe and Mg citrates	Freeze-drying of aqueous solutions and suspensions	Thermal decomposition at 1000 °C, air, 1 h	(a) Forsterite formation during hydrothermal synthesis is accelerated 10 times compared to synthesis from sol–gel precursor. (b) Rate of magnesiowurzite formation at 1000 °C is four times higher than for mixed-oxides ceramic precursor	Finnerty *et al.*, 1978
$3Al_2O_3$*$2SiO_2$ (mullite)	Aqueous sol of $SiO_2 + Al_2(SO_4)_3$	Freeze-drying of aqueous gel	Thermal decomposition at 1200 °C, 1 h	Fast synthesis of mullite powder in the form of spherical granules readily broken into much smaller particles	Wheat *et al.*, 1979
$3Al_2O_3$*2MgO*$5SiO_2$ (cordierite)	Aqueous SiO_2 sol + $MgSO_4$ + $Al_2(SO_4)_3$	Freeze-drying of aqueous sol	Air dehydration at 150 °C, 24 h. Thermal decomposition at 1150 °C, 2 h. Sintering at 1430 °C, 4 h	Preparation of dense (98%) homogeneous cordierite ceramics with no traces of intermediate eutectics formation	Pankov *et al.*, 1982
Na_2O*$6Al_2O_3$	Na_2CO_3 $Al_2(SO_4)_3$	Freeze-drying of aqueous solution	Thermal decomposition; sintering at 1600 °C	Manufacturing of light-weight ceramic insulator with porosity over 90% (density = 0.31 g/cm^3)	Green, 1988
Al_2O_3–ZrO_2–MgO (composite)	Not described	Not described	Sintering at 1650 °C	Density of abrasive composite 96–98% th.d. accompanied by grain size 1–3 μm is obtained at temperature markedly lower than for ceramic precursor	Lopato *et al.*, 1992
(a) Al_2O_3 (corundum) (b) $3Al_2O_3$*$2SiO_2$ (mullite) (c) $(Mg_{1-x}Al_{2/3x})_3Si_4O_{11}$ (steatite)	Corundum, mullite and steatite powders	Freeze-drying of suspensions	Firing of oxide mixture at 900–1000 °C, grinding and mixing with solution of organic binder. Sintering at (a) 1300 °C (b) 850 °C (c) 900 °C	Freeze-drying is used to change the aggregate structure of powders in order to ensure more uniform microstructure of green compact resulted in better sinterability of the last one	Kwiatkowski *et al.*, 1980
$(Zr_{1-x}Sn_x)TiO_4$ $(0.1 < x < 0.3)$ $+0.01$ MnO_2	$ZrO_2 + SnO_2 +$ $TiO_2 + MnO_2$	Freeze-drying of aqueous suspension	Sintering of dried compact at 1350–1425 °C	As above	Yokoyama and Ozeki, 1993

TABLE 3.9 (*Continued*)

Material composition	*Precursors*	*Applied technique*	*Heat treatment*	*Main results*	*References*
SiC	SiC	Freeze-drying of aqueous suspension	Sintering at 1900–2100 °C	As above	Ibigawa Co., 1981; Nissan Motor Co., 1983; Sakai *et al.*, 1985
(a) Asbestos $Mg_2CaSi_4O_{12}$ (b) Montmorillonite	(a) Asbestos powder (b) Montmorillonite suspension	Freeze-drying of aqueous suspensions	Not described	As above	Takeda Ltd, 1977; Yokagawa *et al.*, 1986

the Ag–CdO composition, thick films for microelectronics based on Ag-glass, and fibres for high-intensity light sources based on W + ThO_2. Recently, the technology of composites based on a sialon matrix with silicon carbide reinforcement was improved due to freeze-drying processes. The reported attempts of cryochemical process application in inorganic composite technology are presented in Table 3.11.

3.11 MISCELLANEOUS INORGANIC SUBSTANCES AND MATERIALS

The multifunctional properties of many inorganic substances and materials and rather sparse cryochemical experience in their preparation make it impossible to suggest a universal route and make it reasonable to consider together inorganic products that are miscellaneous in composition, properties and application fields (see Table 3.12). Among them, in the first place, are the target products (ThO_2, UO_2) and waste products of nuclear technology. In the first case, the cryochemical techniques were successful in preparation of free-flowing as well as dust-free and well-pressed powders. In the second case, the cryochemical technology improved the process of separation of radioactive substances. Of special interest is the cryochemical experience in the preparation of powdery Al–Cr phosphate binder with enhanced adhesion ability towards aluminosilicate refractory materials. Among the studies dealing with cryochemical preparation of colloidal materials, the most interesting are those where attempts have been made to develop a fractal description of intermediate states appearing during the evolution of the system on transition from solutions to solid-phase products. The application of cryochemical processes in the technology of carbides (WC, W_2C and SiC) made it possible to optimize the conditions of green compact preparation and to achieve better sintered density. The unique attempt of cryochemical synthesis of sulphides has yielded a high-density $CaLa_2S_4$ ceramic – as promising material for optical windows in the 8–14 µm infrared range.

TABLE 3.10 Application of freeze-drying processes for the preparation of ultrafine powders of metals and alloys.

Metal or alloy	*Precursor*	*Applied technique*	*Heat treatment*	*Main results*	*References*
W W-0.25 Re	$(NH_4)_2WO_4{*}H_2O$; $(NH_4)ReO_4$; hexamethylene tetramine (hexamine)	Freeze-drying of aqueous solutions frozen in trays or at cold metal surface	Thermal decomposition: at: (a) 250–300 °C, 3–6 h. (H_2 flow + hexamine vapour) (b) 200 °C, 30 min + 300 °C, 60 min + H_2 reduction at 800–900 °C, 2 h. Sintering: H_2; 900 °C + 2400 °C, 10 h	First principle works on the freeze-drying synthesis of materials. Synthesis of nanosized W and W–Re powders allowed to achieve 95% density of sintered compact	Landsberg and Campbell, 1965; Landsberg, 1967
W	Ammonium paratungstate	Freeze-drying of aqueous solution	Thermal decomposition at 500 °C followed by H_2 reduction at 650–900 °C	W powder consisting of 0.2–0.4 μm sized particles with reduced tendency of particle coarsening during sintering	Sale, 1977
(a) Ni–Fe (b) Cu–Ni–Fe Cu–Ni–Co	Sulphates	Freeze-drying of aqueous solutions	Thermal decomposition: 850 °C, air. Sintering: (a) 1200 °C, air (b) 1000 °C, air, 14 h. Reduction: (a) 300 °C, H_2, 6 h (b) 1000 °C, H_2, 2 h (c) 1150 °C, H_2, 4 h	Reduction of presintered homogeneous oxide pellets resulted in attainment of 95–98% density of alloy compact	Robbins *et al.*, 1980
Cu, Cu–Ni, Co–Ni, W–Re	Not described	Freeze-drying of aqueous solutions	Thermal decomposition, reduction, hot isostatic pressing	Fully dense fine-grained alloy bodies	Gelles and Roehrig, 1972
Ni	Formate	Freeze-drying of aqueous solutions	Thermal decomposition: 180–200 °C vacuum	Ultrafine Ni powder for diffusion welding insert	Tarlavskii *et al.*, 1986
Cu–Ni	Formates	Freeze-drying of aqueous solutions	Evaporation of formates in vacuo (160–170 °C) and their condensation on cooled substrate, followed by thermal decomposition at 180–200 °C in vacuo	Fine homogeneous Cu–Ni coating on dielectric and/or thermally unstable substrate	Arkhangelskii *et al.*, 1987
Ag	$AgNO_3$	Freeze-drying of suspension followed by cryo-precipitation by K_2CO_3 solutions	Thermal decomposition at 320–340 °C, 30 min	Ultradisperse (0.01 μm) Ag powder with minimum sintering temperature reduced up to 0.35 T_{melt}	Tretyakov *et al.*, 1987
Zn	Zn	Cryogrinding at 77 K	No thermal treatment	Cryogrinding of Zn granules in the field of low-temperature frangibility leads to preparation of submicron (<0.1 μm) Zn powder	Pyagai, 1982; Pyagai *et al.*, 1982a
Fe	Fe	Cryogrinding at 77 K	As above	Milling of carbonyl Fe at low temperatures results in breaking the initial aggregates of 200–1000 particles for single 1–3 μm sized Fe grains	Pyagai, 1982; Pyagai *et al.*, 1982b

TABLE 3.11 Application of cryochemical processes for the preparation of inorganic composites.

Composition	*Precursor*	*Applied technique*	*Heat treatment*	*Main results*	*References*
Co-MgO, $Ni-Al_2O_3$, Pb-MgO, $Fe-ThO_2$, $Co-Ni-ThO_2$	Not described	Freeze-drying of aqueous solutions	Not described	Dispersion-strengthened composites with Me particle size 100–500 Å	Gelles and Roehrig, 1972
(a) $Cu-ThO_2$ (b) Cu-W (c) $Ni-ThO_2$ $Fe-ThO_2$ $Cu-Ni-ThO_2$ $W-SiO_2$ Pb-MgO	(a) $CuSO_4*5H_2O$; $Th(SO_4)_2$ (b) $Cu(OH)_2$; $H_2W_4O_{13}*9H_2O$ (c) Not described	(a) Freeze-drying of aqueous solutions (b) Freeze-drying of suspension (c) Not described	(a) Vacuum dehydration: 120 °C, 8 h. Thermal decomposition: 980 °C, 1 h, air. Reduction: 98 °C, 1 h, H_2 (b, c) Not described	Fine powders of metals with homogeneously distributed 60–90 nm particles of minor component are used in the synthesis of dispersion-strengthened composites	Roehrig and Wright, 1972
$Cu-Al_2O_3$	$Cu(SO_4)$ or $Cu(NO_3)_2$; $Al_2(SO_4)_3$ or $Al(NO_3)_3$	(a) Freeze-drying of aqueous solution of sulphates (b) Cryoprecipitation by KOH at 253 K followed by freeze-drying of suspension (c) Cryoprecipitation of $CuC_2O_4*0.5H_2O$ by water–alcohol solution of $H_2C_2O_4$ at 253 K followed by freeze-drying of product suspension in aqueous solution of $Al_2(C_2O_4)_3$ (d) Co-precipitation by KOH at 20 °C followed by freeze-drying of frozen suspension	Thermal decomposition: 970–1000 °C, air. Reduction: 530–550 °C, H_2 flow	Method of quantitative estimate of conductive composites homogeneity based on determination of percolation threshold and dispersions of EDX intensities allowed to confirm better homogeneity of composite obtained through solution freeze-drying	Mozhaev, 1988; Graboy *et al.*, 1991
Ag-CdO	Nitrates	Cryoprecipitation by K_2CO_3 followed by freeze-drying of suspension	Thermal decomposition 350 °C, 1 h. Air annealing 380–500 °C, 1 h. Sintering: 815–900 °C, 1 h	Higher sintered density (up to 99.5%) and lower grain size of oxide (0.2–0.8 μm) resulted in four times enhancement of wear resistance of electric contacts compared to conventional co-precipitation method	Shabatin, 1986; Shabatin *et al.*, 1983
Ag-(Bi_2O_3:B_2O_3:V_2O_5) Ag-Pt-(Bi_2O_3:B_2O_3:V_2O_5)	$Bi(NO_3)_3$, $VOSO_4$, H_3BO_3, $AgNO_3$	Freeze-drying of suspension of Ag and oxide gless powder followed by (A) and (B): (A) synthesis of Ag powder by cryoprecipitation; (B) synthesis of oxide glass powder by cryoprecipitation and suspension freeze-drying	Thermal decomposition at intermediate stages: (A) 350 °C, 1 h, air (B) RF plasma treatment	Due to better homogeneity of Ag distribution the conductivity of thick films obtained from these powders improved three times compared to ones from standard precursors.	Shabatin, 1986; Shabatin *et al.*, 1987a, 1988
$Si_{6-z}Al_zO_zN_{8-z}$-SiC ($z = 0.3; 0.75$)	β'-sialon, SiC particulates	Freeze-drying of aqueous suspension	Sintering at 1700–1800 °C, 1 h or hot pressing: 1400–1500 °C, 30 MPa, 1 h., or sintering at $P = 10$ atm. 1700 °C, 1 h in the purified N_2 or $N_2/CO = 1/1$ or $Ar/N_2/CO = 90/5/5$	Reduced agglomeration of components in the green compact achieved by freeze-drying processing combined with hot pressing gives a chance for improvement of strength of SiC-reinforced composite.	Hanson *et al.*, 1992

TABLE 3.12 Application of cryochemical processes to the preparation of miscellaneous inorganic substances and materials.

Material compositions	*Precursors*	*Applied techniques*	*Heat treatment*	*Main results*	*References*
ThO_2	$Th(NO_3)_4$	Freeze-drying of aqueous solutions	Thermal decomposition: 235 °C, argon. Air calcination: 800 °C. Sintering: 1800 °C, 12 h, wet H_2	Free-flowing granulated powders of ThO_2 are used in preparation of dense (97–98%) nuclear fuel compacts. Reduced number of unit operations compared to existing process promotes higher safety during synthesis	Clayton *et al.*, 1988
ThO_2, UO_2	$Th(NO_3)_4$; $UO_2(NO_3)_2$	Freeze-drying of aqueous solutions or suspensions obtained by treatment of cryogranules with NH_4OH	Treatment of freeze-dried salts by NH_4OH followed by thermal decomposition	Ability of cryogranules to keep spherical forms during reaction with NH_4OH is used for preparation of spherical UO_2 and ThO_2 particles	Tinkle, 1973
UO_2	$UO_2(NO_3)_2$*$2H_2O$	Freeze-drying of aqueous solutions	Thermal decomposition: by heating at 80 K/h to 300 °C. Reduction at 670 °C, 16 h, He-6%H_2 mixture. Sintering at 1650 °C, 3 h, He-6%H_2	Dense (94%) sintered UO_2 pellets have been obtained from free-flowing, dust-free UO_2 powders while no radioactive waste streams appeared during their synthesis	White *et al.*, 1981
Radioactive wastes and biological objects	Radioactive waste solutions and slurries	Freeze-drying of aqueous solutions and suspensions	No thermal treatment	Concentration of radioactive substance in solid freeze-drying product or in sublimed ice	Takanobu and Kondoh, 1985; Kondo *et al.*, 1985; Wickenden, 1993
Model nuclear fuel storage tank waste sludge	Disodium nickel ferrocyanide, sodium nitrate, sodium nitrite	Freeze-drying of suspensions	As above	Preparation of uniform sample for analysis during model studies	Bajic *et al.*, 1995
Phosphate binder	Al–Cr–phosphate slurry	Freeze-drying of suspension	As above	Two times enhancement of adhesion to aluminosilicate refractories	Tretyakov *et al.*, 1988
$Fe(OH)_{3-x}(NO_3)_x$	$Fe(NO_3)_3$	Freeze-drying of colloidals solutions obtained by anion exchange treatment of nitrate solution	As above	Anion exchange treatment gives a possibility to obtain wide spectrum of hydroxinitrates with different NO_3/OH ratio and fractal dimension	Vertegel *et al.*, 1995; Kalinin *et al.*, 1995
$Fe(OH)_3$	$Fe(OH)_3$	Freeze-drying of aqueous suspension	As above	Atomic structure of $Fe(OH)_3$ around ferric ion has been kept unchanged in the course of freeze drying	Shinoda *et al.*, 1994
$Fe_2(SO_4)_{3-x}(OH)_{2x}$ $Al(NO_3)_{3-x}(OH)_x$	$Fe_2(SO_4)_3$ $Al(NO_3)_3$	Freeze-drying of colloidal solutions and colloidal suspensions obtained by titration of aqueous solutions by ammonia	As above	Preparation of colloidal powders of ferric hydroxosulfate and aluminium hydroxonitrate	Jaeger *et al.*, 1974

TABLE 3.12 (*Continued*)

Material compositions	*Precursors*	*Applied techniques*	*Heat treatment*	*Main results*	*References*
WC W_2C	$(NH_4)_2WO_4$; graphite	Freeze-drying of colloidal graphite suspension in aqueous solution of tungstate	Reduction in H_2; vacuum annealing at 1500 °C for 30 min	Fine-grained mixture of 85% of WC and 15% W_2C consists of 20–50 nm particles aggregated into 700 nm chains	Roehrig and Wright, 1972
SiC	SiC	Freeze-drying of aqueous suspension	Not described	Optimization of green compact preparation leads to micro-uniform grain size distribution and better sintered density	Mueller-Zell and Hahn, 1987
$CaLa_2S_4$	$La(CH_3COO)_3$; $Ca(CH_3COO)_2$	Freeze-drying of aqueous solution	Thermal decomposition: 700 °C, 3 h. Sulphurization: 850 °C, 48 h, H_2S. Sintering: 1200–1400 °C, 8 h, H_2S. Hot isostatic pressing: 1400 °C, 200 mPa, argon. S-annealing: 1400 °C, 2 h, H_2S	The density of pellet attains 95.3% th.d. after sintering and ~ theoretical value after HIPing. Narrow grain size distribution improves transparency up to 40% at $\lambda = 13\ \mu m$, for 0.45 mm thick ceramics	Wang, 1993
(a) CuO (b) Y_2O_3	Nitrates	Freeze-drying of aqueous solutions	Thermal; decomposition: (a) 250–270 °C (b) 570–630 °C	Synthesis of fine powders of CuO ($D = 0.2\ \mu m$, SSA $\leq 6.5\ m^2/g$) and Y_2O_3($D = 0.1\ \mu m$, SSA $\leq 40\ m^2/g$)	Alexeev *et al.*, 1990
NiO	$NiCl_2$ or $Ni(NO_3)_2$ or $NiSO_4$	Freeze-drying of aqueous solutions and suspensions obtained by NaOH precipitation from chloride solutions	Thermal decomposition: 75–425 °C, 5 h, vacuum + 550 °C, 12 h, air	Optimization of thermal decomposition conditions resulted in development of specific surface area up to 150 m^2/g	Gay-Lancermin and Dufour, 1974
NiO	$NiSO_4$	Freeze-drying of aqueous solution	Thermal decomposition at 925 °C, 10 h. Sintering at 1400–1500 °C	Dense (96–97% th.d.) NiO ceramics is used to study the influence of different factors on the mechanism of NiO sintering and grain growth	Harris *et al.*, 1987
$Al_{2-x}Cr_xO_3$	$Al(NO_3)_3$ $Cr(NO_3)_3$	Freeze-drying of suspension of co-precipitated hydroxides	Thermal decomposition at 1330 °C, 10 h	High homogeneity of Cr distribution allowed to study $(Cr^{3+})_x$ clusters formation in $Al_{2-x}Cr_xO_3$ solid solution	Viktorov *et al.*, 1989
$Al_{2-x}Cr_xO_3$	$Al(NO_3)_3$ $Cr(NO_3)_3$	Freeze-drying of gel, obtained by urea hydrolysis of aqueous solutions	Not described	Decrease the temperature of solid solution synthesis and improvement of optimal homogeneity of powder ruby	Loukachine *et al.*, 1995
(a) Fe_2O_3 (b) Al_2O_3	(a) $FeSO_4$ (b) $Al_2(SO_4)_3$	Freeze-drying of aqueous solutions	Thermal decomposition: (a) 600–800 °C, 2–64 h (b) 1100–1200 °C, 8–10 h	300–500 Å oxide particles are joined into the chain-like aggregates	Johnson and Schnettler, 1970; Johnson and Gallagher, 1971a, b
Fe_2O_3	$(NH_4)_4Fe(CN)_6$ $(NH_4)_3Fe(C_2O_4)_3$ $FeSO_4$; $Fe(NO_3)_3$	Freeze-drying of aqueous solutions	Thermal decomposition 475–660 °C, 2–5 h. Sintering: 600–1200 °C, 16 h	Study of the influence of salt precursor on the morphology of Fe_2O_3 powders and their sintering behaviour. Dependence of densification rate on agglomeration is not observed	Gallagher *et al.*, 1976
Fe_2O_3	$(NH_4)_3Fe(C_2O_4)_3$	Freeze-drying of aqueous solution	Thermal decomposition: 300 °C, 1 h, air	Finely dispersed Fe_2O_3 powder with specific surface area up to 160 m^2/g	Mozhaev *et al.*, 1982

TABLE 3.12 *(Continued)*

Material compositions	*Precursors*	*Applied techniques*	*Heat treatment*	*Main results*	*References*
Fe_2O_3	$FeSO_4$	Freeze-drying of aqueous suspension obtained by precipitation by $NaCO_3$ and air bubbling	Thermal decomposition: 125–400 °C, air	Transparent iron oxide pigment	Patil and Netherton, 1977
Na-alumosilicate	Alkali blue pigment powder	Freeze-drying of aqueous suspension	No thermal treatment	Soft-grained alkali blue pigment with high colouring strength	Iyengar, 1982
$Fe_2O_3(Li_2O)$	$(NH_4)_2Fe(SO_4)_2$ $*6H_2O$; Li_2SO_4	Freeze-drying of aqueous solutions	Thermal decomposition: 850 °C, 15 min	Thermoelectric power measurements confirmed better homogeneity of dopant distribution compared to traditional methods	Oleinikov *et al.*, 1974; Pershin, 1977
(a) $Fe_2O_3(Li_2O)$ (b) $Fe_2O_3(SnO_2)$	(a) $(NH_4)_2Fe(SO_4)_2$ $*6H_2O$; Li_2SO_4 (b) $FeSO_4$, $SnSO_4$	Acetone cryo-extraction of frozen aqueous solutions	Thermal decomposition: 850 °C, 2 h	As above	Kuznetsova *et al.*, 1979a, b; Mozhaev, 1988
Surface-doped ZnO	$Zn(NO_3)_2$, $SbCl_3$, $Bi(NO_3)_3$, $Co(NO_3)_2$, $Mn(NO_3)_2$, $Cr(NO_3)_3$, $Ni(NO_3)_2$, Na_4SiO_4	Freeze-drying of suspensions obtained by precipitation from individual solutions by aqueous ammonia and ammonium bicarbonate	Thermal decomposition of separate powders at 300 °C. Mixing of components and sintering at 1300 °C for 2 h	ZnO varistor with excellent non-linearity to voltage, life performances and capability of energy dissipation	Kanai *et al.*, 1983
TiO_2, SiO_2, Al_2O_3, Fe_2O_3, ZrO_2, SnO_2, Sb_2O_3	Nitrates, Chlorides	Freeze-drying of colloidal solutions	—	Preparation of flakes of inorganic oxides	Sato *et al.*, 1985
SiO_2	Aqueous sol of silica	Freeze-drying of suspension	No thermal treatment	Structurization of aqueous gel is avoided	Kolosentsev *et al.*, 1975
Ag_2O	Ag_2O	Freeze-drying of aqueous suspension of Ag_2O in acrylic acid and surfactant-contained solution	As above	Ag_2O-cathode of silver oxide battery	Toshiba Battery Co., 1983
Ta_2O_5, Al_2O_3, TiO_2, Nb_2O_5, HfO_2, ZrO_2, SiO_2, Ge_2O_3	Hydroxides	Freeze-drying of aqueous suspensions	As above	Thick films obtained from synthesized powders are used as component of humiding sensors	Matshushita Electric Co., 1981
Pt- and Sb-doped $SnO_2(TiO_2)$	$SnCl_4$, $TiCl_4$, $SbCl_5$, $H_2PtCl_6*6H_2O$	Freeze-drying of suspensions obtained by co-precipitation by aqueous ammonia and $(NH_4)_2CO_3$	Thermal decomposition: 600 °C, 1 h, N_2. Thick-film preparation by screen printing and annealing at 600–1000 °C for 2 h in N_2 ambient	Homogeneous distribution of TiO_2 and Pt dopants considerably improves the sensitivity of thick film sensor to hydrocarbon gases	Chung *et al.*, 1992; Chung and Lee, 1993
β-MoO_3	Na_2MoO_4	Freeze-drying of aqueous solution of molybdic acid obtained by cation exchange treatment of molybdate solution	Thermal decomposition: 300 °C, 1 h, O_2	Freeze-drying synthesis of recently discovered modification MoO_3 allowed to perform its structural refinement by neutron diffraction	Parise *et al.*, 1991

TABLE 3.12 *(Continued)*

Material compositions	*Precursors*	*Applied techniques*	*Heat treatment*	*Main results*	*References*
SiO_2, ZrO_2, Al_2O_3, $ZrSiO_4$	Oxides	Freeze-drying of aqueous suspension containing organic binder	No thermal treatment	Preparation of fine powders for application in abrasive suspensions	Zhiboedov *et al.*, 1977; Shabatin *et al.*, 1987; Naito *et al.*, 1987; Nippon Mining Co., 1983; Rieser, 1983
Doped oxides, aluminates, ferrites	Not described	Freeze-drying of aqueous solutions and suspensions	Not described	Review of results obtained in Laboratoire des microstructures (CNRS) and by other authors	Paulus, 1976
Individual and multicomponent oxides	Not described	Freeze-drying of aqueous solutions and suspensions	Not described	A review of different methods of oxide synthesis with special attention being paid to titanates and aluminates	Cousin and Ross, 1990
Inorganic oxidizers	$MClO_4$, MNO_3 ($M = NH_4$, Na, K)	Freeze-drying of aqueous solution	No thermal treatment	Fine grained (≤ 1 μm) powders of inorganic oxidizers for propellant applications	McBride, 1974
Ammonium perchlorate	NH_4ClO_4	Freeze-drying of aqueous solution	As above	Needle-like powders with particle size 2–5 μm, for propellant applications	Kohga *et al.*, 1992, 1993
δ-$Cu(HCOO)_2$	$Cu(HCOO)_2$	Freeze-drying of aqueous solution	As above	New metastable form of copper formate easily obtained by freeze-drying method	Arkhangelskii *et al.*, 1987
Complex aluminosilicates	SiO_2, Al_2O_3, TiO_2, $NaNO_3$, KNO_3, $Cu(NO_3)_2$, $Mg(NO_3)_2$, $LiNO_3$	Freeze-drying of multicomponent suspension of oxides in aqueous solution of nitrates	Air decomposition: 200 °C, 2 h, stepwise heating from 400 °C to 750 °C with 12 h, annealings and intermediate grindings. Defoaming at 850 °C, 60–90 min. Vacuum melting at 1100 °C, 2 h and 1300 °C, 1 h, followed by quenching	Synthesis of feldspathic glasses for bioapplication studies	Rizkalla *et al.*, 1991

Among many publications devoted to cryochemical syntheses of individual and complex metal oxides (see Table 3.12) especially interesting are those concerning successful development of high-quality precursors of some advanced materials, including:

- ruby solid solutions $Al_{2-x}Cr_xO_3$ with high optical homogeneity;
- pigments with improved transparency and colouring strength;
- metal oxide varistors with improved characteristics such as non-linearity to voltage, service life and capability of energy dissipation;
- SnO_2-based thick film gas sensors with higher sensitivity to hydrocarbon gases;
- fine abrasive powders for polishing of contact lenses and optical products of water-soluble glasses.

Finally, the successful attempts of cryochemical synthesis of ammonium perchlorate, new metastable form of copper formate, and the preparation of feldspathic glasses should also be mentioned.

3.12 BIOMEDICAL MATERIALS AND DRUGS

Biomedical materials are synthetic polymers, metals, ceramics and natural macromolecules (biopolymers), which are manufactured or processed to be suitable for use in or as medical devices or prostheses (Helmus, 1990). These materials typically come into contact with cells, proteins, tissues, organs and organ systems. They can be implanted for long-term use, e.g. an artificial hip, or for temporary use, e.g. an intravenous catheter. Besides some cases, when a material is used by itself, such as collagen injections for filling soft tissue defects, biomedical materials are used as components in medical devices (Helmus, 1991). Table 3.13 summarizes the most common types of biomedical materials.

It is well known that such materials are used for various purposes – from manufacturing of artificial eye crystalline lenses to artificial heart valves. It cannot be denied that the processes of cryochemical technology could be useful everywhere; however, at present there is a limited range of biomaterials prepared by using low-temperature processes. Some of the most common examples are cited below. Some of them concern preparation of biopolymers. Freeze-drying of polyurethane solutions was used for production of porous materials as an artificial leather substitute (Taguchi, 1991). The same technique was used to prepare filter membranes based on acylated chitin (Kimita *et al.*, 1989). A particular place among biopolymers is occupied by collagenes obtained as powders, sponges and composites. Using freeze-drying techniques, collagen sponges (Takazawa and Morita, 1991), composites (Bastchenko *et al.*, 1992; Berchenko *et al.*, 1992) and water-soluble powders for cosmetic applications (De Carcaradec, 1982) were prepared.

Another series of cryochemical applications relates to bioceramics, mainly to hydroxyapatite $Ca_{10}(PO_4)_6(OH)_2$ (Hattori *et al.*, 1986, 1987; Shabatin *et al.*, 1991). The formation of the latter material comprises a complex topochemical process regardless of the preparation route. Freeze-drying of gel obtained by mixing of calcium acetate $(CH_3COO)_2Ca$ and triethylphosphate $(C_2H_5O)_3PO_4$ solutions with subsequent thermal treatment made it possible to obtain a finely-dispersed product with suppressed aggregation of primary particles.

Of great interest is the experience of preparation of ferments immobilized on sorbents by freeze-drying of sorbent suspension co-precipitated with the enzyme (Kennedy and Kay, 1976).

The most important technique for conservation of biological activity of biopreparations and drugs is the freeze-drying of water and alcohol solutions

TABLE 3.13 Biomedical materials and applications.

Materials	*Application*
Non-degradable synthetic-commodity polymers Polyamides, polycarbonates, polyesters, polyformaldehyde, polyolefins, polyvinyl chloride	Housing materials, vascular grafts, heart valve stents, sutures, mesh for hernia repair, tubing, blood bags
Non-degradable synthetic-value-added polymers Fluorocarbons, hydrogels, polyolefin elastomers, polyurethanes, silicones, ultrahigh molecular weight polyethylene	Vascular grafts, contact lenses, catheter coatings, tubing, artificial heart bladder, catheters, soft-tissue reconstruction, tubing, acetabular cup
Biodegradables Albumin, cross-linked, collagen/gelatin, polyamino acids, poly-anhydrides, polycaprolactones, polyactic/glycolic acid copolymers, polyhydroxybutyrates, polyorthoesters	Vascular grafts coatings, cell encapsulation, soft-tissue reconstruction, controlled release, cell adhesion peptides, bone plates, sutures
Biologically derived materials Bovine carotid artery, ligaments and pericardium, human umbilical vein, porcine heart valves	Vascular grafts, ligaments, pericardial substitute, heart valves
Bioderived macromolecules Chitosans, collagen, elastin, gelatin, hyaluronic acid	Experimental, wound dressing, controlled release, soft tissue injectables, artificial heart bladder coating, surgical non-adhesion
Tissue adhesives Cyanoacrylates, fibrin glue, molluscan glue	Wound closure, microsurgery, vascular graft coating, enhancement of cell adhesion
Metal and metallic alloys Cobalt chrome molybdenum alloys, nitinol (shape memory) alloys, stainless steels, Ti and titanium alloys	Orthopaedic and dental implants, vascular stents, heart valve stents, orthopaedic wire, artificial heart housing
Ceramics, inorganics and glasses Aluminium, calcium and phosphorous oxides, bioglass, glass ceramics, high-density alumina, hydroxyapatites	Degradable bone filler, enhanced bone growth, bioactive phosphorous calcium glass, orthopaedic coating, encapsulation of implantable medical electronic, acetabular cup, ball of hip prosthesis, bioactive ceramic, orthopaedic coating, bone fillers
Carbons Glassy carbons, pyrolitic (low temperature isotropic) carbon, ultralow temperature isotropic carbon	Fibres for orthopaedic composites, heart valves, dental implants, coating on heart-sensitive polymers
Passive coatings Albumin, alkyl chains, fluorocarbons, hydrogels, silica-free silicones	Thromboresistance, absorbs albumin for thromboresistance, reduced drag for catheters, improved wound healing for soft-tissue reconstruction
Bioactive coatings Anticoagulants, antimicrobials, bioactive ceramic and glasses, cell adhesion peptides and proteins, negative surface charge, thrombolytics	Thromboresistance, infection resistance, bone adhesion and formation, soft tissue adhesion, enhanced cell adhesion, epithelium, endothelium

(Hayano *et al.*, 1979; Nippon Chemiphar Co., 1981; Williams *et al.*, 1984). The main features of the process, in this case, consist of using low freezing rates and the use of cryoprotectors (Willemer, 1975) to prevent formation of large ice crystals capable of destroying living tissues. However, in some cases, the crystallization processes of ice under cooling of living organisms are not malicious but beneficial, because the sharp edges of ice crystals can function as a knife cut with respect to malignant tumours.

In some cases freeze-drying techniques make it possible not only to maintain but also to increase the biological activity of powder drugs considerably due to increase of their specific surface and rate of their assimilation by organisms (Leonidov *et al.*, 1984, 1994).

References

Abbaro, S.A., Tseung, A.C.C. and Hibbert, D.B. (1980) Synergism in hydrogen evolution on platinized tungsten trioxide in acid medium, *J. Electrochem. Soc.*, **5**, 1106.

Akinc, M. and Rasmussen, M.D. (1983) Surface characteristics of yttria precursors in relation to their sintering behavior, *Mater. Sci. Res.*, vol. 15 (Advanced Materials Characterisation), 159–70.

Alconchel, S.A., Ulla, M.A. and Lombardo, E.A. (1994) Processing and properties of superconducting $La_{1.85}Sr_{0.15}CuO_4$ powder by double-step calcining, *Materials Science & Engineering B* [Solid-State Materials for Advanced Technology], Vol. B27, Iss: 2–3, 117–27.

Alconchel, S.A., Ulla, M.A. and Lombardo, E.A. (1995) Structural and morphological evolution from a freeze-dried precursor to the $La_{1.85}Sr_{0.15}CuO_4$ superconductor during thermal processing, *Journal of Materials Science*, **30**(7), 1733–41.

Alconchel, S.A., Ulla, M.A. and Lombardo, E.A. (1995) Study of lanthanum-deficient compositions in the systems La–Cu–O and La–Sr–Cu–O, *Journal of Materials Science*, **30**(9), 2227–34.

Alexeev, A.F., Dyakin, V.V., Palekha, K.K. *et al.* (1990) Some features of ultrafine powders of copper and yttrium oxides preparation by cryochemical method, *Powder Metallurgy* (Russ.), **1**, 1.

Alexeev, A.F., Palekha, K.K., Ponomarchuk, S.G. *et al.* (1991) Preparation of powders by cryochemical technology, *Russ. Phys. Chem. Mater. Treatment*, **3**, 121.

Amato, I., Baudrocco, F. and Martorana, D. (1976) Evaluation of freeze drying and spray drying processes for preparing transparent alumina, *Mat. Sci. Eng.*, **26**, 73.

Anastasyuk, N.V. (1972) Studies of effectiveness of chemical methods of ferrites preparation, PhD thesis, Moscow: Moscow State University.

Anderton, D.J. and Sale, F.R. (1979) Production of strontium-doped lanthanum cobaltite conducting oxide powder by freeze-drying technique, *Powder Metallurgy*, **1**, 8.

Andrianov, V.A., Kozin, M.G., Romashkina, I.L. (1991) Reversible and irreversible thermal variations in the emission Mössbauer spectra of $YBa_2Cu_3(^{57}Co)O_{6+x}$, *Superconductivity: Physics, Chemistry, Technique*, **4**(6), 1035–42.

Andrianov, V.A., Kozin, M.G., Romashkina, I.L. *et al.* (1992) Mössbauer studies of the $YBa_2Cu_3(^{57}Co)O_{6+x}$ ceramics under oxygen exchange with a gas ambient, *Physica C*, **192**, 8–12.

Arkhangelskii, I.V., Shabatin, V.P., Tretyakov, Yu.D. *et al.* (1987) The composition for gas phase copper plating of dielectric materials, *Pat. USSR* No. 4254249/26 from 01.06.87.

Avdeev, L.Z., Brandt, N.B., Volkozub, A.V. *et al.* (1987a) Temperature and field dependences of the magnetization of $RBa_2Cu_3O_x$. superconductors (R = Y,Ho), *Pis'ma v Zh. Eksp. Teor. Fiz.*, **46**, Suppl., 82.

Avdeev, L.Z., Brandt, N.B., Volkozub, A.V. *et al.* (1987b) Magnetic properties of bulk and powder samples of the superconductor $YBa_2Cu_3O_x$, *Pis'ma v Zh. Eksp. Teor. Fiz.*, **46**, 86.

Bajic, S.J., Luo Siquan, Jones, R.W. and McClelland, J.F. (1995) Analysis of underground storage tank waste simulants by Fourier transform infrared photoacoustic spectroscopy, *Appl. Spectroscopy*, **49**(7), 1000–5.

Barboux, P., Tarascon, J.M., Greene, L.H. *et al.* (1988) Bulk and thick films of the superconducting phase $YBa_2Cu_3O_{7-\delta}$ made by controlled precipitation and sol–gel process, *J. Appl. Phys.*, **63**(8), 2725–9.

Basargin, O.V., Rudnitskii, L.A., Moshchalkov, V.V. *et al.* (1988) Anomalies of the thermal expansion coefficient in $YBa_2Cu_3O_x$ ceramics, *Solid State Physics*, **30**, 877.

Bashkuev, A.E. (1985) Influence of dopants on the resistivity and microstructure of Ni–Zn ferrite, MSc thesis, Moscow: MSU.

Bastchenko, Y.V., Barinov, S.M., Berchenko, G.N. and Shabatin, V.P. (1992) Hydroxylapatite–collagen composite granules, preparation and behaviour in rats' femoral muscle tissue, *4th World Congress on Biomaterials*, Berlin, April, 536.

Beltran-Porter, D., Martinez-Tamayo, E., Ibanez, R. *et al.* (1989) Comparative study of synthetic procedures for YBaCuO-type oxides, *Solid State Ionics*, **32/33**, 1160–6.

Bentzen, J.J. and Nicholson, P.S. (1980) The preparation and characterization of dense, highly conductive $Na_5GdSi_4O_{12}$ Nasicon (NGS), *Mat.Res.Bull.*, **15**, 1737–45.

Berchenko, G., Bruskina, G., Desyan, G. *et al.* (1992) Use of composition materials based on collagen, hydroxylapatite and tricalcium phosphate for the substitution of bone defects, *III European East–West Conference on Materials and Processes*, Strasbourg, 3–6 Nov., D-1U/P1.

Bermejo, E., Quarton, M. and Lacour, C. (1994) Solid state synthesis and sintering below 1000 degrees C of lithium ferrite freeze-dried powders, *Materials Research Bulletin*, **29**(9), 965–72.

Bermejo, E., Lacour, C. and Quarton, M. (1995) A cryochemical procedure for preparing ultrafine Li–Zn ferrite powder, *Journal of Materials Science Letters*, **14**(19), 1346–8.

Bevan, H.L. and Tseung, A.C.C. (1974) The electrochemical reduction of oxygen on high surface area lithium doped nickel oxides, *Electrochimica Acta*, **9**, 201.

Beyer, L.A., Kalnas, Ch.E., Roy, D. and Lloyd, I.K. (1987) Characterization of freeze-dried alumina powders prepared with and without a heat lamp to speed sublimation, *Am. Ceram. Soc. Bull.*, **66**(11), 1647.

Bich, V.M., Mozhaev, A.P., Kuznetsova, M.A. *et al.* (1983) Physico-chemical studies of Li–Cu ferrite preparation, *Russ. J. Inorg. Mater.*, **19**(5), 807–10.

Bonsdorf, G., Langbein, H. and Knese, K. (1995) Investigations into phase formation of $LiFe_5O_8$ from decomposed freeze-dried Li–Fe-formates, *Mat. Res. Bull.*, **30**(2), 175.

Borisova, N.M., Golubenko, Z.V., Kuz'micheva, T.G. *et al.* (1992) Optimization principles for preparation methods and properties of fine ferrite materials, *J. Magn. Magn. Mater.*, **114**, 317–28.

Brandt, N.B., Moshchalkov, V.V., Gippius, A.A. *et al.* (1988) ESR and magnetic properties of high Tc superconductors $RBa_2Cu_3O_7$ (R = Ho, Sm, Eu, Y), *Solid State Physics*, **30**(4), 1210–14.

Brosha, E.L., Sanchez, E., Davies, P.K. *et al.* (1991) Preparation of pure and chemically substituted $XBa_2Cu_4O_8$ via freeze-dried nitrates, *Physica C*, **184**, 353–61.

Brosha, E.L., Garzon, F.H., Raistrick, I.D. and Davies, P.K. (1994) Low-temperature synthesis and phase equilibria in the Y–Cu–O binary system, *Journal of the American Ceramic Society*, **77**(5), 1139–42.

Brosha, E.L., Garzon, F.H., Raistrick, I.D. and Davies, P.K. (1995) Low-temperature phase equilibria in the Y–Ba–Cu–O system, *Journal of the American Ceramic Society*, **78**(7), 1745–52.

Brosha, E.L., Garzon, F.H., Raistrick, I.D. and Davies, P.K. (1996) Low-temperature synthesis, structure and stability of $Ba_{2+y}Cu_3O_6$, *J. Solid State Chem.*, **122**, 176–85.

De Carcaradec, N. (1982) Lyophilized native collagen preparations for cosmetics, *Fr. Pat.* 82/8.308 from 7 May.

Chang, H.L. and Lloyd, I.K. (1993) Alumina prepared from freeze-dried aluminum isopropoxide, *J. Am. Ceram. Soc.*, **76**(5), 1357–60.

Christen, S., Langbein, H. and Jaenicke-Rössler, K. (1992) The thermal decomposition of crystalline and freeze-dried *m*-oxo-acetato-complexes of manganese and iron, *Thermochimica Acta*, **209**, 253.

Chung, W.Y., Lee, D.D. and Sohn, B.K. (1992) Effects of added TiO_2 on characteristics of SnO_2-based thick film–gas sensors, *Thin Solid Films*, **221**, 304.

Chung, W.Y. and Lee, D.D. (1993) Effects of phase transition of added TiO_2 on characteristics of SnO_2-based hydrocarbon-gas sensors, *Sensors and Actuators B*, **13–14**, 517.

Clayton, J.C., Burke, T.J. and Hutchinson, C.R. (1987) Preparation, characterization, and processing of thoria-based fuels for light water breeder reactors, *Advances in Ceramics*, Vol. 21: Ceramic Powder Science, 69.

Clayton, P.R., Jantsky, J.G., Sigalas, I. and McLachan, D.S. (1988) The influence of chemical preparation routes on the properties and microstructures of $YBa_2Cu_3O_{7-x}$ superconducting ceramics, *Proc. Brit. Ceram. Soc.*, **40**, 19–29.

Coppa, N.V., Myer, G.H., Salomon, R.E. *et al.* (1992) Preparation, thermal processing behavior and characterization of YBCO from freeze-dried nitrate precursors, *J. Mater. Res.*, **7**(8), 2017–26.

Coppa, N.V., Hults, W.L., Smith, J.L. and Brynestad, J. (1994) Synthesis, processing behavior and characterization of bismuth superconductors using freeze-dried nitrate precursors, *J. Mater. Res.*, **9**(10), 2510–18.

Cousin, P. and Ross, R.A. (1990) Preparation of mixed oxides: a review, *Mat. Sci. Eng.*, **A130**, 119.

Davidson, C.R., Kissel, G. and Srinivasan, S. (1982) Electrode kinetics of the oxygen evolution reaction at $NiCo_2O_4$ from 30% KOH. Dependence on temperature, *J. Electroanal. Chem.*, **132**, 129.

Dierickx, D., Van der Biest, O., Rosseel, K. *et al.* (1994a) Melt processing and morphology of freeze-dried $YBa_2Cu_3O_{7-x}$ high T_c superconductors, *Physica C*, **235–40**, 451–2.

Dierickx, D., Van der Biest, O. and Boon, W. (1994b) Controlled precipitation of pinning centres during melt growth of freeze-dried precursors, *Electroceramics IV*, **2**, 979–82, Aachen.

Dierickx, D. and Van der Biest, O. (1995a) Intermediate phases during processing of freeze-dried YBaCuO precursors, *European Journal of Solid State and Inorganic Chemistry*, **32**(7–8), 711–18.

Dierickx, D., Rosseel, K., Boon, W. *et al.* (1995b) Influence of precursor preparation on the Y_2BaCuO_5 size distribution throughout melt processed bulk $YBa_2Cu_3O_x$ superconductors, *Applied Superconductivity 1995, Proceedings of EUCAS 1995*, Institute of Physics, **1**, 175–8.

Dogan, F., Roosen, A. and Hausner, H. (1987) Influence of hydroxide-precursor processing on the densification of yttrium oxide powders, *Advances in Ceramics*, Vol. 21: Ceramic Powder Science, 681–9.

Dou, S.X., Liu, H.K., Apperley, M.H. *et al.* (1990) Critical current density in superconducting Bi–Pb–Sr–Ca–Cu–O wires and coils, *Supercond. Sci. Technol.*, **3**, 138–42.

Dou, S.X., Liu, H.K., Guo, Y.C., Sorrell, C.C. (1991a) Improved Jc-H characteristics and flexibility in superconducting Ag-clad $(Bi,Pb)_2Sr_2Ca_2Cu_3O_{10+y}$ wires, *International Workshop MSU-HTSC II, Chemistry and Technology of High-Temperature Superconductors*, Moscow, October 14-18, *Proceedings*, **1**, 73–82.

Dou, S.X., Liu, H.K., Wang, J. and Bian, W.M. (1991b) Transmission electron microscope investigation and magnetic properties of HIPed Bi–Pb–Sr–Ca–Cu–O, *Supercond. Sci. Technol.*, **4**, 21–6.

Dudnik, E.V., Konchakovskaya, L.D., Karpikov, I.I. and Pavlenko, N.P. (1988) Properties of cryochemical precursor of Mg–Al spinel, *Powder Metallurgy (Russ.)*, **7**, 18.

Elashkin, M.V., Zhukov, A.A., Shabatin, V.P. *et al.* (1989) Superconductivity in $YBa_2Cu_3O_{7-x}$ composites, *Physica C*, **162–4**, 1173–4.

Elashkin, M.V., Shabatin, V.P., Mosolov, A.B. and Tretyakov, Yu.D. (1990) Comparison of the composites Ag–$YBa_2Cu_3O_{7-x}$ by several techniques, *MatTech'90, The first East–West symposium on materials and processes*, 10–18 June, Finland, sec. 4, 5.

Eror, N.G. (1993) Synthesis and microstructure of $Bi_{1.8}Pb_{0.4}Sr_2Ca_{2.2}Cu_3O_x$ obtained from freeze-dried precursors, *International Workshop MSU-HTSC III, Chemistry and Technology of High-Temperature Superconductors*, Moscow, 20–23 Sept., 1–24.

Ezrahni, M., Brohan, I., Marie, A.M. *et al.* (1993) Investigations on structural configurations and texture of $YBa_2Cu_3O_{7-x}$, *Eur. J. Solid State Inorg. Chem.*, **30**, 19–34.

Fadeeva, V.I., Voinov, I.D. and Tretyakov, Yu.D. (1978) Fine crystalline structure of cobalt ferrite with different chemical and thermal prehistory, *Russ. J. Inorg. Mater.*, **14**(2), 312–16.

Fedotov, G.N. (1984) Synthesis and properties of structurized piezocomposites PZT-polymer obtained by cryochemical methods, PhD thesis, Moscow: Moscow State University.

Fedotov, G.N., Metlin, Yu.G. and Tretyakov, Yu.D. (1983) The way of formation of porous ceramic PZT framework, *Pat. USSR* No. 3674676/29-33 from 19.12.83.

Finnerty, T.A., Waychunas, G.A. and Thomas, W.M. (1978) The preparation of starting mixes for mineral synthesis by a freeze-dry technique, *American Mineralogist*, **63**, 415.

Fokin, V.A. (1977) Cryochemical synthesis of ferroelectric materials bazed on PZT, PhD thesis, Moscow: Moscow State University.

Fokin, V.A., Belov, V.V., Zhabotinskii, V.A. *et al.* (1979) The technique to prepare PZT-based polycrystalline oxide material, *Pat. USSR* No. 2824814/29-26 from 28.09.79.

Gallagher, P.K. and Schrey, F. (1970) The thermal decomposition of freeze-dried tantalum and mixed lithium–niobium oxalate, *Thermochim. Acta*, **1**, 465–76.

Gallagher, P.K., Johnson Jr., D.W., Remeika, J.P. *et al.* (1975) The activity of $La_{0.7}Sr_{0.3}MnO_3$ without Pt and $La_{0.7}Sr_{0.3}MnO_3$ with varying Pt contents for the calalytic oxidation of Co, *Mat. Res. Bull.*, **10**, 529.

Gallagher, P.K., Johnson Jr., D.W. and Schrey F. (1976) Some effects of the source and calcination of iron oxide on its sintering behavior, *Am. Ceram. Soc. Bull.*, **55**(6), 589–93.

Gallagher, P.K., Johnson Jr., D.W. and Vogel, E.M. (1977) Preparation, structure, and selected catalytic properties of the system $LaMn_{1-x}Cu_xO_{3-y}$, *J. Am. Ceram. Soc.*, **60**(1–2), 28–31.

Gay-Lancermin, C. and Dufour, L.C. (1974) Freeze drying and reactivity of mineral powders, *Powder Technol.*, **10**(6), 289–93.

Gazza, G.E. (1972) Hot pressing of $LiAl_5O_8$, *J. Am. Ceram. Soc.*, **55**(3), 172–3.

Gelles, S.H. and Roehrig, F.K. (1972) Freeze-drying metals and ceramics, *Journal of Metals*, June, 23.

Goldshtein, R.V., Elashkin, M.V., Klimov, D.M. *et al.* (1990) Density and mechanical properties of composite $YBa_2Cu_3O_{7-x}$–Ag materials, *Reports of the USSR Academy of Sciences*, **313**(5), 1121–4.

Goodilin, E.A., Popov, G.Yu., Oleinikov, N.N. *et al.* (1995) The formation of metastable superconducting phase $YBa_2Cu_3O_{7-x}$, *Reports of the Russian Academy of Sciences*, **344**(6), 773–5.

Gorst, J., Do, D.D. and Desai, N.J. (1989) Removal of nitric oxide from fuel gas using mixed oxide sorbents, *Fuel Processing Technology*, **22**, 5.

Graboy, I.E., Mozhaev, A.P. and Tretyakov, Yu.D. (1991) Comparison of chemical homogeneity of composites obtained by different methods via specific electro-conductivity measurements, *Russ. J. Inorg. Materials*, **27**(4), 870–2.

Green, D.J. (1988) Preparation of freeze-dried powder for the manufacture of light weight ceramic thermal insulators, *US Pat.* Appl. 213, 392 from 30 June.

Green, D.J. and Hutchison, S. (1980) Solution spray-dried and freeze-dried sodium beta-alumina powders: preparation and hot-pressing, *Mater. Sci. Monogr.*, **6**, 964–80.

M'Hamdi, E.M. and Lacour, C. (1992) Synthèse du compose $Bi_2Sr_2Ca_2Cu_3O_{10}$ sans Pb à partir d'une poudre lyophilise, *Ann.Chim. Fr.*, **17**, 421–32.

M'Hamdi, E.M., Morineau, R. and Lacour, C. (1993) Etude du mecanisme de formation et de development de $Bi_2Sr_2Ca_2Cu_3O_{10}$ a partir d'un precurseur lyophilise, *Ann. Chim. Fr.*, **18**, 139–57.

Hanson, W.B., Bower, R.M. and Hendry, A. (1992) Atmospheres for sintering ceramic matrix composites containing silicon carbide, *Key Engineering Materials*, **72–4**, 127.

Harris, L.B., Taylor, R.I. and Atkinson, A. (1987) Preparation of polycrystalline nickel oxide for diffusion studies, *J. Mater. Sci.*, **22**, 1993.

Hattori, T., Tatsumoto, H. and Hiraya, A. (1986) Cryochemically prepared Al–*i*-propoxide and its thermal decomposition, *Yogyo Kyokaishi*, **94**, 922–9.

Hattori, T., Iwadate, Y., Inai, H. *et al.* (1987) Preparation of hydroxylapatite powder using freeze drying method, *Yogyo Kyokaishi*, **95**(8), 825–7.

Hausner, H. and Roosen, R. (1983) Ceramic powders of high sintering activity, *Germ. Pat.* No. 3, 310, 924 from 25 March.

Hayano, K., Yoshikura, J. and Tsuru, S. (1979) Effects of drying and freeze drying on *b*-glucosidase in soil, *Toketsu oyobi kanso kenkyukai kaishi*, **25**, 36–9 (CA 94-152345v).

Helmus, M.N. (1990) Outlook for biomedical materials, *DR reports*, Decision Resources, Inc.

Helmus, M.N. (1991) Overview of biomedical materials, *MRS Bulletin*, **16**(9), 33–8.

Hibbert, D.B. and Tseung, A.C.C. (1979) A critical examination of a cryochemical method for the preparation of high surface area semiconducting powders. Part 2. Non-aqueous solvents. The behaviour of solid solutions in freeze-drying, *J. Mater. Sci.*, **14**, 2665.

Hibbert, D.B., Lovegrove, J. and Tseung, A.C.C. (1987) A critical examination of a cryochemical method for the preparation of high surface area semiconducting powders. Part 3. Factors which determine surface area, *J. Mater. Sci.*, **22**, 3755.

Ho Jang-Cheng, Liu Kuo-Shung and Lin I-Nan (1995) Synthesis of $Pb(Mg_{1/3}Nb_{2/3})O_3$ perovskite by an alkoxide method, *Journal of Materials Science*, **30**(15), 3936–43.

Hobbs, B.S. and Tseung, A.C.C. (1972) The anodic oxidation of hydrogen on platinized tungsten oxides. I. Composition of tungsten blue in platinized WO_3 hydrogen electrocatalyst, *J. Electrochem. Soc.*, **119**(5), 580–3.

Hodge, J.D. (1983) Powder processing and crystallization of β- and β''-aluminas, *Ceram. Bull.*, **62**(2), 244–8.

Horn, J., Borner, M., Semmelhack, H.C. *et al.* (1991) On the preparation of $YBa_2Cu_4O_8$, *Solid State Comm.*, **79**(6), 483–6.

Horowitz, H.S., Bordia, R.K., Torardi, C.C. *et al.* (1989) The effect of synthesis method on the processing and properties of $YBa_2Cu_3O_{6+x}$, *Solid State Ionics*, **32/33**(2), 1087–9.

Ibigawa Electric Industry Co. Ltd. (1981) Silicon carbide sinters, *Jap. Pat.* No. 57, 145, 077 from 5 May.

Ito, T., Kimura, Y. and Hiraki A. (1991) High-quality $YBa_2Cu_3O_{7-x}$ ceramics prepared from freeze-dried nitrates, *Jap. J. Appl. Phys.*, **30**(7B), L1253–5.

Iwaida, S., Itatani, K., Howell, F.S. *et al.* (1994) Sinterability of yttria-stabilized zirconia powder prepared by ultrasonic spray freeze-drying technique, *Materials and Manufacturing Processes*, **9**(4), 665–79.

Iyengar, D.R. (1982) Soft grained alkali blue pigments with high colouring strength, *Ger. Pat.* 3, 340, 488 from 12 Nov.

Jaeger, R.E., Miller, T.J. and Williams, J.C. (1974) Effects of ammonium hydroxide on phase separation in the cryochemical processing of salt solutions, *Ceramic Bulletin*, **53**(12), 850.

Johnson Jr., D.W. and Schnettler, F.J. (1970) Characterization of freeze-dried Al_2O_3 and Fe_2O_3, *J. Am. Ceram. Soc.*, **53**(8), 440–4.

Johnson Jr., D.W. and Gallagher, P.K. (1971) Kinetics of the decomposition of freeze-dried aluminum sulfate and ammonium aluminum sulfate, *J. Am. Ceram. Soc.*, **75**(9), 461–5.

Johnson Jr., D.W. and Gallagher, P.K. (1971) The kinetics of the decomposition of freeze-dried iron(II) sulfate, *J. Phys. Chem.*, **75**(9), 1179–85.

Johnson Jr., D.W., Gallagher, P.K., Nitti, D.J. *et al.* (1974) Effect of preparation technique and calcination temperature on the densification of lithium ferrites, *Am. Ceram. Soc. Bull.*, **53**(2), 163–7.

Johnson Jr., D.W., Gallagher, P.K., Schrey, F. and Rhodes, W.W. (1976) Preparation of high surface area substituted $LaMnO_3$ catalysts, *Ceramic Bulletin*, **55**(5), 520.

Johnson Jr., D.W., Gallagher, P.K., Schnettler, F.J. and Vogel, E.M. (1977) Novel preparative techniques for supported oxide catalysts, *Ceramic Bulletin*, **56**(9), 785.

Johnson, S.M., Gusman, M.T. and Rowcliffe, D.J. (1987) Preparation of superconducting powders by freeze-drying, *Adv. Ceram. Materials*, **2**(3B), 337–42.

Kalinin, S.V., Vertegel, A.A., Oleynikov, N.N. and Tretyakov, Yu.D. (1995) Synthesis and properties of fractal particles of iron(III) hydroxonitrate, *35th IUPAC Congress*, 14–19 Aug., Istanbul, Abstracts II, sections 4–6, 1385.

Kanai, H., Takahashi, T., Imai, M. *et al.* (1983) Metal oxide varistor made by a co-precipitation process and freeze-dried, *U.S. Pat.* No. 4, 540, 971 from 22 June.

Kawakami, T., Miura, H., Hashiba, I. and Hattori, Y. (1990) Manufacture of hydrogen catalyst, *Jap. Pat.* No. 04, 200, 639 from 29 Nov.

Kazin, P.E., Shabatin, V.P. and Pershin, V.I. (1992) Cryochemical synthesis and magnetic properties of fine powders of $YBa_2Cu_3O_7$ superconductor, *Russ. J. Inorg. Materials*, **28**(9), 1887–91.

Kazin, P.E., Shlyakhtin, O.A., Putlyaev, V.I. *et al.* (1995) Preparation of Bi-2212 with submicron alien phase precipitates for flux pinning improvement, *International Workshop MSU-HTSC IV, Chemistry and Technology of High-Temperature Superconductors*, Moscow, 7–12 Oct., OR-14.

Kemnitz, E., Scheurell, S., Naumov, S.W. *et al.* (1993) New aspects on fluorination of Bi(Pb)–Sr–Ca–Cu–O – reasons for the improved superconducting properties, *Eur. J. Solid State Inorg. Chem.*, **30**, 701–12.

Kennedy, J.F. and Kay, I.M. (1976) Hydrous titanium oxides – new supports for the simple immobilization of enzymes, *J.C.S. Perkin I*, **3**, 329.

Kim, Y.S. and Monforte, F.R. (1971) Theoretically dense (99.9%) polycrystalline alumina prepared from cryochemically processed powders, *Ceramic Bulletin*, **50**(6), 532.

Kimita, M., Jitzuzauki, K. and Takagi, S. (1989) Porous chithin-based materials and manufacture, *Jap. Pat.* No. 02, 261, 802 from 3 Apr. (CA 114-45363).

Kimura, Y., Ito, T., Yoshikawa, H. *et al.* (1990) Growth and characterization of homogeneous $YBa_2Cu_3O_{7-x}$ powders prepared by a freeze-drying method, *Jap. J. Appl. Phys.*, **29**(8), L1409–11.

Kimura, Y., Ito, T., Yoshikawa, H. and Hiraki, A. (1991) Superconducting $YBa_2Cu_3O_{7-x}$ particles prepared from freeze-dried nitrates, *Jap. J. Appl. Phys.*, **30**(5A), L798–801.

King, H.W., Payzant, E.A. and Bhatnagar, A. (1989) Effect of powder processing on normal state properties of the superconductors $YBa_2Cu_3O_{7-x}$ and $ErBa_2Cu_3O_{7-x}$, *J. Canad. Ceram. Soc.*, **58**(3), 48–54.

King, W.J.and Tseung, A.C.C. (1974a) The reduction of oxygen on nickel–cobalt oxides – I. The influence of composition and preparation method on the activity of nickel–cobalt oxides, *Electrochimica Acta*, **19**, 485.

King, W.J. and Tseung, A.C.C. (1974b) The reduction of oxygen on nickel–cobalt oxides – II. Correlation between crystal structure and activity of Co_2NiO_4 and related oxides, *Electrochimica Acta*, **19**, 493.

Kirchnerova, J. and Hibbert, B. (1990) Vacancy ordering in $La_{1-x}Sr_xCoO_{3-y}$ prepared by calcination of freeze-dried powders of nitrates, *Mat. Res. Bull.*, **25**, 585.

Kirchnerova, J. and Hibbert, D.B. (1993) Structures and properties of $La_{1-x}Sr_xCoO_{3-y}$ prepared by freeze drying, *J. Mat. Sci.*, **28**, 5800.

Kirchnerova, J., Klvana, D., Vaillancourt, J. and Chaouki, J. (1993) Evaluation of some cobalt and nickel based perovskites prepared by freeze-drying as combustion catalysts, *Catalysis Letters*, **21**, 77–87.

Kirchnerova, J. and Klvana, D. (1994) Preparation and characterization of high surface perovskite electrocatalysts, *Int. J. Hydrogen Energy*, **19**(6), 501.

Kirchnerova, J., Klvana, D., Hinatsu, J. *et al.* (1995) New Ni–Co based perovskites as electrocatalysts in alkaline fuel cells, *New Materials for Fuel Cell Systems I. Proceedings of the First International Symposium on New Materials for Fuel Cell Systems*, Montreal, 190–200.

Klvana, D., Chaouki, J., Repellin-Lacroix, M. and Pajonk, G.M. (1989) A new method of preparation of aerogel-like materials using a freeze-drying process, *Revue de physique applique*, Colloque C4, t.24, pp. C4–29.

Klvana, D., Chaouki, J., Perras, L. and Belanger, G. (1992) Powder catalyst for a new hydrogenation process for aromatic hydrocarbons, *Progress in Catalysis*, Elsevier Sci. Publ., 239–46.

Klvana, D., Vaillancourt, J., Kirchnerova, J. and Chaouki, J. (1994) Combustion of methane over $La_{0.66}Sr_{0.34}Ni_{0.3}Co_{0.7}O_3$ and $La_{0.4}Sr_{0.6}Fe_{0.4}Co_{0.6}O_3$ prepared by freeze-drying, *Applied Catalysis A: General*, **109**, 181.

Kohga, M., Suzuki, M. and Hagihara, Y. (1992) Ultrafine ammonium perchlorate prepared by freeze-drying method, *Kogyo Kagaku*, **53**(6), 295–300 (CA 118-215911p).

Kohga, M. and Hagihara, Y. (1993) AP-HTPB composite propellant using ultrafine ammonium perchlorate prepared by freeze-dry method, *Kogyo Kagaku*, **54**(2), 64–9 (CA 119-229461).

Kolosentsev, S.D., Belotserkovskii, G.M. and Plachenov, T.G. (1975) Structure of silica gels obtained by freezing-out of silica sols, *Russ. J. Appl. Chem.*, **48**(1), 252.

Kolotyrkin, V.I., Metlin, Yu.G. and Tretyakov, Yu.D. (1978) Lithium titanate with ramsdellite structure as a cation-conducted solid electrolyte in reversible galvanic cells, *Courier of Moscow State University*, **19**(5), 595–9 (Russ.).

Kolotyrkin, V.I. (1980) Synthesis and properties of some Li-conducting solid state ionics, PhD thesis, Moscow: Moscow State University.

Komm, T.Z., Kaul, A.R., Plokhih, A.A. *et al.* (1981a) The technique to prepare beta-alumina, *Pat. USSR* No. 3232871/22-02 from 04.01.81.

Komm, T.Z., Kaul, A.R., Kalinina, L.A. and Tretyakov, Yu.D. (1981b) The technique to prepare beta-alumina doped by yttrium oxide, *Pat. USSR* No. 3232496/22-02 from 04.01.81.

Kondo, I., Takanobu, O., Mukai, O. and Hamaya K. (1985) Treatment of radioactive wastewater, *Jap. Pat.* No. 62, 71, 899 from 25 Sep. (CA 107-123056).

Kopelev, N.S., Zalishchansky, M.E., Luzhkov, A.V. and Kiselev, Yu.M. (1990) The influence of thallium content on the superconductive properties of thallium-based and related ceramics, *Proc. 3rd German–Soviet Bilateral Seminar on High Temperature Superconductivity*, Karlsruhe, 8–12 Oct., 617–24.

Krishnaraj, P., Lelovic, M., Eror, N.G. and Balachandran U. (1993) Synthesis and microstructure of $Bi_{1.8}Pb_{0.4}Sr_2Ca_{2.2}Cu_3O_x$ obtained from freeze-dried precursors, *Physica C*, **215**, 305–12.

Krishnaraj, P., Lelovic, M., Eror, N.G. and Balachandran, U. (1994) Synthesis and oxygen stoichiometry of $Bi_2Sr_2CaCu_2O_{8+x}$, *Physica C*, **234**(3–4), 318–22.

Krishnaraj, P., Lelovic, M., Eror, N.G. and Balachandran, U. (1995) Oxygen stoichiometry, structure and superconductivity in $Bi_2Sr_2CaCu_2O_{8+x}$, *Physica C*, **246**(3–4), 271–6.

Kulakov, A.B. (1988) Thermal decomposition of nitrates in the synthesis of Y–Ba cuprates, PhD thesis, Moscow: Moscow State University.

Kulakov, A.B., Shabatin, V.P., Gordeev, I.V. *et al.* (1987) The method to prepare powders of high temperature superconductors, *Pat. USSR* No. 4299016/31-02 from 16.06.87.

Kulakov, A.B., Tesker, A.M. and Zalishchansky, M.E. *et al.* (1989) The processes leading to Y–Ba cuprates formation during synthesis by means of nitrates, *Russ. J. Inorg. Materials*, **25**(9), 1499–503.

Kuribayashi, K. and Nicholson, P.S. (1980) Dense conductive β-Al_2O_3 ceramics by spray/freeze-drying, *Mat. Res. Bull.*, **15**, 1595–601.

Kuznetsova, M.A., Mozhaev, A.P., Oleinikov, N.N. *et al.* (1979a) Investigation of cryoextraction conditions during synthesis of multicomponent oxide materials by cryochemical method, *Chemistry and Technology of Oxide Magnetic Materials*, Volgograd, 140–5 (Russ.).

Kuznetsova, M.A., Mozhaev, A.P., Tretyakov, Yu.D. (1979b) Influence of the nature of precursor salts on the chemical homogeneity and sintering processes of doped hematite obtained by cryochemical method, *Russ. J. Inorg. Materials*, **15**(11), 2014–18.

Kwiatkowski, A., Reszka, K. and Szymanski, A. (1980) Preparation of corundum and steatite ceramics by the freeze-drying method, *Ceramurgia Int.*, **6**(2), 79–82.

Lacour, C. (1980) Relationship between thermochemical treatment and transparency of $MgAl_2O_4$ prepared from freeze-dried solutions, *Sci. Sintering*, **12**, 39–48.

Lacour, C. and Paulus, M. (1979) Lyophilisation parameters of ceramic compounds, *Science of Sintering*, **11**(3), 193–202.

Lacour, C., Laher-Lacour, F., Dubon, A. *et al.* (1990) Freeze drying preparation of YBaCuO, *Physica C*, **167**, 287–90.

Lacour, C., Laher-Lacour, F., Dubon, A. *et al.* (1991) Electrical properties correlated to the chemical and granular microstructures in YBaCuO ceramics, *Phys. Stat. Sol. (a)*, **123**, 241–54.

Landsberg, A. (1967) Preparation of homogeneous powders composed of ultra-fine particles, *U.S. Pat.* No. 3, 357, 819 from 12 Dec.

Landsberg, A. and Campbell, T.T. (1965) Freeze-dry technique for making ultra-fine metal powder, *Journal of Metals*, 856.

Langbein, H. and Fischer, S. (1991) Investigation of the formation of nickel–zinc ferrite from coprecipitated oxalates, *Thermochimica Acta*, **182**, 39.

Langbein, H., Michalk, C., Knese, K. and Eichhorn, P. (1991) Thermal decomposition of freeze-dried complex acetates of iron, nickel and zinc, and structural characterisation of products, *J. Europ. Ceram. Soc.*, **8**, 171.

Larbalestier, D. (1988) High temperature oxide superconductors – their potential for high field magnets use, *IEEE Trans. Magn.*, **24**, 711.

De Lau, J.G.M., Bongaerts, P.F.G., Wijgergans, J.L.H.M. and Stuijts, A.L. (1975) Untersuchung chemischer Inhomogenitaten in Ferrit-Pulvern unter Verwendung des Heisspressens, *Ber. Dt. Keram. Ges.*, **52**(7), 207.

Lee, S.R. (1992) Specific features of real structure of high T_c materials based on Y–Ba cuprates, PhD thesis, Moscow State University, Moscow.

Lee, S.R., Oleinikov, N.N. and Gas'kov, A.M. (1992) Influence of cooling rate on the structure and properties of $YBa_2Cu_3O_7$ thick films, *Russ. J. Inorg. Materials*, **28**(10/11), 2141–6.

Lee, S.R., Shlyakhtin, O.A., Oleinikov, N.N. and Lee Sung-Ik (1995a) Some aspects of Hg-based superconductors preparation, *International Workshop MSU-HTSC IV, Chemistry and Technology of High-Temperature Superconductors*, Moscow, 7–12 Oct., 42.

Lee, S.R., Shlyakhtin, O.A., Mun Mi-Ock *et al.* (1995b) A freeze-drying approach to the preparation of $HgBa_2Ca_2Cu_3O_{8+x}$ superconductor, *Supercond. Sci. Technol.*, **8**, 60–4.

Lelovic, M., Krishnaraj, P., Eror, N.G. and Balachandran, U. (1995a) Minimum critical current density of 10^5 A/cm^2 at 77 K in the thin layer of $Bi_{1.8}Pb_{0.4}Sr_{2.0}Ca_{2.2}Cu_{3.0}O_y$ superconductor near the Ag in the Ag-sheathed tapes, *Physica C*, **242**, 246–50.

Lelovic, M., Krishnaraj, P., Eror, N.G. and Balachandran, U. (1995b) Magnetization current of a thin layer of $Bi_{1.8}Pb_{0.4}Sr_{2.0}Ca_{2..2}Cu_{3.0}O_y$ near silver in Ag-sheathed BSCCO-2223 tapes, *Supercond. Sci. Technol.*, **8**, 336–40.

Lelovic, M., Krishnaraj, P., Eror, N.G. *et al.* (1996) Transport critical current density above 10^5 A cm^{-2} at 77 K in $Bi_{1.8}Pb_{0.4}Sr_{2.0}Ca_{2.2}Cu_{3.0}O_y$ superconducting tapes made by the Ag wire-in-tube method, *Supercond. Sci. Technol.*, **9**(3), 201–4.

Leonidov, N.B., Shabatin, V.P. and Tretyakov, Yu.D. (1984) The technique to prepare medicinal medium in the powder form, *Pat. USSR* No. 3696036/28-13 from 19.01.84.

Leonidov, N.B., Uspenskaya, S.I., Pershin, V.I. *et al.* (1994) Cryochemical synthesis of drugs and bioactive substances, *Intern. Conf. on Low Temp. Chemistry*, Moscow, 5–9 Sept., 107.

Levina, M.E., Savchenko, L.A., Tretyakov, Yu.D. and Fokin, V.A. (1978) Influence of the freeze drying conditions on the properties of materials obtained by cryochemical method, *Russ. J. Inorg. Materials*, **14**(7), 1352–4.

Levina, M.E., Marton, I., Kolbeneva, G.I. and Tretyakov Yu.D. (1980) Synthesis of $Pb(Zr_{0.53}Ti_{0.47})$ solid solution by cryocoprecipitation, *Russ. J. Inorg. Materials*, **16**(11), 2005–8.

Lloyd, I.K. and Kovel, R.J. (1988) Pore morphology in cryochemically prepared alumina powders and their sulfate precursors, *J. Mater. Sci.*, **23**, 185–8.

Lloyd, I.K., Blachere, J.R., Beyer, L.A. *et al.* (1989) Characterization of freeze-dried alumina powders using electron microscopy, *Ultramicroscopy*, **29**, 153.

Lopato, L.M., Dudnik, E.V., Zaitseva, Z.A. *et al.* (1992) Application of cryochemical method for powders preparation in the system Al_2O_3–ZrO_2–MgO, *Powder Metallurgy* (Russ.), **6**, 51.

Loukachine, A.V., Vertegel, A.A., Oleynikov, N.N. *et al.* (1995) Synthesis of ruby using sol–gel method, *35th IUPAC Congress*, 14–19 Aug., Istanbul, Abstracts II, Sections 4–6, 1344.

Luccini, E., Meriani, S. and Sbalzero, O. (1989) Preparation of zirconia–ceria powders by coprecipitation of a mixed zirconium carbonate in water with urea, *Int. J. of Materials and Product Technology*, **4**(2), 167.

Matsuda, M., Ogawa, Y., Aihara, Y. *et al.* (1993) Synthesis of $Ba_2RCu_4O_8$ (R = Y, Sm) superconductors at oxygen pressure of 1 atm by spray-frozen/freeze-drying method, *J. Am. Ceram. Soc.*, **76**(6), 1618–20.

Matsushita Electric Industrial Co. Ltd. (1981) Humidity sensor preparation, *Jap. Pat.* No. 58, 25, 216 from 7 Aug.

McBride, D.A. (1974) The method to prepare finely dipersed inorganic oxidizers, *Pat. USA* No. 3.888.017 from 25 March.

Medelius, H. and Rowcliffe, D.J. (1989) Solution routes to synthesize superconducting oxides, *Mat. Sci. Eng.*, **A109**, 289–92.

Metlin, Yu.G. and Tretyakov, Yu.D. (1994) Chemical routes for preparation of oxide high-temperature superconducting powders and precursors for superconductive ceramics, coatings and composites, *J. Mater. Chem.*, **4**(11), 1659–65.

Mikhailenko, A.A. and Fomina, G.A. (1990) Structure, properties and application of Al_2O_3-based powders obtained by cryochemical method, *Russ. Phys. Chem. Mater. Treatment*, **6**, 119.

Mizuno, N., Fujiwara, Y. and Misono, M. (1989) Pronounced synergetic effect in the catalytic properties of $LaMn_{1-x}Cu_xO_3$, *J. Chem. Soc., Chem. Commun.*, 316.

Moshchalkov, V.V., Muttik, I.G., Samarin, N.A. *et al.* (1988) Superconductivity and localization in $YBa_2Cu_3O_x$. oxides, *Fizika Nizkih Temperatur*, **14**(9), 988–92.

Mozhaev, A.P., Pamyatnyh, A.Yu. and Tretyakov, Yu.D. (1980) Investigation of lithium titanate with ramsdelite structure, *Russ. J. Inorg. Mater.*, **16**(12), 2193–9.

Mozhaev, A.P., Shabatin, V.P., Sheredeko, M.V. and Tretyakov, Yu.D. (1982a) The technique to prepare polycrystalline oxide material, *Pat. USSR* No. 3445579/23-26 from 24.05.82.

Mozhaev, A.P., Telegina, L.I., Tretyakov, Yu.D., Golovchansky, A.V. (1982b) Cryochemical technology of finely dispersed salt and oxide powders, *Fine Powders and Powder Based Materials*, Kiev: Naukova Dumka, 57–62 (Russ.).

Mozhaev, A.P. (1988) Low temperature processes in the synthesis of some solid state materials, *Dr. Sci. Thesis*, Moscow, MSU.

Mueller-Zell, A. and Hahn, C. (1987) Improvement of sintered silicon carbide SSiC through process optimization, *Keram. Z.*, **39**(5), 315–17.

Nagai, M. and Nicholson, P.S. (1982) Relative stability of spray-frozen/freeze-dried potassium β-alumina powders, *Mat. Res. Bull.*, **17**, 1131.

Nagai, M., Kuribayashi, K. and Nicholson, P.S. (1983) Spray freeze/freezing drying and sintering of β-Al_2O_3 ceramics, *Ceramic Powders*, 461–8.

Nagai, M., Kushida, T. and Nishino, T. (1989) Fabrication and evaluation of porous β/β''-Al_2O_3 ceramics prepared by the sol–gel process, *Solid State Ionics*, **35**, 213.

Nagai, M., Kushida, T. and Nishino, T. (1990) Effect of water on the conductivity of porous Na β/β''-Al_2O_3 ceramics, *Solid State Ionics*, **40/41**, 107.

Naito, M., Sakai, K. and Okada, M. (1987) Manufacture of spherical silica glass powder, *Jap. Pat.* No. 63, 277, 527 from 11 May.

Nakajima, K., Shimada, S. and Inagaki, M. (1995) Preparation of zirconium oxide gel using electrochemical method, *Journal of the Ceramic Society of Japan*, **103**(3), 304–6.

Nikulin, A.D., Shikov, A.K., Khlebova, N.E. *et al.* (1993) The effect of the ceramic core initial phase composition on the Ag-sheathed Bi-2223 tapes critical properties. In H.C. Freyhardt, *Applied Superconductivity. Papers of the 1st European Conference*, vol. 1, Oberursel: DGM Informationsges., Germany, 217–20.

Nikulin, A.D., Shikov, A.K., Khlebova, N.E. *et al.* (1994) Ag-2223 (Bi) tapes structure and superconducting properties in dependence of initial powder type, *IEEE Transactions on Magnetics*, **30**(4: Pt 2), 2368–71.

Nippon Chemiphar Co. (1981) Preparation of pharmaceutical crystalline powder by freeze-drying, *Jap. Pat.* No. 82, 127, 406 from 29 Jan.

Nippon Mining Co. Ltd (1983) Ultrafine zirconia, *Jap. Pat.* No. 60 96,528 from 1 Nov.

Nissan Motor Co. Ltd (1983) Silicon carbide powder for atmospheric sintering, *Jap. Pat.* No. 60.96.574 from 26 Oct.

Oleinikov, N.N., Pershin, V.I., Tretyakov, Yu.D. *et al.* (1974) The technique for powder preparation, *Pat. USSR* No. 2053978/33 from 09.08.74.

Oleinikov, N.N., Tretyakov, Yu.D. and Anastasyuk, N.V. (1978) The technique to prepare ferrite powder, *Pat. USSR* No. 2634446/22-02 from 27.06.78.

Oleinikov, N.N. (1988) Features of ferrites formation during synthesis of magnetic materials with controlled microstructure and properties, *Dr. Sci. Thesis*, Moscow: Moscow State University.

Oleinikov, N.N., Lee, S.R., Kazin, P.E. *et al.* (1992) Actual structure of $YBa_2Cu_3O_{7-x}$ high T_c powder and possibilities for identifying it, *Superconductivity: Physics, Chemistry, Technique*, **5**(1), 105–12.

Oleinikov, N.N., Lee, S.R., Goodilin, E.A. and Kazin, P.E. (1993) A modified QMG method of high J_c YBCO preparation, *J. Alloys and Compounds*, **195**, 27–30.

Olesch, T., Kemnitz, E. and Wilde, W. (1992) Synthesis and reaction route of bulk superconducting $YBa_2Cu_4O_8$ and $Y_{0.9}Ca_{0.1}Ba_2Cu_4O_8$, *Eur. J. Solid State Inorg. Chem.*, **29**, 249–55.

Orehotsky, J., Huang, H., Davidson, C.R. and Srinivasan, S. (1979) Oxygen evolution on $Ni_xFe_{3-x}O_4$ electrodes, *J. Electroanal. Chem.*, **95**, 233.

Os'kina, T.E., Tretyakov, Yu.D., Shapiro, Ya.A. and Soldatov, E.A. (1990) Influence of environmental components on the high density $YBa_2Cu_3O_{6.6}$ ceramics, *Russ. J. Inorg. Mater.*, **26**(4), 849–52.

Pajonk, G.M., Repellin-Lacroix, M., Abouarnadasse, S. *et al.* (1990) From sol–gel to aerogels and cryogels, *J. Non-Crystalline Solids*, **121**, 66–7.

Pankov, V.V., Paulus, M. and Dubon, A. (1982) Powder production for rapid sintering of oxides suspended in liquid solution of other cations by the combined suspension-solution freeze-drying technique, *Ann. Chim. Fr.*, **7**, 377.

Pankov, V.V., Paulus, M. and Dubon, A. (1983) Powder preparation for rapid sintering by the combined suspension-solution freeze drying technique, *Proc. Brit. Ceram. Soc.*, **33**, 17.

Parise, J.B., McCarron III, E.M., Von Dreele, R. and Goldstone, J.A. (1991) β-MoO_3 produced from a novel freeze drying route, *J. Solid State Chem.*, **93**, 193.

Patil, A.S. and Netherton, L.E. (1977) Transparent iron oxide pigment prepared by freeze drying, *Can. Pat.* No. 1, 102, 506 from 2 June.

Paulus, M. (1976) Dispersion par lyophilisation en vue du frittage à basses temperatures, *Ann. Chim.*, **1**(2–4), 187.

Payzant, E.A. and King, H.W. (1990) Structure and normal state physical properties for transition metal substitutions in $La_{1.85}Sr_{0.15}CuO_{4-\delta}$, *J. Canad. Ceram. Soc.*, **59**(1), 68–74.

Pekarsky, A. and Nicholson, P.S. (1980) The relative stability of spray-frozen/freeze-dried β-Al_2O_3 powders, *Mat. Res. Bull.*, **15**, 1517.

Pershin, V.I. (1977) Studies of physico-chemical processes during preparation of ferrites and ferrite forming oxides by cryochemical method, PhD thesis, Moscow: Moscow State University.

Pershin, V.I., Mozhaev, A.P., Naumov, S.V. and Lyashchenko, A.K. (1991) Water–salt compositions and synthesis of $YBa_2Cu_3O_{7-x}$ from nitrate solutions with additives of acetates and urea, *International Workshop MSU-HTSC II, Chemistry and Technology of High-Temperature Superconductors*, Moscow, 14–18 Oct., Proceedings, 297–301.

Pershin, V.I., Naumov, S.V., Mozhaev, A.P. *et al.* (1993) Synthesis of $YBa_2Cu_3O_7$ using nitrate–acetate solutions, *Russ. J. Inorg. Materials*, **29**(7), 967–70.

Pershin, V.I., Naumov, S.V., Mozhaev, A.P. *et al.* (1994) Synthesis of $YBa_2Cu_3O_{7-y}$ from urea doped nitrate solutions, *Russ. J. Inorg. Chem.*, **39**(6), 867–70.

Pham Thi, M., Kormann, R. and Morell, A. (1989) Synthèse par lyophilisation de poudres pour ceramiques supraconductrices, *L'Industrie Ceramique*, **836**(3/89), 197–200.

Plokhih, A.A. (1985) Fundamentals and practical solutions of the cryochemical synthesis of β-alumina, PhD thesis, Moscow: Moscow State University.

Poillerat, G. (1994) Electrocatalyse sur oxydes metalliques, *Journal de physique IV*, **4**, C1–107.

Primo, V., Sapina, F., Sanchis, M.J. *et al.* (1992) A new improved synthesis of the 110 K bismuth superconducting phase: freeze drying of acetic solutions, *Mater. Letters*, **15**, 149–55.

Primo, V., Sapina, F., Sanchis, M.J. *et al.* (1993) Fast synthesis of single-phased 110 K bismuth superconductor by freeze drying of acetic precursors. Kinetic role of calcium and copper oxides, *Solid State Ionics*, **63–5**, 872–82.

Pupysheva, O.V., Shlyakhtin, O.A., Putlyaev, V.I. and Tretyakov, Yu.D. (1995) Melt casting synthesis of $Bi_2Sr_2CaCu_2O_y$–$SrTiO_3$ composites from freeze-dried precursors, *International Workshop MSU-HTSC IV, Chemistry and Technology of High-Temperature Superconductors*, Moscow, 7–12 Oct, 32.

Pupysheva, O.V., Shlyakhtin, O.A., Lennikov, V.V. *et al.* (1999) Melt synthesis of $Bi_2Sr_2CaCu_2O_x$–$SrZrO_3$ composites from cryochemical presursors, *Russ. J. Inorg. Mater.*, **33**(9), 1–6.

Putilin, S.N., Antipov, E.V., Chmaissem, O. and Marezio, M. (1993) Superconductivity at 94 K in $HgBa_2CuO_{4+\delta}$, *Nature*, **362**(6417), 226.

Pyagai, R.N. (1982) Influence of cryotreatment on the reactivity of solids, PhD thesis, Moscow: Moscow State University.

Pyagai, R.N., Oleinikov, N.N. and Tretyakov, Yu. D. (1982) Influence of cryomilling on acid dissolution of Zn metal, *Powder Metallurgy* (Russ.), **3**, 6–12.

Pyagai, R.N., Oleynikov, N.N., Tretyakov, Yu.D. and Anisin, S.I. (1982) Disaggregation of Fe powders at low temperatures, *Powder Metallurgy* (Russ.), **5**, 9–12.

Ramme, R. and Hausner, H. (1987) Mechanical properties of ZrO_2 (2% Y_2O_3) derived from freeze-dried coprecipitated hydroxides, *Ber. DKG*, Bd.64, Nr.1/2, S.12.

Rao, C.N.R., Nagarajan, R. and Vijayaraghavan, R. (1993) Synthesis of cuprate superconductors, *Supercond. Sci. Technol.*, **6**(1), 1.

Rasmussen, M.D., Akinc, M. and Berard, M.F. (1984) Effect of precursor freeze-drying conditions on the sinterability of hydroxide-derived Y_2O_3 powders, *Ceram. Int.*, **10**(3), 99–104.

Reetz, T. and Moritz, T. (1991) Spherical sinterable ceramic spraying granulate and its manufacture, *Germ. Pat.*, No. 4, 118, 752 from 6 June.

Rieser, U. (1983) Sintered ceramic articles, *Ger. Pat.*, No. 3, 322, 889 from 24 June.

Rigterink, M.D. (1972) Advances in technology of the cryochemical process, *Ceramic Bulletin*, **51**(2), 158.

Rizkalla, A.S., Jones, D.W., Hall, G.C. *et al.* (1991) The influence of spray- and freeze-drying preparation methods on the chemical compositions of six feldspathic glass formulations, *Br. Ceram. Trans. J.*, **90**(1).

Robbins, M., Staudinger, A. and Shabaka, S. (1980) Alloy pre-form preparation through the reduction of sintered oxides, *J. Mater. Sci.*, **15**, 527.

Roehrig, F.K. and Wright, T.R. (1972) Freeze drying: a unique approach to the synthesis of ultrafine powders, *J. Vac. Sci. Technol.*, **9**(6), 1368.

Roehrig, F.K. and Wright, T.R. (1972) Carbide synthesis by freeze drying, *J. Am. Ceram. Soc.*, **55**(1), 58.

Roosen, A. and Hausner, H. (1983) The influence of processing conditions on the sintering behaviour of coprecipitated calcia-stabilized zirconia powders, *Mater. Sci. Monogr.*, vol. 16 (Ceramic Powders), 773–82.

Rudnitskii, L.A., Moshchalkov, V.V., Kaul, A.R. *et al.* (1988) Dilatometry and thermomicrogravimetry of $YBa_2Cu_3O_x$ superconductor in hydrogen flow, *Solid State Physics*, **30**(10), 3132–4.

Rudolph, D. and Reetz, T. (1988) Herstellung feindisperser anorganischer Materialen durch Gefriertrocknung, *Chem. Techn.*, **40,** Jg., Heft 10, s. 434.

Ryabov, N.P., Kornilov, V.P., Makarov, A.I. and Shapovalov, Yu.I. (1978) Installation for freeze drying of powdered ceramic materials, *Glass and Ceramics*, **7**, 24–5 (Russ.).

Sakai, M., Naka, K. and Otani, K. (1985) Preparation of ceramic mixtures for injection molding, *Jap. Pat.* No. 61, 219, 760 from 22 May.

Sale, F.R. (1977) Applications of freeze-drying in metallurgy and materials science, *Metall. Mater. Tech.*, **9**, 439–43.

Sanchis, M.J., Burgos, M., Carrasco, R. *et al.* (1993) Stability and synthetic pathways: novel routes to $CaCuO_2$, *Solid State Ionics*, **66**, 27.

Sarkar, S.K., Sharma, M.L. and Bhaskar, L.H. (1984) Effect of the processing parameters on the surface area of freeze-dried and spray-dried alumina, *J. Am. Ceram. Soc.*, **67**(9), C172–3.

Sato, G., Komatsu, M., Tanaka, H. and Fujii, S. (1985) Formation of inorganic oxide flakes, *Jap. Pat.* No. 62, 03, 003 from 28 June.

Sato, T. (1985) Magnetic recording medium, *Jap. Pat.* No. 61, 182, 631 from 7 Feb.

Savchenko, L.A. (1982) Specific features of the freeze drying in the fast frozen water–salt systems, PhD thesis, Moscow: Moscow State University.

Schaeffer, R.W., Macho, J., Salomon, R.E. *et al.* (1991a) Synthesis and characterisation of YBaCuO superconductors from solutions in liquid ammonia, *J. of Superconductivity*, **4**(5), 365–74.

Schaeffer, R.W., Macho, J., Salomon, R.E. *et al.* (1991b) Synthesis and characterisation of cuprate superconductors from solutions in liquid ammonia, *Ceram. Trans.*, **18** (Supercond. Ceram. 2), 201–14.

Scheurell, S., Olesch, T., Kemnitz, E. *et al.* (1993) Halogenation of Y–Ba–Cu–O and Bi(Pb)–Sr–Ca–Cu–O phases and its implication for superconductivity, *Proceedings of the Beijing International Conference: High Temperature Superconductivity (BHTSC '92)*, Singapore, 156–60.

Schlenkrich, F., Fischer, K., Wilde, W. *et al.* (1990) The preparation of $YBa_2Cu_3O_{7x}$ powders using solution techniques, *Mat. Sci. Forum*, **62–4**, 63–4.

Shabatin, V.P. (1986) The development of solid state materials synthesis methods using cryocoprecipitation of hydroxides, oxalates and carbonates, PhD thesis, Moscow: Moscow State University.

Shabatin, V.P., Mozhaev, A.P., Skazin, V.E. *et al.* (1983a) The technique to prepare polycrystalline ferrite material with garnet structure, *Pat. USSR* No. 3654908/23-26 from 20.10.83.

Shabatin, V.P., Barkovskii, A.I., Mozhaev, A.P. *et al.* (1983b) The technique to prepare precursor for composite metal–ceramic electric contacts, *Pat. USSR* No. 3680469/22-02 from 23.12.83.

Shabatin, V.P., Mozhaev, A.P., Tretyakov, Yu.D. *et al.* (1984) The technique to prepare doped nickel–zinc ferrites, *Pat. USSR* No. 3810877/22-02 from 02.11.84.

Shabatin, V.P., Tretyakov, Yu.D., Sokolovsky, V.R. *et al.* (1987a) The technique to prepare powder compositions, *Pat. USSR* No. 4213515/31-02 from 26.01.87.

Shabatin, V.P., Fomina, Ya.V., Metlin, Yu.G. (1987b) The technique to prepare high T_c ceramic superconductors, *Pat. USSR* No. 4244655/31-33 from 12.05.87.

Shabatin, V.P., Myshlevskaya, A.L., Karpinskaya, V.M. *et al.* (1987c) The technique to prepare polishing suspension, *Pat. USSR* No. 4260437/23-05 from 11.06.87.

Shabatin, V.P., Sokolovsky, V.R., Mozhaev, A.P. *et al.* (1988) The technique to prepare glass powders, *Pat. USSR* No. 4366362/33 from 19.01.88.

Shabatin, V.P., Venkateshvara, Reddy Bommareddy, Pershin, V.I. *et al.* (1989) The technique to prepare Bi-containing high T_c superconducting material, *Pat. USSR* No. 4749380/33 from 17.10.89.

Shabatin, V.P., Kazin, P.E. and Pershin, V.I. (1991a) Cryochemical preparation and magnetic properties of the superconducting $YBa_2Cu_3O_x$ ultrafine powders, *Waseda International Symposium on Superconducting Materials*, 23–25 Oct., Tokyo, 33.

Shabatin, V.P., Baschenko, Yu.V., Maximov, P.V. and Morozov, V.I. (1991b) The method of hydroxylapatite powders preparation, *Pat. USSR* No. 5016313/26 from 12 Dec.

Sharikov, F.Yu. (1991) Cryochemical synthesis of finely dispersed oxide powders using ion exchange, PhD thesis, Moscow: Moscow State University.

Shinoda, K., Matsubara, E., Muramatsu, A. and Waseda, Y. (1994) Local structure of ferric hydroxide $Fe(OH)_3$ in aqueous solution by the anomalous X-ray scattering and EXAFS methods, *Materials Transactions, JIM*, **35**(6), 394–8.

Shishov, V.V. (1978) Specific features of synthesis and properties of Li, K, Na metaniobates and their solid solutions, PhD thesis, Moscow: Moscow State University.

Shlyakhtin, O.A. (1981) Influence of doping and sintering conditions on some structure-sensitive properties of cryochemically prepared Ni–Zn ferrites, MSc thesis, Moscow: Moscow State University.

Shlyakhtin, O.A. (1985) The role of the water in topochemical processes during cryochemical synthesis of Ni–Zn ferrites, PhD thesis, Moscow: Moscow State University.

Shlyakhtin, O.A., Pokholok, K.V., Oleinikov, N.N. and Tretyakov, Yu.D. (1989) Use of the Mössbauer marker in the analysis of the structure of $YBa_2Cu_3O_x$-based compounds, *Superconductivity: Physics, Chemistry, Technique*, **2**(2), 38–41.

Shlyakhtin, O.A., Moysa, V.S., Korsakov, I.E. and Mozhaev, A.P. (1992) Introduction of $BaZrO_3$ fine particles in melt grown $YBa_2Cu_3O_x$ through cryochemical processing, *EMRS 1992 Fall Meeting*, Strasbourg, A-V.10.

Shlyakhtin, O.A., Kulakov, A.B., Badun, Yu.V. *et al.* (1994) Freeze drying preparation of $YBa_2Cu_3O_x$ powders: chemical study of the process, *Trans. Mat. Res. Soc. Japan*, **14A**, 7–10.

Shlyakhtin, O.A., Putlyaev, V.I., Tretyakov, Yu.D. and Jansen, M. (1995) Topochemical factors in Bi-2223 synthesis, *International Workshop MSU-HTSC IV, Chemistry and Technology of High-Temperature Superconductors*, Moscow, 7–12 Oct., 33.

Simon, R. (1989) High temperature superconductors for microelectronics, *Solid State Technol.*, **32**, 141.

Sokolov, I.S. (1978) Influence of chemical and thermal prehistory on the properties of ferrite powders obtained by cryochemical method, PhD thesis, Moscow: Moscow State University.

Song Kang-Ho, Liu Hua-Kun, Dou Shi-Xue and Sorrell, C.C. (1990) Rapid formation of the 110 K phase in Bi–Pb–Sr–Ca–Cu–O through freeze-drying powder processing, *J. Am. Ceram. Soc.*, **73**(6), 1771–3.

Song, K.-H., Liu, H.K., Dou, S.X. and Sorrell, C.C. (1991a) Freeze drying of $(Bi,Pb)_2Sr_2Ca_2Cu_3O_{10+y}$, *Key Engineering Materials*, **53–55**, 662–5.

Song, K.-H., Sorrell, C.C., Dou, S.X. and Liu, H.K. (1991b) Effect of the hot-pressing on the bismuth–lead–strontium–calcium–copper–oxygen system, *J. Am. Ceram. Soc.*, **74**(10), 2577–82.

Startseva, I.V., Levina, M.E., Tretyakov, Yu.D. and Fokin, V.A. (1977) The technique to prepare polycrystalline oxide material, *Pat. USSR* No. 2503211/29-33 from 23.06.77.

Startseva, I.V. (1981) Effectiveness of some preparation methods of PZT based polycrystalline materials, PhD thesis, Moscow: Moscow State University.

Strasik, M., Luhman, T.S. and Eror, N.G. (1988) A novel preparation technique for high-temperature superconducting $YBa_2Cu_3O_{7-x}$, *SAMPE Quarterly*, **20**(1), 11–14.

Strelkov, A.V., Zhilin, A.B., Gaskov, A.M. and Tretyakov, Yu.D. (1993) Characterization of high-T_c superconductors by mass spectral methods of surface analysis (LIMS, SNMS), *J. Alloys and Compounds*, **195**, 145–8.

Sugano, T. and Karita, Y. (1985a) Concentration of water-soluble magnetic fluids, *Jap. Pat.* No. 61, 622, 591 from 29 March.

Sugano, T. and Karita, Y. (1985b) Manufacture of magnetic ultrafine particles by freeze-drying of suspensions, *Jap. Pat.* No. 61, 220, 719 from 27 March.

Tachivaki, T., Kimura, Y., Ito, T. and Hiraki, A. (1991a) Characterization of Y–Ba–Cu–O superconductor prepared from freeze-dried fine powders, *Supercond. Sci. Technol.*, **4**, 591–4.

Tachivaki, T., Kimura, Y., Ito, T. and Hiraki, A. (1991b) Characterization of Y–Ba–Cu–O superconductor prepared from freeze-dried fine powders, *Proc. ISEC'91*, Glasgow, Scotland, 272–5.

Tachivaki, T., Koizumi, S., Kimura, Y. *et al.* (1991c) Preparation and characterization of Y–Ba–Cu–O superconductor by using freeze-drying of the suspensions from coprecipitation and alkoxide route, *International Workshop MSU-HTSC II, Chemistry and Technology of High-Temperature Superconductors*, Moscow, 14–18 Oct., Proceedings, **1**, 214–23.

Tachivaki, T., Suzuki, M., Okajima, H. *et al.* (1993) Preparation of fine powders for electronic materials by freeze-drying, *Applied Surface Science*, **70/71**, 751–4.

Taguchi, Y. (1991) Manufacture of synthetic leather from dry-method polyurethane resin, *Jap. Pat.* No. 04, 308, 282 from 5 Apr. (CA 118-256282c).

Takahashi, K., Ito, T., Yoshikawa, H. and Hiraki, A. (1993) Normal atmospheric preparation of $YBa_2Cu_4O_8$ superconductor from nitrate solution by freeze-drying method, *Jap. J. Appl. Phys.*, **32**, L1211–14.

Takanobu, O. and Kondoh, I. (1985) Method and apparatus for treatment of liquid radioactive waste, *Ger. Pat.* No. 3, 625, 602 from 29 July (CA 106-109806a).

Takazawa, H. and Morita, S. (1991) Manufacture of collagen sponge, *Jap. Pat.* No. 05, 43, 734 from 21 Aug. (CA 119-119962f).

Takeda Chemical Industries Ltd. (1977) Freeze drying of ceramic moldings, *Jpn. Pat.* No. 60, 22, 667 from 5 Dec. 1977.

Tang, Y.Q., Sheng, Z.Z., Chen, Z.Y. *et al.* (1992) Fabrication of highly oriented $Tl_2Ba_2Ca_2Cu_3O_{10}$ films with T_c over 120 K by chemical deposition and Tl-diffusion, *Supercond. Sci. Technol.*, **5**(9), 538.

Tarlavskii, V.E., Rozanov, S.D., Shabatin, V.P. *et al.* (1986) The technique of diffusion welding through intermediate insert, *Pat. USSR* No. 4052887/25-27 from 10.04.86.

Terentiev, A.N., Shlyakhtin, O.A., Kuznetzov, V.M. and Mozhaev, A.P. (1993) Generation of new pinning centers by doping melt grown $YBa_2Cu_3O_x$ by $BaZrO_3$, *IUMRS-ICAM-93, The 3rd IUMRS International Conference on Advanced Materials*, Tokyo, Japan, 31 Aug.–4 Sept., HH-36.

Tesker, A.M. (1986) Principles of diagnostics of multistage synthesis of Ni–Zn and Mn–Zn ferrite powders, PhD thesis, Moscow: Moscow State University.

Tesker, A.M., Shapiro, Ya.A., Zalyaletdinov, R.K. *et al.* (1989) Influence of microstructure on some properties of Y–Ba–Cu–O ceramics, *Russ. J. Inorg. Mater.*, **25**(1), 172–4.

Thampi, K.R., Subba Rao, M., Schwartz, W. *et al.* (1988) Preparation of $SrTiO_3$ by sol–gel techniques for the photoinduced production of H_2 and surface peroxides from water, *J. Chem. Soc. Faraday I*, **84**(5), 1703–12.

Thierauf, A., Egger, Ch., Sporn, D. and Keck, K. (1992) Crystallization behaviour of freeze-dried $YBa_2Cu_3O_{7-x}$ powders at low temperatures, *Supercond. Sci. Technol.*, **5**, 123–9.

Thomson Jr., J. (1974) Chemical preparation of PLZT powders from aqueous nitrate solutions, *Ceramic Bulletin*, **53**(5), 421–4, 433.

Tinkle, M.C. (1973) Cryochemical method of forming spherical metal oxide particles from metal salt solutions, *U.S. Pat.* No. 3, 776, 988 from 4 Dec. (Hollow microspheres of actinide oxides).

Toshiba Battery Co. Ltd (1983) Silver oxide battery, *Jap. Pat.* No. 58, 94, 756 from 30 Nov.

Trambouze, Y. (1970) Application de la lyophilisation preparation des adsorbants, catalyseurs et poudres metalliques, *Science et technique. Chimie et industrie – Genie chimique*, **103**(18), 2338.

Tretyakov, Yu.D., Oleinikov, N.N., Anastasyuk, N.V. and Pershin, V.I. (1973) The technique to prepare polycrystalline materials, *Pat. USSR* No. 1937660/29-33 from 14.06.73.

Tretyakov, Yu.D., Levina, M.E., Prisedsky, V.V. *et al.* (1977) Synthesis of lead titanate–zirconate from cryochemical precursors, *Russ. J. Inorg. Materials*, **13**(10), 1825–9.

Tretyakov, Yu.D. and Oleinikov, N.N. (1980) Cryochemical processes in ferrite preparation, *The 3rd International Conference on Ferrites*, Kyoto, 29 Sept.–2 Oct., Summaries, 8.

Tretyakov, Yu.D. and Fokin, V.A. (1982) Development of cryochemical processes in the technology of oxide ferroelectrics, *Russ. J. Inorg. Materials*, **18**(10), 1645–9.

Tretyakov, Yu.D., Shabatin, V.P., Sokolovskii, V.R. *et al.* (1987) The technique to prepare silver powder, *Pat. USSR* No. 4186083/31-02 from 26.01.87.

Tretyakov, Yu.D., Shabatin, V.P., Faktorovich, G.S. *et al.* (1988) The technique to prepare alumochromphosphate binder, *Pat. USSR* No. 4438346/23-33 from 07.06.88.

Tsai, M.T. and Shih, H.C. (1993a) A modified powder route for the preparation of sol–gel process-synthesized magnesia and alumina, *J. Mater. Sci. Letters*, **12**, 1025.

Tsai, M.T. and Shih, H.C. (1993b) Effects of powder processing on the characterization of magnesia derived from alkoxide precursors. Part. II. Compaction and sintering behaviour, *J. Mater. Sci.*, **28**(16), 4530–5.

Tseung, A.C.C. and Bevan, H.L. (1970) Preparation and characterization of high surface area semiconducting oxides, *J. Mater. Sci.*, **5**, 604.

Tseung, A.C.C. and Yeung, K.L.K. (1978) Reduction of oxygen on Teflon-bonded $NiCo_2O_4$/graphite electrodes, *J. Electrochem. Soc.*, **126**(6), 1003.

Vertegel, A.A., Kalinin, S.V., Oleynikov, N.N. and Tretyakov, Yu.D. (1995) The fractal particles of iron(III) hydroxonitrate: from solution to solid state, *J. Non-Crystalline Solids*, **181**, 146–50.

Vertegel, A.A., Lukashin, A.V., Oleinikov, N.N. and Tretyakov, Yu.D. (1997) Specific features of cryochemical synthesis of yttrium–iron garnet from colloidal solutions, to be published in *Reports of Russ. Acad. Sci.*

Viktorov, V.V., Evdokimov, I.V. and Shabatin, V.P. (1989) The formation of the structure of $(Cr_xAl_{1-x})_2O_3$ solid solutions, *Repts. Acad. Scu. USSR*, **308**(5), 1138–41.

Visco, S.J. and Kennedy, J.H. (1983) Investigation of $Na_5GdSi_4O_{12}$ (NGS) nasicon and highly doped NGS nasicon prepared by spray-freeze/freeze-dry methods using complex plane analysis, *Solid State Ionics*, **9/10**, 885.

Vogel, E.M., Johnson Jr., D.W. and Gallagher, P.K. (1977) Oxygen stoichiometry in $LaMn_{1-x}Cu_xO_{3-y}$ by thermogravimetry, *J. Am. Ceram. Soc.*, **60**(1–2), 31–3.

Voinov, I.D. (1976) Influence of chemical and thermal prehistory on the substructure and catalytic activity of cobalt ferrite, PhD thesis, Moscow: Moscow State University.

Volfkovitch, S.I., Tretyakov, Yu.D., Kubasova, L.V. *et al.* (1979a) The technique to prepare borophosphate catalyst for *n*-butyl alcohol dehydration, *Pat. USSR* No. 2790685/23-04 from 24.05.79.

Volfkovitch, S.I., Mitichenko, M.G. and Savchenko, L.A. (1979b) The technique to prepare porous alumophosphate, *Pat. USSR* No. 2815234/23-26 from 23.08.79.

Wachowski, L., Zielinski, S. and Burewicz, A. (1980) Explosion method for the synthesis of perovskite-type binary oxides, *Acta Chim. Acad. Sci. Hung*, **104**(2), 141.

Wang, Ch.T., Lin, L.Sh. and Yang, Sh.J. (1992) Preparation of $MgAl_2O_4$ spinel powders via freeze-drying of alcoxide precursors, *J. Am. Ceram. Soc.*, **75**(8), 2240.

Wang, J., Bultitude, J., Stevens, R. *et al.* (1989) Structure and properties of the $YBa_2Cu_3O_{7-x}$ superconducting ceramics, *Proc. Brit. Ceram. Soc.*, **41**, 281–8.

Wang, L. and Lloyd, I.K. (1991) Sinterability of calcined freeze-dried alumina powders, *J. Am. Ceram. Soc.*, **74**(11), 2934.

Wang, L.H. (1993) Processing and densification of $CaLa_2S_4$ powder by the acetates route, *J. Mater. Sci. Lett.*, **12**, 212.

West, A.R. (1984) *Solid State Chemistry and Its Applications*, New York: Wiley & Sons, Pt 2, Chapter 13.2.3.

Wheat, T.A., Sallam, E.M.H. and Chaklader, A.C.D. (1979) Synthesis of mullite by a freeze-dry process, *Ceram. Intern.*, **5**, 42–4.

White, G.D., Lavender, J.C. and Hart, P.E. (1981) UO_2 pellets from freeze-dried oxide monospheres, *Ceramic Bulletin*, **60**(2), 260–1.

Wickenden, D.A. (1993) Determination of tritiated water in biological matrixes, *Sci. Total Environ.*, **130–1**, 121–8 (CA 118-229581t).

Willemer, H. (1975) Low temperature freeze drying of aqueous and non-aqueous solutions using liquid nitrogen. In S.A. Goldblith, L. Rey and W.W. Rothmayne (eds), *Freeze Drying and Advanced Food Technology*, London–New York–San Francisco: Academic Press, 461–77.

Williams, N.A., Polli, G.P. and Parenter, J. (1984) The lyophilization of pharmaceuticals: a literature review, *Sci. Technol.*, **38**(2), 48–59.

Xue, L.A., Riley, F.L. and Brook, R.J. (1986) Tert-butyl alcohol – a medium for freeze-drying: application to barium titanate, *Br. Ceram. Trans. J.*, **85**(2), 47–8.

Yamashita, K., Ramanujachary, K.V. and Greenblatt, M. (1995) Hydrothermal synthesis and low temperature conduction properties of substituted ceria ceramics, *Solid State Ionics, Diffusion & Reactions* **81**(1–2), 53–60.

Yan, M.F., Huo, T.C.D. and Ling, H.C. (1987) Preparation of $Y_3Al_5O_{12}$-based phosphor powders, *J. Electrochem. Soc.*, **134**(2), 493.

Yanson, I.K., Rybalchenko, L.F., Fisun, V.V. *et al.* (1988) Point-contact studies of high-temperature superconductor $YBa_2Cu_3O_{7-\delta}$, *Fizika Nizkih Temperatur*, **14**(7), 732–7.

Yasuda, H., Fujiwara, Y., Mizuno, N. and Misono, M. (1994) Oxidation of carbon monoxide on $LaMn_{1-x}Cu_xO_3$ perovskite-type mixed oxides, *J. Chem. Soc. Faraday Trans.*, **90**(8), 1183.

Yokogawa, H., Hirao, S., Yokoyama, M. *et al.* (1986) Manufacture of porous inorganic articles for thermal insulation, *Jap. Pat.* No. 65, 159, 270 from 23 Dec.

Yokoyama, Y. and Ozeki, H. (1993) Microwave dielectric ceramic compositions and their manufacture, *Jap. Pat.* No. 06, 215, 626 from 18 Jan.

Zalishchanskii, M.E., Shabatin, V.P., Kulakov, A.B. *et al.* (1988) The method for preparation of high T_c superconductors, *Pat. USSR* No. 4384238/31-33 from 25.02.88.

Zalischansky, M.E., Shabatin, V.P. and Kopelev, N.S. (1989) The employment of freeze-drying method for the synthesis of mixed oxides in the system of 'Tl–Ba–Ca–Cu–O', *Progress in High Temperature Superconductivity*, **22**, 183 (Beijing International Conf. on High Temperature Superconductivity, Beijing, China, 4–8 Sept.).

Zhiboedov, V.G., Makarov, V.M., Skotnikov, A.S. *et al.* (1976) The technique to prepare granular polycrystalline oxides of group III and IV elements of Periodic Table, *Pat. USSR* No. 2434366 from 27.12.76.

Zhiboedov, V.G., Fokin, V.A., Tretyakov, Yu.D. and Skotnikov, A.S. (1977) The technique to prepare polycrystalline materials, *Pat. USSR* No. 2511848/29-33 from 21.07.77.

Zhukov, A.A., Komarkov, D.A., Moschalkov, V.V. *et al.* (1989) Magnetic field dependence of the critical current density in $YBa_2Cu_3O_{7-x}$ ceramics, *Physica C*, **162–4**, 1623–4.

Zhukov, A.A., Komarkov, D.A. and Moschalkov, V.V. (1990) The effect of magnetic field on the critical current of $YBa_2Cu_3O_{7-x}$ ceramics, *Jap. J. Appl. Phys.*, **29**(5), 4760–2.

Zhukov, A.A., Komarkov, D.A., Mircovic, J. *et al.* (1991) Critical currents in granular bi-based superconductors, *International Workshop MSU-HTSC II, Chemistry and Technology of High-Temperature Superconductors*, Moscow, 14–18 Oct., Proceedings, 334–8.

Zubov, I.V., Ilyushin, A.S., Novakova, A.A. *et al.* (1988) Moessbauer spectroscopy studies of crystal fields in $YBa_2(Cu_{1-x}Fe_x)_3O_{7-y}$, *J. Magn. Magn. Mater.*, **76/77**, 623–5.

4

Safety and hazards in cryochemical technology of advanced materials

The cryochemical technology of advanced materials deals with various physical and chemical processes which are rarely ecologically safe. Ecological problems arising can be divided into two main groups: the problems brought about by cryochemical processes themselves and those connected with the chemical nature of synthesized materials.

The largest source of pollution in cryochemical technology is refrigerating units using Freons as a cooling agent. The Montreal Protocol (UN Environment Program, 1991) signed by many nations of the world appeals to restriction in Freons' use because of their negative influence on the Earth's ozone layer. Complete renouncement of Freons (chlorofluorocarbons, CFCs) seems hardly to be possible, because besides a proper combination of thermophysical parameters these compounds are non-flammable and do not maintain combustion; they also cause no corrosion, are well compatible with lubricants and are non-toxic. For these reasons the majority of modern cooling facilities are designed for the use of CFC refrigerants.

A possible solution of the problem connected with CFC pollution could be the modification of their composition. Recent studies (World Meteorological Organization, 1990) have shown that the degradation of ozone is catalysed mainly by organic compounds containing chlorine and bromine. The greatest effect is caused by long-lived compounds stable in higher atmospheric layers. This suggests two ways for the problem solution: the substitution of CFCs with less stable compounds containing hydrogen (HCFC) and the use of chlorine-free fluorine derivatives (HFC) (Fischer and McFarland, 1992). Thus, for example, HFC-134a (tetrafluoroethane, CH_2FCF_3) has similar thermodynamic properties to CFC-12 (dichlorodifluoromethane, CCl_2F_2) and has attracted the most attention as a substitute refrigerant for CFC-12. One major drawback is that it is not miscible with mineral oils that have been used in the past, and the system energy requirements are strongly dependent on the lubricant used. Extensive efforts have gone into identifying the best oils for use with HFC-134a to enhance compressor reliability and system energy deficiency. The components and composition of the ternary blend have been carefully selected in order to approximate the thermal properties of CFC-12. Only small and relatively inexpensive design changes would be needed to incorporate the blend into refrigerators/freezers originally designed for the use with CFC-12.

HFC-152a (difluoroethane, CH_3CHF_2) has been proposed because its theoretical vapour compression efficiency is better than CFC-12 or HFC-134a. However, a compensating factor including operating temperature, heat capacity and thermal conductivity may allow HFC-134a system efficiencies comparable to those that can be achieved with HFC-152a. Also, because HFC-152a is flammable some people have raised concerns over its use in an application where a non-flammable refrigerant has been the standard. The plus side of each of these alternative refrigerants is that lubricants have finally been identified and compressors and heat exchangers can be redesigned so that systems using any of these refrigerants could be as efficient or even slightly more efficient than current refrigerators using CFC-12.

Another aspect of the problem connected with CFC use, up to the present, has been paid rather little attention: it implies the influence of CFCs on the global climate warming. The high stability of CFCs, and their high reflecting ability in the IR range, significantly contribute to the global warming despite their much lower concentrations compared to CO_2 (*Climate Change*, 1990). Nevertheless, in this case, the problem can also be resolved in the same approach. Thus, while HCFC and HFC possess significant reflecting ability as well, their stability in higher atmosphere layers is tenfold lower than that of CFCs, so their contribution in the global warming is obviously much lower (Fischer *et al.*, 1991).

However, all these considerations mainly concern the large consumers of refrigeration, which are household systems and food industry. Thus, the capacities of the latter, applying cold for production of foods by using sublimation drying, several orders exceed the capacities of the sublimators used for preparation of advanced materials. Taking into account that cryochemical processes are much more labour-consuming than traditional ceramic technologies, it can be expected that in the near future the amount of advanced materials produced by cryochemical technology will not exceed several tens or hundreds of kilograms of the final product. This implies that the ecological problems connected with the generation of cold for cryochemical production of advanced materials will still be insignificant compared to the problems brought about by household systems and food industry.

Since cryochemical technology is applied on a limited scale, it can be stated that its hazardous wastes, including gases (SO_2, NO, NO_2) and condenser defrosting liquids containing various organic and inorganic chemicals, would not considerably change the total amount of hazardous wastes produced by industry (Dorfman *et al.*, 1992). For the USA only, the latter is estimated to be $8*10^8$ tons. At the same time, the possibility of local pollution connected with certain cryochemical production in quite a small region cannot be precluded.

Next to the problem of local pollution is the problem of staff safety in various technological processes. One of the main features of cryochemical technique is the use of materials and cooling agents at very low, cryogenic temperatures ($T < 204\,K$); first of all, the use of liquid nitrogen in cryocrystallization may be mentioned. The low heat of evaporation of N_2 ($\Delta H = 5.6\,kJ/mole$) enables a part of nitrogen to be rapidly evaporated, forming a gas blanket over the surface of the warmer object. Due to this, a short-time (several seconds) contact with liquid nitrogen – in particular, the contact of its drops with the skin – is not dangerous. The exception is the contact of liquid nitrogen drops with the eyes, so they must be protected with goggles or masks. The probability of such contact is much lower if special closed facilities are applied, for example as described by Coppa (1992) – during spraying the pulverization device and cooling agents are isolated from the operator. A more radical solution to the problem seems to be the freezing of solutions by spraying them into a vacuum chamber (Brazhnikov, 1983), when the contact of the operator with cooling agents is excluded in principle.

If isolated spraying devices are applied, care should be taken to provide communication between the gas in the chamber and the atmosphere, to prevent spontaneous closing-up; it should be kept in mind that the large volume of gases eliminated during evaporation can result in a dangerous increase of pressure.

Additional precautions are quite necessary if organic solvents are used in contact with liquid nitrogen or cooled surfaces; for example, in sublimation drying of non-aqueous solutions. The condensate formed on the cooled surfaces may contain a significant amount of liquid oxygen, forming explosive mixtures with organic solvents.

Special attention should be paid to the material of the vessels and units being in contact with liquid nitrogen (Edeskuty and Stewart, 1996). At cryogenic temperatures many kinds of laboratory glasses and plastics become friable or can be destroyed at fast cooling due to internal strains. It is more convenient to use stainless steel, copper, bronzes, aluminium or nickel alloys. Among plastics, Teflon, Dacron and Nylon are the most stable.

Not only in the case of contact with cooling agents should precautions be observed. In some cases, especially in large-scale syntheses, significant amounts of acids and organic solvents, more or less dangerous for human organism (Sarokin *et al.*, 1985), are used (Table 4.1). The precautions which should be taken in this case are much more similar to those usually accepted in the chemical laboratory (Marshall, 1987).

Considerably more serious are the problems arising due to chemical

TABLE 4.1 Overview of known health effects of some laboratory solvents and liquids (Kenworthy, 1990).

Chemical name	*Carcino-genicity*	*Develop-mental toxicity*	*Repro-ductive toxicity*	*Acute toxicity*	*Chronic (system) toxicity*	*Neuro-toxicity*	*Environ-mental toxicity*	*Bioaccum-ulation*
Acetone					X		X	
Ammonia				X	X			X
Benzene	X	X	X		X		X	
n-Butanol					X			
tert-Butyl alcohol								
Freon 113					X			
Hydro-chloric acid				X	X			
Isopropyl alcohol	X				X	X		
Methanol						X		
Nitric acid				X				
Sodium hydroxide (solution)							X	
Sulphuric acid				X	X		X	
Toluene		X	X				X	

peculiarities of synthesized materials. These problems are still aggravated by using the processes including the preparation of water–salt solutions, their dispersion and producing of fine aerosols, as well as various highly dispersed salt and oxide products. Taking into account an extreme diversity of advanced materials in their chemical composition, we preferred to confine the discussion to the problems arising in cryochemical production of superconducting ceramic materials. There are at least two reasons for this choice: superconducting ceramics are multicomponent materials; they usually contain highly dangerous oxides of thallium, mercury, lead, barium, copper and rare-earth elements. The toxicity values for various oxides (nitrates) used in synthesis of superconducting materials are presented in Table 4.2. These and other available data on substance toxicity allow the following conclusions to be drawn (Kiselev *et al.*, 1989):

1. All studied substances were found to be more toxic while penetrating via the respiratory tract (i.e. by breathing) rather than introduced through the alimentary canal. This notion is especially important because under conditions of cryochemical technology the delivery of substances in the organism by

TABLE 4.2 Experimental data on toxicity of some oxides and nitrates applied in the synthesis of high temperature superconductors.

Elements	*Toxicity* LD_{50} (mg/kg)						*Acute toxicosis threshold*	
	Oxide		*Nitrate*		*Oxide*		*Oxide*	
	Mice	*Rats*	*Mice*	*Rats*	*Mice*	*Rats*	*Mice*	*Rats*
	Alimentary introduction				*Peritoneal introduction*		*Peritoneal introduction,* mg/kg	*Respiratory introduction,* mg/m^3
Cu		470		940	273		45.4	
Ca					3059		259	
Sr			1028	1892	667		50	74*
Ba				390	300	293*	25	
Tl		39	20–30		72		0.001	
Cd	72		100					
Y	9480	15 340			425	230		92
Gd–Lu	11 400				2420			95
La–Eu	14 400		1125		3080 590*			
Pb					217 291		36.1 48.5	
Bi	10 000	5000	3710	4042				22.5 14.7*

* – data for nitrates.

respiration is the most common. According to decreasing toxicity; the substances can be arranged in the following row:

Tl–Cd–Ba–Cu–Y–Sr–(Gd–Lu)–La–Eu

2. Thallium is considerably more toxic than cadmium, which, in turn, is 100 times more toxic than any other element involved.
3. At peritoneal introduction the toxicity decreases in the following row (data for cadmium are missing):

Tl–Pb–Cu–Ba–Y–Sr–Ca–(La–Lu)

4. According to the limit of acute peritoneal toxicosis, the toxicity decreases in the following row:

Tl–Ba–Pb–Cu–Sr–Ca

5. Nitrates are more toxic than oxides and carbonates of corresponding metals and possess more intense irritating action on the skin and mucosa.

The introduction of oxides and other compounds used in HTSC technology into the human organism via the respiratory tract is the most probable, while their introduction via alimentary canal and, for some substances (thallium, lead) through the skin is often negligible.

Thallium is a polytrophic poison retained in the organism for a long time. Its concentration in the blood is insignificant even immediately after toxicosis because the blood transports thallium towards the organs, intensively bonding and accumulating it in large amounts: the thyroid, salivary glands and hairs. The average lethal dose depends on individual sensibility and is estimated as 6.0 to 40.0 mg per kg of body weight (Weissflog, 1969).

Copper Acute toxicoses with copper oxides and salts are provoked by their interaction with tissue proteins, leading to acute irritation of mucosa of the respiratory tract and alimentary canal. Chronic toxicosis with copper compounds may cause functional disorders of the nervous system and diseases of the liver and kidneys.

Calcium Inhalations of lime dust may result in pneumonia. Limestone (calcium carbonate) itself causes various malfunctions of the trachea (rhynites, laringites, pharingites), olfactory disorders, as well as chronic bronchite. Pathogenic effect of lime dust results from its irritatory and fibrogenic properties; occupational bronchial diseases are very often of a toxico-chemical nature.

Strontium Strontium may induce some alterations in the lungs and cause diseases of the respiratory organs and the nervous system; it may be responsible for heart pains as well. Strontium nitrate was found to have irritatory action on the skin. With the excessive introduction of strontium into the organism, the bone tissues, liver and blood become affected first.

Barium Water-soluble barium salts (nitrate, acetate) are very toxic: 0.2–0.5 g of the carbonate at intestinal introduction causes toxicosis, while a 2–4 g dose is lethal. With acute toxicosis, the myocardium, nervous system and blood vessels are affected. With chronic toxicosis, the diseases of bone tissues (barium displaces calcium and phosphor from bones), marrow, gonads and liver are observed. With chronic toxicoses with barium salts in aerosols, irritations of the trachea, eyes and skin can occur.

Lead Lead, similar to thallium, is a polythropic poison affecting all organs and systems of human and animal organism. The most significant disorders occur in the blood, nervous and cardiovascular systems, as well as in the liver and alimentary canal. Lead is classed among cumulative poisons and is accumulated in the bones, liver, kidneys and pancreas. The toxicity of its compounds and their pathogenic action vary, depending on solubility of the salts in the body liquids. Lead possesses slight mutagenic action, resting among heavy metals at the last place (Hg $>$ Zn $>$ Ag $>$ Al $>$ Cd $>$ Pb).

Rare-earth elements Acute toxicoses with rare-earth element (REE) oxides in industrial technological processes are less probable because of their rather low toxicity (Table 4.2). However, the limit of inhalatory action of yttrium oxides (92–95 mg/m^3) proves them to be quite dangerous upon respiratory introduction. With long-time inhalation of REE oxides, the alterations in the blood composition and coagulability, neural disorders, metabolic malfunctions, the atrophy of liver, kidneys and spleen are observed. Among REE the most toxic are oxides of the yttrium group, as well as yttrium oxide itself. REE oxides possess rather low fibrinogenic activity. Their action on the skin and mucosa is not so harsh; corresponding nitrates may provoke considerably greater reaction.

People working with REE compounds often suffer from the diseases of respiratory organs, caused by the combined action of REE aerosols with acid and alkaline vapours.

Bismuth There is no information in the literature concerning professional diseases caused by bismuth and its derivatives. However, available data on its influence on animals point to the possibility of acute and chronic toxicoses under technological conditions. Established for animals the limits of acute respiratory action for oxide and nitrate are quite low. The limit of long-time action of bismuth nitrate at its inhalation was found to be 2.0 mg/m^3. The ability of bismuth to accumulate is estimated as weak for the nitrate and chloride, and high for the sulphate. The accumulation of bismuth occurs mainly in the kidneys, liver, brain and spleen. Bismuth and its compounds are able to cause functional disorders in cardiovascular and central nervous systems, as well as in carbohydrate, lipide and protein metabolism. The nitrate has an irritating action on the skin, while for the oxide this is not the case.

Cadmium Cadmium oxide is classed among the substances with direct toxic action, causing acute toxicoses like toxic pneumonia and death from emphysema or bronchopneumonia in 3–5 days. The symptoms are observed after a 1–8 h latent period from the moment of oxide inhalation.

At acute oral cadmium toxicosis the following symptoms are usually observed: burning sensations in the mouth extended to the epigastral area, and further along all abdominal area, nausea, diarrhoea.

Nitrates At introduction of metal nitrates in the organism, the increase of met- and sulphohaemoglobin in the blood is observed. The alteration in external breath like enhanced frequency of respiration (reflectory increase of the frequency due to oxygen deficit) may occur. Nitrates possess carcinogenic action towards human and animal organisms, determined by their possible transformation into mutagenic nitrosamines.

To summarize, it is necessary to emphasize once more that all the components used for the preparation of HTSC ceramics, as well as the substances which can form during technological processes (especially in emergency situations), should not be treated carelessly.

In industrial technology these poisons can also possess various combined actions; accordingly, it seems reasonable for an experimenter to choose those combinations which are less harmful for the human body. However, the data on such combined actions are very scarce in the literature. It seems that the combination of copper, bismuth and lead is undesirable because copper and lead favour retention of bismuth in the organism. At low calcium level (0.3%) in food, the toxicity of lead considerably increases; this give reason to suppose that the combination of lead with calcium is to some extent favourable. The combination of thallium, lead and cadmium seems undesirable because they are the most toxic substances with similar cumulative action.

Thus, the cryochemical technique, as well as all the other techniques of inorganic materials synthesis, in its present state does not completely exclude undesirable wastes and often is connected with some hazards in the course of certain processes. However, full comprehension of the reasons and nature of the presented problems, as well as taking the necessary precautions, allow considerable reduction in the negative effects and, at the same time, maintain the principal advantages of the technique. Further improvement of cryochemical techniques and, first of all, of the applied equipment, will allow it to meet ecological requirements, which are becoming increasingly stringent with the increase of societal needs.

References

Brazhnikov, S.M. (1983) Heat and mass transfer during evaporative freezing in vacuum, PhD thesis, Moscow: Moscow Institute of Chemical Machinery (MIChM).

Climate Change (1990) The IPCC Scientific Assessment, Intergovernmental Panel on Climate Change.

Coppa, N.V., Myer, G.H., Salomon, R.E. *et al.* (1992) Preparation, thermal processing behavior and characterization of YBCO from freeze-dried nitrate precursors, *J. Mater. Res.*, **7**(8), 2017–26.

Dorfman, M.H., Muir, W.R. and Miller, C.G. (1992) *Environmental Dividends: Cutting More Chemical Wastes*, New York: INFORM Inc.

Fischer, S.K., Hughes, P.J., Fairchild, P.D. *et al.* (1991) *Energy and Global Warming Impacts of CFC Alternative Technologies*, New York: Arthur D. Little.

Fischer, S.K. and McFarland, M. (1992) Alternatives to CFCs and global warming: a systems approach to evaluating net contributions, *MRS Bulletin*, **17**(3), 39–42.

Edeskuty, F.J. and Stewart, W.F. (1996) *Safety in the Handling of Cryogenic Fluids*, New York: Plenum Publ. Co.

Kenworthy, L. (1990) *A Citizen's Guide to Promoting Toxic Waste Reduction*, New York: INFORM Inc.

Kiselev, Yu.M., Shabalina, L.P. and Spiridonova, V.S. (1989) Industrial safety aspects during HTSC ceramics preparation, *J. Mend. Chem. Soc.*, **34**(4), 544–50 (Russ.).

Marshall, V.C. (1987) *Major Chemical Hazards*, New York: Halsted Press.

Sarokin, D.J., Muir, W.R., Miller, C.G. and Sperber, S.R. (1985) *Cutting Chemical Wastes. An INFORM Report*, New York: INFORM Inc.

United Nations Environment Program (1991) *Handbook for the Montreal Protocol on Substances that Deplete the Ozone Layer.*

Weissflog, A. (1969) *Beitrag zur Klinik der Thalliumvergiftung*, Bonn.

World Meteorological Organization. Scientific Assessment of Stratospheric Ozone, Geneva: Global Ozone Research and Monitoring Project, 1990.

5

Conclusion

There are many reasons to affirm that in the near future the low-temperature processes will become as important as high-temperature ones. Moreover, a well-planned combination of high-temperature and low-temperature processes can ensure the greatest effectiveness in material technology. Cold, for the most part, supplements traditional technological procedures and makes them more controllable and efficient. This concerns the situations wherein instead of chemical precipitation at room temperature cryoprecipitation is used, or common salting-out is replaced by cryoextraction; pulverization processes can also be carried out at low temperature (cryodispersion) and drying processes can take place at low temperatures under low gas pressures (freeze-drying).

However, cold itself does not ensure the desired effects if physico-chemical peculiarities of the precursors are not taken into account. Very often, in an attempt to modify the technology of advanced materials, processes of freezing and sublimation drying are carried out in a similar manner as in the food industry in the production of instant products (coffee, tea, milk, meat, fish etc.). Moreover, in the materials technology the freezing and sublimation drying equipment designed for the food industry is traditionally used. However, such an approach seems not to be reasonable. The freezing of solutions by direct contact with liquid nitrogen used in many cases is not efficient and, regarding cooling rate, is inferior to crystallization on a cooled surface. In the latter case, the product structure changes abruptly from the dendrite to fine granular with higher chemical homogeneity. Up to the present time, only little attention was given to these differences, while such a transition can be decisive for some physico-chemical systems, especially for the multicomponent ones.

The majority of products of cryochemical synthesis are highly dispersed stochastic systems, while the principal aim of this kind of synthesis is bulk dense material with considerably higher ordering. The most successful approach to the quantitative description of system evolution from the stochastic state to an ordered one is the application of fractal concepts, which make it possible to explain many observed phenomena. Besides, by improving cryochemical (and other) technologies it should be taken into account that highly homogeneous precursors do not always ensure successful results in obtaining the desired product.

Almost all systems that are attractive for materials engineering demonstrate some features of **deterministic chaos systems**, i.e. infinitely small deviations of initial conditions leading to drastic change in the properties of the analysed system.

In terms of **deterministic chaos** there are no absolutely reproducible processes in materials engineering at all so both structure-sensitive and physico-chemical properties of materials (like chemical degradation ability upon storage and application) are also hardly reproducible parameters from a theoretical point of view.

In principle in materials science there are many empirical methods allowing an increase in the reproducibility of obtained product properties. In **deterministic chaos** theory all of these methods can be considered as the following principal decisions:

1. **Chaos 'prevention'** by means of system simplification. One possible novel method of accomplishing this is based on the approach of **decomponization** to control the development of structurally complicated phases. This approach subdivides the whole process into a number of deterministic stages.
2. Achieving high '**accuracy**' of initial precursor parameters to make the system predictable for a reasonable period of time as well as acceleration of the interaction rate to keep the interaction time within the predictable duration interval of the process.
3. Intensive **suppression** of chaos by counteraction based on feedback relations or other **map-based algorithms**. In the frame of this branch of interest it is necesssary to know the details of chaotic behaviour of the system so that a unique method of chaos control, involving both computer simulation and experimental study of topochemical reactions occurring kinetically, diffusionally, or in mixed regime, can be undertaken.

Further investigations in this direction followed by revision of many traditional concepts can lead to considerable progress in the branch of material syntheses, especially those involving a great number of chemically dissimilar components.

As was shown earlier, in the ceramic technology of advanced materials the best results were achieved in the preparation of magnetic ceramics, ceramic electrolytes, piezoceramics transparent in optical range and porous ceramics based on heavy metal dioxides. However, these are only the first steps. For example, in the preparation of magnetic ceramics the cryochemical technology can be especially effective in obtaining materials with high permeability, magnetostriction and high coercive force, including materials for radiation-absorbing systems.

Success achieved in cryochemical technology of polyaluminate ceramics with the β-alumina structure makes it reasonable to apply this technology for preparation of ceramic electrolytes based on zirconium dioxide, as well as for production of other multicomponent and heat-resisting materials with unipolar ionic conductivity. Up to the present, very little was made for development of cryochemical technologies for ceramics transparent in the optical range, although the positive experience with piezoceramics based on lead titanate–zirconate doped with lanthanum oxide raises hopes for successful preparation of other materials with high transparence in the visual and far infrared spectral range, especially when cryochemical synthesis of powders is combined with hot pressing.

Very little work has been done so far on the development of cryochemical technology of non-oxide ceramics in spite of positive experience in the preparation of carbides and borides of some heavy metals. Much can be expected from the application of cryochemical techniques for preparing the oxide and, perhaps, chalcogenide luminophores, although cryochemical syntheses of chalcogenides were not always successful. The patents on cryochemical synthesis of composite materials positively reflect the prospects of this approach. Besides traditional metal–oxide composites with required density, rigidity and durability at heat stresses, the cryochemical technology can lead to the successful production of metalloceramic electrodes, light sources of high output and new kinds of solid electrolytes on the basis of dispersoids.

Cryochemical technology of solid-state materials will be more flexible if, besides aqueous solutions, non-aqueous ones of the components involved are used as starting objects for cryocrystallization.

Especially promising from the technological point of view is the combination of cooling with other novel processes. The close interdependence of physical, chemical and biological phenomena gives reasons to affirm that the investigation of cryoeffects in any particular branch can lead to fundamental discoveries, which in turn will be necessary for the development of new advanced technologies.

Further reading

CHAPTER 2

Brazhnikov, S.M., Mozhaev, A.P., Kiryushin, N.V. and Volynets, A.Z. (1991) Kinetics and mass transfer in the process of sublimation in vacuum, *Ultrapure Substances*, **6**, 76–83 (Russ.).

Golovchansky, A.V., Naumov, S.V., Mozhaev, A.P. and Tretyakov, Yu.D. (1982) Studies of freezing of aqueous solutions of electrolytes, *Russ. J. Phys. Chem.*, **56**(9), 2337.

Khomutov, V.V. and Lyashchenko, A.K. (1995) Properties of water–salt compositions for preparing high-T_c superconductors containing yttrium, barium and copper, *Russ. J. Inorg. Chem.*, **40**(2), 80–5 (Engl.).

Kuleshov, I.V. (1984) Calorimetric studies of processes and products of freeze drying synthesis, PhD thesis, Moscow: Moscow State University.

Kuleshov, I.V., Gordeev, I.V., Kesler, Ya.A. and Tretyakov, Yu.D. (1984) Thermochemical studies of schoenites like double salts, *Courier of Moscow State University, Chemistry*, **4**, 523–4.

Kuleshov, I.V., Gordeev, I.V., Kesler, Ya.A. and Mozhaev, A.P. (1987) Thermochemical study of the system 'cobalt chloride–water' in the field of concentrated solutions, *Russ. J. Phys. Chem.*, **61**(2), 341–4.

Kuznetsov, V.V., Lyashchenko, A.K. and Trostin, V.N. (1993) X-ray studies of binary and multicomponent nitrate solutions of Y, Ba and Cu, *Russ. J. Inorg. Chem.*, **38**(1), 159–61.

Lyashchenko, A.K., Mozhaev, A.P., Naumov, S.V. and Golovchansky, A.V. (1983) Cryocrystallization of water–salt systems and specific features of structures of superconcentrated solutions, *Russ. J. Struct. Chem.*, **24**(2), 48–53.

Mirkovich, V.V. and Wheat, T.A. (1970) Use of liquid nitrogen in spray-freezing, *Am. Ceram. Soc. Bull.*, **49**(8), 724–5.

Naumov, S.V., Mozhaev, A.P., Tretyakov, Yu.D. and Topor, N.D. (1982) Investigation of crystallization of aqueous sulphate solutions by differential scanning calorimetry, *Russ. J. Phys. Chem.*, **56**(1), 232–3.

Swamy, C.S. (1974) Freeze-drying as a cryochemical technique in material synthesis, *Indian Chem. J.*, **8**, 17–20.

Telegina, L.I., Mozhaev, A.P. and Tretyakov, Yu.D. (1980) Studies of physico-chemical transformations in fast frozen salt solutions, *Russ. J. Inorg. Materials*, **16**(10), 1853–6.
Telegina, L.I., Mozhaev, A.P. and Tretyakov, Yu.D. (1983) Kinetics of granulated ice dissolution in organic solvents, *Russ. J. Phys. Chem.*, **57**(2), 386–90.
Volynets, A.Z. (1980) Heat and mass transfer in the vacuum freeze-drying technology, *Dr. Sci. Thesis*, Moscow: Moscow Academy of Chemical Machinery.

CHAPTER 3

Phosphor materials

Inoue, K., Nakada, Y. and Asada, M. (1986) Preparation of pigmented phosphors by freeze-drying a water-soluble polymer-containing slurry, *Jap. Pat.* No. 6, 313, 5480 A2 from 28 Nov.
Takada, Y., Hashimoto, N., Sawada, T. and Nagao, C. (1990) Rare earth silicate fluorescent materials and manufacture thereof, *Jap. Pat.* No. 3, 207, 787 from 10 Jan.

Magnetic materials

Buettner, H.J. and Wiegand, A.U. (1977) Study of the freeze-drying process for manufacturing ferrite powders with high specific surfaces, *Hermsdorfer Techn. Mitt.*, **17**(47), 491–5.
Gallagher, P.K., Johnson, D.W., Vogel, E.M. and Schrey, F. (1977) Microstructure of some freeze-dried ferrites, *Ceramic Microstructures'76*, Boulder, CO: Westview Press, 423–32.
Kuzmicheva, T.G., Olkhovik, L.P. and Shabatin, V.P. (1990) The method of hexagonal barium ferrite preparation, *Pat. USSR* No. 4843538/33 from 28 June.
Suchoff, L.A. (1971) Decomposing freeze-dried oxalates in vacuum to uniformly submicron ferrite particles, *US Pat.* No. 3, 607, 753 from 21 Sept.
Suenaga, W. and Yamanashi, K. (1989) Treatment of magnetic powders, *Jap. Pat.* No. 2, 307, 202 from 23 May.
Sugano, T. and Karita, Y. (1985) Concentration of water-soluble magnetic fluids, *Jap. Pat.* No. 61, 222, 591 from 29 Mar. (R).
Sugano, T. and Karita, Y. (1986) Manufacture of ferrofluids, *Jap. Pat.* No. 6, 311, 6407 A2 from 5 Nov.

Ferroelectric materials

Akbas, M.A. and Lee, W.F. (1995) Synthesis and sintering of PLZT powder made by freeze/alcohol drying of composition citrate solutions, *J. Eur. Ceram. Soc.*, **15**(1), 57.
Kondou, S., Kakegawa, K. and Sasaki, Y. (1990) Synthesis of $Pb(Zr_xTi_{1-x})O_3$ by freeze drying method, *Nippon Kagaku Kaishi*, **7**, 75–8.
McHale, J.M., McIntyre, P.C. and Coppa, N.V. (1996) Nanocrystalline $BaTiO_3$ from freeze-dried nitrate solutions, *J. Mat. Res.*, **11**(5), 1199.

Solid-state ionics

Chowdhry, U. and Cannon, R.M. (1978) Microstructural evolution during the processing of sodium β-alumina. In H. Palmour III, R.F. Davis and T.M. Hare (eds), *Processing of Polycrystalline Ceramics*, New York: Plenum Press.

Ida, T. (1990) Manufacture of solid electrolyte capacitors, *Jap. Pat.* No. 3, 215, 918 A2 from 18 Jan.

Li, F., Zhang, J. and Pang, H. (1992) Preparation of NASICON and lithium titanium chromium phosphorus oxide ($Li_{1.8}Ti_{1.2}Cr_{0.8}P_3O_{12}$) fast ionic conductor by spray (freeze) drying, *Huadong Huagong Xueyan Xuebao*, **18**, 113–19.

Nicholson, P.S. (1992) Synthesis of polycrystalline hydronium- and ammonium-β/β''-alumina and potential application in steam-electrolysis/fuel-cells, *Chem. Solid State Mater.*, **2** (Proton Conductors), 499–510.

Superconducting ceramic materials

Badica, P., Aldica, G. and Morozov, V.V. (1996) Influence of pyrolization temperature on final characteristics of BSCCO ceramics produced by spray-frozen, freeze-drying method, *J. Mat. Sci. Letters*, **15**(3), 187.

Badun, Ju.V., Shlyahtin, O.A., Kuzmenkov, A.I. and Mozhaev, A.P. (1991) Cation stoichiometry deflections during HTSC nitrate synthesis, *Waseda International Symposium on Superconducting Materials. Abstracts*, Tokyo, 31.

Badun, Ju.V., Shlyakhtin, O.A., Kuzmenkov, A.I. *et al.* (1993) Cation stoichiometry deviations during high-temperature superconductor nitrate synthesis, *Phase Transitions*, **41**(1–4), 157–8.

Barboux, P. and Tarascon, J.M. (1987) Method of making high T_c oxide superconductor particles by adjusting pH value and cationic ratio of admixed solution followed by drying, heating and firming, *US Pat.* No. 4, 962, 082 A from 19 Nov.

Van der Biest, O. and Kwarciak, J. (1991) Ceramic superconductors synthesized by sol–gel methods, *J. Australas. Ceram. Soc.*, **27**(1–2), 51–6.

Chen, D.H., Sheen, S.R. and Chang, C.T. (1992) Chemical method for preparation of copper oxide superconductors, *Huaxue*, **50**, 158–67.

Churagulov, B.R., Lyubimov, S.L., Kravchuk, K.G. and Zagnit'ko, E.V. (1993) The use of high temperature concentrated solutions of yttrium, barium and copper nitrates in the cryochemical preparation of high temperature superconductors, *Russ. J. Inorg. Chem.*, **38**(7), 1149–52.

Dierickx, D. and Van der Biest, O. (1995) Intermediate phases during processing of freeze-dried YBaCuO precursors, *Eur. J. Solid State Inorg. Chem.*, **32**(7), 711.

Garvie, D.A., Finlayson, T.R. and Przelozny, Z. (1991) Flux pinning study of $Y_xRE_{1-x}Ba_2Cu_3O_{7-\delta}$ superconductors, *Adv. Supercond. IV, Proc. 4th Intern. Symp. Supercond.*, Tokyo, 327–30.

Hiraki, A., Ito, T., Kimura, Y. and Watabe, H. (1991) Manufacture of superconductor powder material, *Jap. Pat.* No. 4, 295, 003 from 25 Mar.

Johnson, S.M., Gusman, M.I., Hildebrand, D.L. *et al.* (1988a) Critical current density in bulk yttrium barium copper oxide $YBa_2Cu_3O_{7-\delta}$, *Ceram. Supercond. 2, Res. Update*, Westerville: Am. Ceram. Soc., 300–7.

Johnson, S.M., Gusman, M.I., Hildebrand, D.L. and Eom, C.B. (1988) Synthesis of superconducting powders by freeze-drying, *Mater. Res. Soc. Symp. Proc.*, **121** (Better Ceramics through Chemistry 3), 413–20.

Martynenko, L.I., Kupriyanova, G.N., Kovaleva, I.B. *et al.* (1990) Preparation of high T_c superconductors by complexonates, *International Symposium on High-T_c Superconductivity*, Hanoi, 29 Oct., OC-3.

Mikhailov, B.P., Kadyrbaev, A.R., Burkhanov, G.S. *et al.* (1990) Influence of Ag_2O on the structure and superconducting properties of $YBa_2Cu_3O_{7-x}$-conductors, *Superconductivity: Physics, Chemistry, Technique*, **3**(8: Pt 2), 1926–31.

Mozhaev, A.P., Pershin, V.I. and Shabatin, V.P. (1989) Methods of synthesis of high-T_c superconductors, *J. Russ. Mendeleev Chem. Soc.*, **34**, 504–9.

Nakamura, N., Nakano, T. and Shimotomai, M. (1987) Oxide superconductor source suspension and manufacture of oxide superconductor source, *Jap. Pat.* No. 1, 096, 016 A2 from 9 Oct.

Shinho, Y., Kosuge, S., Yoshikawa, H. and Hiratani, A. (1987) Manufacturing of yttrium barium copper oxide superconductor powder, *Jap. Pat.* No. 1, 093, 422 A2 from 5 Oct.

Sipachev, V.I., Arkhangel'skii, I.V., Shabatin, V.P. and Reshetova, L.N. (1991) The use of liquid ammonia in the synthesis of high-T_c materials, *Waseda International Symposium on Superconducting Materials*, 23–25 Oct., Tokyo, 34.

Sipachev, V.I., Arkhangel'skii, I.V., Shabatin, V.P. *et al.* (1991) Synthesis of intercalates of high-T_c ceramics, *Waseda International Symposium on Superconducting Materials*, 23–5 Oct., Tokyo, 35.

Tachivaki, T., Sugimoto, J. and Hiraki, A. (1996) Characterization of freeze-dried powders prepared by alkoxide route for YBCO superconductors, *Appl. Surf. Sci.*, **100/101**, 272.

Tretyakov, Yu.D. (1988) Preparation and physico-chemical properties of superconductive oxides, *Progress in High Temperature Superconductivity*, **18** (The Science of Superconductivity and New Materials), World Scientific, 152–60.

Tretyakov, Yu.D., Oleinikov, N.N., Shlyakhtin, O.A. and Kazin, P.E. (1993) Recent development of melt technology of high-T_c ceramic superconductors, *3rd International Workshop on Chemistry and Technology of High Temperature Superconductors (MSU-HTSC-III)*, Moscow, 20–23 Sept., Programme and Abstracts, O-1.

Wang, C.T., Lin, L.S., Lin, J.H. *et al.* (1988) $YBa_2Cu_3O_{7-\delta}$ superconductors produced by hot pressing of freeze-dried, carbonate- and oxalate-coprecipitated powders, *Mater. Res. Soc. Symp. Proc.*, **99** (High-Temperature Superconductors), 257–60.

Weyl, C., Zhao, J., Lacour, C., *et al.* (1995) Investigation of a nanocrystalline $YBa_2Cu_3O_{7-\delta}$ powder by neutron and electron diffractions, *J. Phys. Chem. Solids*, **56**(9), 1211–21.

Electrode materials

Boskovic, S. and Stevanovic, M. (1978) Cryogenic synthesis of lithium nickel oxide solid solution powder, *Hem. Ind.*, **32**, 450–2.

Matsushita Electric Industrial Co. (1979) Nonaqueous electrolyte battery, *Jap. Pat.* No. 5, 510, 4073 from 2 Feb.

Tseung, A.C.C. and Bevan, H.L. (1973) A reversible oxygen electrode, *Electroanal. Chem., Interfac. Electrochem.*, **45**, 429–38.

Tseung, A.C.C., Rasiyah, N.P. and Hibbert, D.B. (1980) Electrocatalyst, *Eur. Pat.* No. 34447 from 11 Feb.

Tseung, A.C.C., Botejue, J.C. and Rasiyah, P.N. (1984) Optimization of Teflon-bonded gas evolving electrodes, *Rep. Comm. Eur. Comm.*, EUR 9528, 145.

Materials for catalysts and sorbents

Berentsveig, V.V. and Shabatin, V.P. (1992) The catalytic method of cleaning waste gases from carbon monoxide, *Pat. USSR* No. 1837948 from 2 Mar.

Chen, X., Zhong, T., Li, Y., Freeze-drying method of preparing perovskite-like lanthanum cobalt oxide ($LaCoO_3$) catalyst, *Huaxue Tongbao*, **3**, 20–2.
Gorst, J., Eales, A., Desai, N.J. and Do, D.D. (1987) High capacity mixed oxide sorbents for nitrogen oxide (NO_x) removal from flue gas, *Recent Trends Chem. React. Eng., Proc. 2nd Intern. Chem. React. Eng. Conf.*, New Delhi: Wiley East, 369–82.
Gortsema, F.P., Hatwell, H. and Johnson, G.K. (1972) Reforming catalyst on a high surface area, fibrous ceramic support, *Germ. Pat.* No. 2, 303, 884 from 27 Jan.
Kennedy, J.F., Humphreys, J.D. and Barker, S.A. (1981) Further facile immobilization of enzymes on hydrous metal oxides and use of their immobilization reversibility phenomena for the recovery of peptide antibiotics, *Enzyme Microb. Technol.*, **3**(2), 129–36.
Kim, H. (1980) The study on the supported oxide catalysts for reducing carbon monoxide gas in automotive exhaust gas, *Yo Op Hoe Chi*, **17**(4), 179–87.
Ladavos, A.K. and Pomonis, P.J. (1991) Intercalation of lanthana and lanthana–nickel oxide oxidic species into montmorillonite layered structure, *Stud. Surf. Sci. Catal.*, **63** (Prep. Catal. 5), 319–28.
Sermon, P.A. and Knight, S.J. (1975) Improvements in and relating to gas purification catalysts, *Brit. Pat.* No. 1, 558, 167 from 16 May.

Refractory oxides

Baer, D., Foerthmann, R., Naoumidis, A. and Nickel, H. (1992) Studies for the production of aluminium oxide ceramics with defined microstructure parameters using colloidal-chemical processing, *Ber. Forschungszentrum Juelich*, Juel-2612.
Berrin, L., Johnson, D.W. and Nitti, D.J. (1972) High purity reactive alumina powders, *Amer. Ceram. Soc. Bull.*, **51**(10), 840–4, 896–900.
Gu, Y., Zhao, S., Hu, L. and Chen, M. (1993) Preparation of aluminium oxide ultrafine particles via a precipitation process in aqueous solution, *Huagong Yejin*, **14**, 14–21.
Hattori, T. and Mohri, J. (1979) Preparation of alumina by a cryochemical method, *Chiba Daigaku Kogakubu Kenkyu Hokoku*, **31**, 45–9.
Jeon, S.B. and Pyun, Su-Il (1979) Preparation and characterization of stabilized zirconium (IV) oxide by wet chemical methods, *Yo Op Hoe Chi*, **16**, 155–63.
Livage, J., Vivien, D. and Mazieres, C. (1969) Lyophilization of precipitated zirconium oxide and other group VIa oxides, *Rev. Inst. Hautes Temp. Refract.*, **6**(1), 71–3.
Lloyd, I.K., Blachere, J.R., Beyer, L.A. *et al.* (1990) Sintered alumina prepared from spray frozen sulfates, *Ceramic Transactions, Vol. 7: Sintering in Advanced Ceramics*, Am. Ceram. Soc., Westerville, OH, 371–80.
Nakanishi, N., Takano, M. and Kumito, H. (1977) Preparation of fine powders of magnesium oxide by freeze-drying processing, *Funtai Oyobi Funmatsuyakin*, **24**(1), 1–5.
Rakotoson, L. and Paulus, M. (1984) Sintering of a freeze-dried 10 mol.% yttria stabilized zirconia, *Adv. Ceram.*, **12** (Sci. Technol. Zirconia 2), 727–32.
Ramme, R. and Hausner, H. (1986) On the freeze-drying of zirconium hydroxide and its effects on the strength of sintered zirconium dioxide, *Ceramic Forum Int.*, **63**(7–8), 373–7.
Rasmussen, M.D., Akinc, M. and Hunter Jr., O. (1985) Processing of yttria powders derived from hydroxide precursors, *Ceram. Int.*, **11**, 51 5.
Real, M.W. (1986) Freeze drying alumina powders, *Br. Ceram. Proc.*, **38**, 59–66.
Roosen, A. and Hausner, H. (1984) Sintering kinetics of zirconia powder, *Adv. Ceram.*, **12** (Sci. Technol. Zirconia 2), 714–26.

Sipitanos, K.M., Bakoyannakis, D.N. and Kaliakos, A.T. (1979) Preparation of aluminum oxides by freeze drying and conventional drying techniques, and their differences in relation to the values of certain physical characteristics, *Chem. Chron.*, **8**(1), 67–76.

Varela, J.A. (1991) Sol–gel process for the manufacture of doped and undoped zirconia using zirconium nitrate as starting material, *Port. Pat.* No. 9, 102, 212 from 29 May.

Wheat, T.A. (1974) Fired microstructure of freeze-dried lime stabilized zirconia, *J. Can. Ceram. Soc.*, **43**(1), 1–14.

Ultrafine metal and metal alloy powders

Amato, G. and Brunetto, N. (1996) Porous silicon via freeze drying, *Mater. Letters*, **26**(6), 295.

Bhatia, I.N. and Jape, K.B. (1976) Freeze-drying technique for the production of metal powders, *Trans. Powder. Metal. Assoc. India*, **3**, 43–6.

Ferrante, M.J., Lowery, R.R. and Robidart, G.B. (1971) Tungsten and dispersion-strengthened tungsten made by freeze drying, *U.S. At. Energy Comm.*, BM-RI-7485.

Kusai, R. (1972) Freeze drying of metal powders, *Met. Metal. Form.*, **39**, 377–8.

Lenel, F.V., Ansell, G.S. and Nazmy, M.Y. (1971) Creep properties of dispersion-strengthened silver–gallium oxide alloys, *High Temp. – High Pressures*, **3**(4), 439–44.

McCarron, E.M. and Parise, J.B. (1990) Process for preparation of metal oxides of molybdenum or molybdenum and tungsten, *US Pat.* No. 5, 055, 441 from 10 Feb.

Nakamura, K., Asai, T. and Kanzawa, R. (1974) Fine silver powder for electric resistor pastes, *Jap. Pat.* No. 4, 911, 8661 from 13 Nov.

Sumitomo Chemical Co. (1982) Heat-resistant dispersion-strengthened sintered nickel alloys, *Jap. Pat.* No. 5, 820, 7340 from 28 May.

White, G.D. and Gurwell, W.E. (1989) Freeze-dried tungsten heavy alloys, *Adv. Powder Metall.*, **2**, 355–69.

Zeng, Z., Zhao, Q., Xu, Y. and Li, J. (1983) Effect of original ammonium paratungstate on the particle size of reduced tungsten powder, *Xiyou Jinshu*, **2**, 155–62.

Inorganic composites

Anderton, D.J. and Sale, F.R. (1978) The application of the freeze-drying process to the production of silver–cadmium oxide contact material, *Electr. Contacts*, **24**, 183–9.

Joshi, P.B., Murti, N.S.S., Gadgeel, V.L. *et al.* (1995) Preparation and characterization of Ag–ZnO powders for applications in electrical contact materials, *Journal of Materials Science Letters*, **14**(16), 1099–1101.

Nakanishi, N., Takano, M., Yasuda, C. *et al.* (1979) Preparation of fine powders of copper–aluminium oxide dispersion-strengthened alloy by freeze-drying processing, *Funtai Oyobi Funmatsuyakin*, **26**, 79–83.

Nakanishi, N., Takano, M., Chyo, A. *et al.* (1982) Preparation of fine powder of copper–aluminium oxide alloy by the freeze-dry method of molten salts, *Funtai Oyobi Funmatsuyakin*, **29**(5), 182–6.

Wei, W.C. and Novich, B.E. (1987) Manufacture of ceramic or metallic composites from a binderless slip, *US Pat.* No. 8, 904, 735 A1 from 25 Nov.

Miscellaneous inorganic substances and materials

Allardyce, K.E. and Edmondson, B.J. (1972) Tungstic and vanadic acids, *Brit. Pat.* No. 1, 434, 523 from 25 Feb.

Anderton, D.J. and Sale, F.R. (1976) Thermal decomposition of conventionally crystallized and freeze-dried silver nitrate and cadmium nitrate, *Proc. 1st European Symp. Therm. Analysis*, 278–9.

Aoki, F., Yokochi, K. and Umada, G. (1989) Thin film formation on the surface of ceramic powders by freeze-drying process, *Jap. Pat.* No. 3, 126, 653 from 12 Oct.

Barret, P. (1970) Application des techniques de pulverisation electrochimique par etincelles et de lyophilisation à l'obtention de solides de grandes surfaces ou de depots, *Silicates industriels, Colloque 'Proprietes des petites particules'*, Mons, Oct.

Bermejo, E., Dantas, T., Lacour, C. and Quarton, M. (1995) Mechanism of formation of nanocrystalline hematite prepared by freeze-drying, *Materials Research Bulletin*, **30**(5), 645–52.

Bones, R.J. and Teagle, D.A. (1992) Manufacture of sintered artefacts from sinterable powder, and the sintered artefacts obtained, *Germ. Pat.* No. 4, 313, 380 A1 from 24 Apr.

Bucerius, K.M. and Eisenreich, N. (1986) Improving the burning behaviour of solid propellants and tube weapon propellants, the propellants and their uses, agents for improving propellants, and use of the agents, *Germ. Pat.* No. 3, 641, 788 A1 from 6 Dec.

Bucerius, K.M. (1987) Increasing the effectiveness of burning modifiers in double-base solid propellants, *Int. Jahrestag.- Fraunhofer Inst. Treib-Exposivst.*, **18** (Technol. Energ. Mater.), 75/1–75/16.

Chen, Z. and Nie, J. (1988) Preparation of ultrafine oxide powders by freeze-drying and their uses in ceramic materials, *Guisuanyan Tongbao*, **7**(2), 31–9.

Chen, Z., Wan, Y., Rong, J. *et al.* (1989) Preparation of ultrafine powder for complex oxides by freeze drying, *Wuji Cailiao Xuebao*, **4**(2), 157–63.

Coblenz, W.S. and Kavanaugh, M.D. (1992) Ceramic materials, their manufacture and use, *Brit. Pat.* No. 5, 85, 761 A1 from 21 Aug.

Colson, J.C., Cointot, A. and Barret, P. (1969) Freeze drying in the preparation of finely powdered solids, *Bull. Inst. Int. Froid*, Annexe, 1969, No. 9, pp. 167–77.

Daurova, L.I., Mozhaev, A.P., Rogacheva, L.F. *et al.*, (1986) Cryochemical method of synthesis of ceramics based on ferroyttrium garnet, *Russ. Electron. Ind.*, **10**, 64–5.

Deki, S., Nagata, K., Domura, H. and Shimada, Y. (1990) Preparation and properties of non-crystalline alkali metal metatungstate hydrates, *Nippon Kagaku Kaishi*, **6**, 615–20.

Dogan, F., Roosen, A. and Hausner, H. (1985a) Chemical processing and agglomerate characteristics of oxide powders, *Bull. Am. Ceram. Soc.*, **64**(3), 445.

Dogan, F., Roosen, A. and Hausner, H. (1985b) Agglomerate formation, compact characteristics and sintering behavior of ceramic powders, *Science of Ceramics*, **13**(C1), 231–5.

Dogan, F. and Hausner, H. (1988) The role of freeze drying in ceramic powder processing, *Ceramic Trans.*, **1** (Ceram Powder Sci. **2**(Pt. A)), 127.

Felten, E.J. and Stonehart, P. (1976) Tungsten carbide, *Germ. Pat.*, No. 2, 553, 373 from 16 June.

Fujiwara, H., Kudo, S., Uehara, K. *et al.* (1976) Micro powder of transition metal oxide or oxide hydroxide, *Jap. Pat.*, No. 5, 307, 9798 from 24 Dec.

Gallagher, P.K., Johnson, D.W. and Schrey, F. (1970) Thermal decomposition of iron(II) sulfates, *J. Am. Ceram. Soc.*, **53**, 666.

Gallagher, P.K., Johnson, D.W., Schrey, F. and Nitti, D.J. (1973) Preparation and characterization of iron oxides, *Ceram. Bull.*, **52**(12), 842–9.

Hitachi Ltd. (1981) Gas detector, *Jap. Pat.* No. 5, 715, 8547 from 27 Mar.

Horowitz, H.S. and Lewandowski, J.T. (1982) High surface area unagglomerated noble metal pyrochlore compounds, *US Pat.* No. 4, 440, 670 from 30 Sept.

Inward, P.W. (1986) Preparation of basic aluminium halides for antiperspirants, *Euro. Pat.* No. EP 256832 A2 from 11 Aug.

Ischenko, V.V., Shlyakhtin, O.A., Oleinikov, N.N. and Tretyakov, Yu.D. (1995) Morphological evolution of powders during thermal decomposition, *International Workshop MSU-HTSC IV, Chemistry and Technology of High-Temperature Superconductors*, Moscow, 7–12 Oct., 96.

Iwata, T., Tajima, Y. and Ishii, T. (1991) Manufacture of gas-permeable porous ceramics for fuel cells, *Jap. Pat.* No. 5, 51, 279 A2 from 23 Aug.

Jaeger, R.E. and Miller, T.J. (1971) Particulate material by freeze drying, *US Pat.* No. 3, 916, 532 from 14 Sept.

Johnson, D.W. and Gallagher, P.K. (1972) Kinetics of the thermal decomposition of $BeSO_4$, *J. Am. Ceram. Soc.*, **55**(2), 232–3.

Johnson, Jr, D.W. and Gallagher, P.K. (1976) Reactive powders from solution, *Ceramic Processing before Firing*, New York: Wiley & Sons, 125–39.

Kharlanov, A.L., Antipov, E.V., Bryntse, I. *et al.* (1992) Structure, properties and substitutional chemistry of lanthanum strontium copper gallium oxide ($LaSrCuGaO_5$), *Eur. J. Solid State Inorg. Chem.*, **29**(6), 1041–54.

Kubiak, A. and Wolska, E. (1970) Physico-chemical properties of lyophilized aluminium hydroxide, *Rocz. Wyzsz. Szk. Roln. Poznaniu*, **47**, 149–60.

Mahler, W. and Chowdhry, U. (1984) Morphological consequences of freezing gels, *Ultrastructure Processing of Ceramics, Glasses and Composites*, New York: Wiley & Sons, 207–18.

Maki, T., Kokubo, T. and Sakka, S. (1986) Formation of oxide fibers by indirectional freezing of gel, *Bull. Inst. Chem. Res.*, Kyoto Univ., **64**(4), 292–305.

Mamsurova, L.G., Shabatin, V.P., Shlyakhtina, A.V. and Shcherbakova, L.G. (1989) Specific features of cryochemical synthesis of rare-earth titanates, *Russ. J. Inorg. Materials*, **25**(4), 637–41.

Marino, L., Gurzoni, E.T. and Castro, J.H.C. (1977) Sintering of magnesium aluminate $MgAl_2O_4$ obtained by the process of lyophilization, *Ceramica*, **23**(91), 21–30.

Marino, L. and Pereira, L.R. (1982) Preparation of ultrafine powders by the freeze-drying process, *37th Congr. Anu. ABM*, **1**, 727–43.

Moritz, T. (1994) Spray freeze drying of non-aqueous Si_3N_4-slips, *Key Engineering Materials*, **89/91**, 705.

Morris, L.B., Taylor, R.I. and Atkinson, A. (1987) Preparation of polycrystalline nickel oxide for diffusion studies, *J. Mat. Sci.*, **22**(6), 1993–8.

Nakagawa, Z., Hamano, K., Sakaguchi, M. and Kanzaki, S. (1982) Characterization and sinterability of Mg–Al spinel powders prepared with a thermal decomposition of a freeze-dried sulphate, *Yogyo Kyokaishi*, **90**(6), 313–19.

Nakano, T., Kumagai, M. and Uchimura, R. (1986) Manufacture of fine inorganic powders, *Jap. Pat.* No. 6, 307, 4904 A2 from 17 Sept.

Occhionero, M.A., Novich, B.E. and Sundback, C.A. (1987) Cryogenic slip-casting or injection-molding of complex ceramic and metallic articles having uniformly porous defect-free microstructure, *US Pat.* No. 8, 807, 902 from 9 Apr.

Ohtsuka, K., Kondoh, I. and Suzuki, T. (1988) Spent fuel treatment method, *Euro. Pat.* No. 358431 A1 from 5 Sept.

Panitz, J.K.G., Voigt, J.A., Greulich, F.A. *et al.* (1989) The morphology of freeze-dried rubidium chloride powder, *Mater. Res. Soc. Symp. Proc.*, **155**, 37–44.

Paschke, H., Daimer, J. and Haering, R. (1991) Preparation of very fine glass powders of high purity, *US Pat.* No. 5, 340, 776 A from 11 Jan.

Peuckert, M. (1987) Manufacture of free-flowing sintering aid-coated silicon nitride powder and of silicon nitride ceramics from this powder, *Germ. Pat.* No. 3, 711, 191 A1 from 2 Apr.

Reczka, K.R. and Reszka, J. (1987) Freeze-dried ceramics from alcoholic solutions, *Mater. Sci. Monogr.*, **38A** (High Tech Ceramics, Part A), 707–15.

Reetz, T. and Haase, I. (1989) Influence of the freezing process on the properties of the freeze-dried powders, *Ceram. Powder Process. Sci., Proc. 2nd Intern. Conf.*, Dtsch. Keram. Ges., 641–8.

Sakura Color Products Corp. (1976) Solid water colors, *Jap. Pat.* No. 5, 903, 1546 from 14 May.

Sasamoto, T. and Torikai, N. (1984) Preparation of ultrafine oxide powders by freeze drying technique, *Hyomen*, **22**, 708–18.

Strasser, J., McCambridge, T.H., Schnettler, F.J. *et al.* (1970) Feasibility study: freeze-dried ceramics, *Ceramic Age*, **8,** 22–4.

Sundback, C.A., Novich, B.E. and Karas, A.E. (1987) Manufacture of complex ceramic and metallic articles by low pressure injection molding at cryogenic temperatures, *US Pat.* No. 8, 807, 903 A1 from 9 Apr.

Torikai, N., Meguro, T., Sasamoto, T. *et al.* (1984) Preparation of fine particles of spinel-type manganese–cobalt–nickel oxide by freeze drying, *Nippon Kagaku Kaishi*, **6**, 785–91.

Tretyakov, Yu.D. (1991) Cryochemical processes to make ceramics, *Mater. Sci. Monogr.*, **66B** (Ceram. Today – Tomorrow's Ceram., Pt. B), 981–6.

Tretyakov, Yu.D. (1995) Cryochemical technology of advanced materials, *3 Slovenska Konferenca o Materialih in Tehnologijah.* 4–6 Oct., Portorozh, 87.

Walton, K. (1987) Methods of making a molecular oxygen-releasing substance, *Brit. Pat.* No. 8, 806, 374 A from 27 Oct.

Winnik, F.M., Breton, M.P. and Riske, W. (1991) Thermal jet printing inks, *Brit. Pat.* No. EP 497522 A1 from 28 Jan.

White, G.D., Lavender, J.C. and Hart, P.E. (1981) UO_2 pellets from freeze-dried oxide microspheres, *Ceramic Bulletin*, **60**(2), 260.

Yokota, T., Sugawara, T. and Ishikawa Y. (1992) Preparation of porous ceramic particles by freeze-dry processing, *Kagaku Kogaku Ronbunshu*, **18,** 752–6.

Yoshida, I., Watanabe, M., Kyodo, T. and Hatanaka, S. (1984) Manufacture of glasses, *Jap. Pat.* No. 61, 086, 428 A2 from 5 Oct.

Biomedical materials and drugs

Mladenov, D.A., Tsvetkov, Ts. and Vulchanov, N.L. (1993) Freeze drying of biomaterials for the medical practice, *Cryobiology*, **30**(3), 335.

Index

GPSR Compliance

The European Union's (EU) General Product Safety Regulation (GPSR) is a set of rules that requires consumer products to be safe and our obligations to ensure this.

If you have any concerns about our products, you can contact us on ProductSafety@springernature.com

In case Publisher is established outside the EU, the EU authorized representative is:

Springer Nature Customer Service Center GmbH
Europaplatz 3
69115 Heidelberg, Germany

Batch number: 10755284

Printed by Printforce, the Netherlands